Physical Metallurgy for Engineers

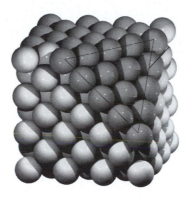

M. Tisza

ASM International
Materials Park, Ohio
and
Freund Publishing House Ltd.
London – Tel Aviv

Physical Metallurgy for Engineers

M. Tisza

ISBN 0-87170-725-X

ASM International
Materials Park, Ohio 44073-0002 USA
and
Freund Publishing House Ltd.
London – Tel Aviv

PREFACE

This book has evolved from an undergraduate course on *Physical Metallurgy* that I have lectured for more than twenty years at the University of Miskolc for students of Mechanical Engineering. The major goal of this book is to provide a solid theoretical ground for understanding the structure and properties of various materials and the basic principles of how their structure and properties can be controlled and modified.

Materials, and particularly metals, have always played a dominant role in the development of mankind. This is clearly illustrated by the fact that significant periods of history refer to certain materials as having played a very important role at that time (for example, *stone age*, *bronze age* and *iron age*).

In recent years, the types and application of materials has changed drastically. A very dynamic development can also be experienced in the field of metallic materials. Let us consider, for example, the various types of high strength metals, superalloys subjected to extreme loads under extreme conditions, the developments in high temperature superconductors, or shape memory alloys. Naturally, this list could be continued endlessly with other non-ferrous metals (e.g. aluminum, titanium, beryllium, zirconium and their alloys), too. These developments were mainly promoted by the leading branches of industry, such as the electronic and automobile industry, as well as aerospace and aircraft engineering.

In recent years, besides metallic materials (and in many cases, in place of them), non-metallic materials (polymers, ceramics and composites) gained ever-increasing application. Thus, our age cannot be characterized by the application of some specific materials, but rather a wide range of new, advanced materials.

This book was primarily written to introduce the fundamentals of *Physical Metallurgy* to students of a Mechanical Engineering Faculty. Although it is entitled *Physical Metallurgy*, it has a broader aim. Besides metals and metallic materials, the book frequently refers to non-metallic materials. Basic principles are considered generically for all materials rather than categorically for certain material groups.

Obviously, due to the limited extent of this book, it is not aimed at dealing with all kinds of materials, in detail. Instead, the theoretical principles and the materials science approach are emphasized to provide engineers with a thorough knowledge that would enable them to solve the emerging practical problems by applying up-to-date, advanced materials to meet the increasing functional and technological requirements.

In the elaboration of the content of this book, I also aimed at enforcing the correct ratio among the various materials, keeping in mind the present and the near future application trends. The book attempts to introduce Physical Metallurgy in a more structured manner than previous books. The setup of the book primarily reflects the structure of the lectures delivered by the author in undergraduate courses for mechanical engineering students at the University of Miskolc during recent years. The educational and research experiences are combined with recent developments and achievements in materials sciences.

Thus, the first half of the book (Chapters 1-7) is devoted to the basic principles upon which physical metallurgy builds. Therefore, the first few chapters summarize the basic idea of the atomic and the crystalline structures (including both ideal and real crystals). This is followed by the analysis of crystallization processes based on the rules of thermodynamics. A substantial part deals with the theoretical basis of the mechanical properties of metallic materials. This general introduction of physical metallurgy is concluded with a comprehensive overview of ideal phase diagrams.

The second half of the book (Chapters 8-13) may be termed *Applied Physical Metallurgy*. Since ferrous alloys still dominate in many industrial fields, a thorough analysis is given on the equilibrium and non-equilibrium phase transformations of various iron-carbon alloys. Both metastable and stable crystallization of iron-carbon alloys is analyzed in detail, applying a special technique of crystallization trees elaborated at the Department of Mechanical Engineering. The most important properties and characteristics of plain carbon steels and cast irons, as well as the most widely applied alloyed steels are introduced mainly from the theoretical point of view of physical metallurgy. Finally, the various groups of non-ferrous metals and their alloys of industrial significance are introduced.

I would like to emphasize that throughout the book the units of the *International System of Units (Système International d'Unités)* are used, commonly referred to as SI Units. Though widely applied all over the world, a brief summary of the most commonly applied units in Physical Metallurgy will be given in the Appendix.

Acknowledgements

I am particularly grateful to the University of Miskolc where I have been delivering lectures for undergraduate students for many years. During these years, I have gained much useful experience in teaching young engineering students, hopefully utilized in this book, also. This opportunity led to the evolvement of this book.

In the compilation of this book, both from the contextual and the didactical point of view, the scholarly activities of Professor Béla Zorkóczy (the former Head of the Department of Mechanical Engineering) had a decisive role. In the elaboration of the curriculum and the contents of the lectures forming the bases of this book, I have also received many useful comments from my colleagues at the Department of Mechanical Engineering in reviewing the various chapters. I would particularly like to name two of them, Ferenc Kovács and József Sárvári. Special thanks are also due to Miklós Tisza Jr. and Tibor Fülöp for the careful preparation of figures.

I would like to express my deepest gratitude to Academician János Prohászka, Professor of Materials Sciences and Engineering at the Technical University of Budapest, not only for his valuable comments and suggestions throughout this work, but also for his encouragement in our joint professional activities in the field of Materials Sciences.

Special thanks are also due to Chris Kirk, Senior Researcher at the University of Bath (United Kingdom) who assisted me by reading and correcting the final manuscript and by giving many useful comments and suggestions.

Finally, I am especially grateful to my family, and particularly to my wife Ann for her understanding and encouragement.

Miskolc, May 2000.

Miklós Tisza

ABOUT THE AUTHOR

Miklós Tisza is a Professor of Engineering at the Department of Mechanical Engineering of the University of Miskolc in Hungary. He was graduated at the Technical University of Heavy Industry in 1972. He acquired his Ph.D. degree at the University of Miskolc in 1977 and was awarded a D.Sc. degree in materials sciences by the Hungarian Academy of Sciences in 1994. Miklós Tisza has been teaching undergraduate and postgraduate materials science and engineering courses for many years. He has written several textbooks including the *Introduction to Materials Sciences*, the *Fundamental Principles of Physical Metallurgy*, and the *Theory and Practice of Metal Forming*.

CONTENTS

1. INTRODUCTION

1.1. The Role of Materials in Technical Development

Materials and energy have always played a dominant role in man's life since the beginning of human civilization. Objects and tools around us are made of various materials; some of them from well-known "*everyday*" materials, like wood, steel and glass. However, there are several other objects consisting of such new materials that were unknown even a few decades ago. Among these, we can mention many new materials, like super-alloys, ceramics, composites.

Most products and tools are designed by mechanical engineers. During production, engineers control manufacturing processes and create production systems as well. This is why it is essential for them to know the various types of materials, their properties and internal structure, the relationships between structure and properties and the principles constituting the basis for property changes. Therefore, it is of utmost importance for students of mechanical engineering – and for practicing engineers, as well - to be familiar with the fundamentals of material sciences.

Hence, the main objective of this book is to introduce students of Mechanical Engineering to the basic knowledge of material sciences, to teach them the fundamentals of physical metallurgy and to give a general overview of materials and materials science related aspects of mechanical engineering.

1.2. The Definition and Scope of Physical Metallurgy

Physical Metallurgy is a science concerned with the physical, chemical and mechanical characteristics of metals (pure metals and metallic alloys), the relations between their structures and properties, as well as the theoretical basis and practical methods of property changes.

This book – as its title indicates – deals mainly with metallic materials. However, in recent years, significant changes have taken place in the field of materials applied in practical engineering. Therefore, beside metallic materials, the book refers in certain chapters to new non-metallic materials of our age, like plastics (polymers), ceramics and composites, as well.

As this book was primarily written to complete the fundamental engineering studies of mechanical engineering students, it is not aimed at dealing with specific materials in detail. Instead, the author wishes to direct the readers' attention to the theoretical basis of physical metallurgy and to train them to be able to understand the *materials science approach and attitude*.

In recent years, besides the dynamic internal development of materials science, integrated development of natural and engineering sciences can be observed. This interdisciplinary approach is becoming increasingly common in engineering studies, as well. This integrated approach in engineering and materials science can be seen all over the world: many new institutions integrating various sciences have recently come into existence in many countries (*Departments of Materials Science and Engineering*). The aim of these institutions is to deal with materials sciences and engineering more effectively using an interdisciplinary approach. All these facts can be considered natural because the main objective of engineering is the utilization of mechanical-scientific knowledge, acquirement and experience while applying natural materials and resources for the benefit of mankind.

The co-operation between engineering and materials sciences has a long history. This interdisciplinary approach must serve as a model for the co-operation of mechanical engineering and materials sciences to a wider extent. In this co-operation, the new and dynamically developing materials play an increasingly significant role beside the traditional materials of practical engineering.

In accordance with the above-mentioned facts, the first part of this book deals with the fundamentals of materials science (atomic structure, crystal structure, the rules of crystallization, the basic rules of elastic and plastic deformation, etc.). This fundamental knowledge can help us to understand the basic principles of the most important characteristics of materials applied in practical engineering. In the following part, those applied metallographic methods – mainly based on the principles of thermodynamics – are introduced which constitute the fundamentals of property changes of different metals.

The dynamic developments of the past decades in materials science must naturally be reflected in university education, as well. The syllabus of introductory materials science subjects, as Physical Metallurgy for Mechanical Engineers, has also been modified according to these changes. Therefore, in the second part of the book, iron-based alloys are described in detail, as they still play a decisive role in industrial practice. However, due to the increasing use and importance of light and non-ferrous metals, as well as non-metallic materials, fundamentals relating to these types of materials will also be discussed in certain chapters.

Introducing these changes into the curriculum and into the content of this book was self-evident since the subject of Physical Metallurgy has always been based on such solid grounds of materials sciences as atomic structure, crystallography, solid-state physics, fundamentals of thermodynamics. These sciences formed the basis for materials science development in the recent years, as well.

1.3. Main Types of Materials

Every material can be classified into one of the three basic types of materials: *metals, polymers* and *ceramics*. Due to the increasing application, nowadays, composites are becoming increasingly significant besides the three main classes of materials: therefore, composites will also be introduced in short. Composites form a combination of the previously mentioned basic materials.

Due to the limited volume of this book, obviously, we can only give a short summary of fundamental properties of these materials and some application examples in order to provide the students of Mechanical Engineering with a comprehensive view of materials sciences.

1.3.1. Metals, metallic materials

The term *metals* includes pure metals and metallic alloys. Pure metal contains only one metallic component, e.g. iron, aluminum, copper, etc. Metallic alloys consist of two or more materials: the basic component (i.e. the base-material) is some kind of metal, however other materials (metalloids, sometimes non-metallic materials) can be also found among the components.

Metals have a crystalline structure in which the atoms are arranged in a regular manner. Metallic materials generally have a lustrous appearance, high thermal and electrical conductivity, relatively high rigidity and certain formability. These properties of metals and metallic alloys are directly connected with their atomic and crystal structure.

Metals can be divided into two large groups. One of them is the group of so-called *iron-based alloys* (ferrous metals). In these materials, the basic component is iron: this group contains steel and cast iron which are still the most important materials in practical engineering.

The second main group includes the so-called *non-ferrous metals* and their alloys: e.g. aluminum, copper, titanium, several other metallic materials and their alloys.

Metals can be considered as materials most widely applied in engineering devices and equipment up to the present day. We can mention household appliances, various machines and devices, car components or several items of equipment for the aircraft industry and astronautics. These components characteristically consist of metals and in many cases they are made from special metallic alloys.

As an example, an aircraft turbine engine is shown in Figure 1.1. In the so-called jet engine applied at steadily increasing temperatures, "*super-alloys*" are required to increase the power and efficiency. Therefore, in the engine a nickel-based super-alloy is applied which is able to withstand high temperatures up to 1000°C.

Figure 1.1
F117-PW2000 Pratt & Whitney aircraft turbine engine
(Courtesy of Pratt & Whitney: A United Technologies Company)

1.3.2. Polymers, plastics

Most polymers are organic compounds that are chemically based on carbon with a very large molecular structure (containing molecular chains or networks). Structurally, most polymer materials are non-crystalline, but there are a few polymers with crystal structure (only polymers with regular structure are able to crystallize) whilst there are others containing both crystalline and non-crystalline domains. According to their inner (molecular) structure, polymers are poor conductors and good insulators. These properties form the basis for several fields of application.

The electronic industry applies polymers increasingly: as an example, an integrated circuit unit of a computer motherboard is shown in Figure 1.2.

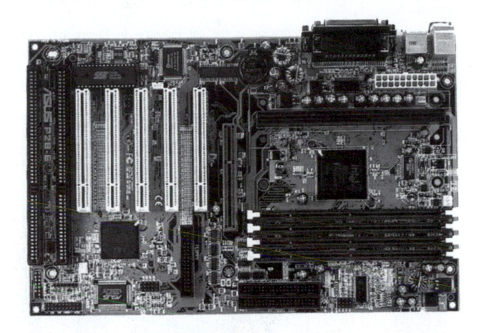

Figure 1.2
A motherboard of a personal computer: application example for polymers

Polymers are generally characterized by a low density and a low melting point (as an example thermoplastic polymer can be mentioned here) but there are so-called thermosets (thermoreactive polymers) as well.

Polymers (plastics) can be found in many fields of today's life. In homes and offices, a great number of products and tools are made from plastics for everyday use. Their low density and relatively high ratio of rigidity/density resulted in wider application in many fields where only metals were previously used. (For example, in the car industry the use of polymers is continually growing: at present, about 25-30% of car components are polymers).

1.3.3. Ceramics

Ceramics are inorganic compounds that contain metallic and non-metallic elements. Ceramic materials may have a crystalline structure and similarly to polymers there are ceramics with both crystalline and non-crystalline domains.

High rigidity and hardness enabling operation at high temperatures characterizes most ceramic materials, though most of them are quite brittle as well. Besides these characteristics, low density, good insulation, good heat and wear resistance are the main advantages of ceramics. These special properties are the main reasons for the rapidly growing industrial application of ceramics in the past few decades.

A promising field of application is the car industry. In Figure 1.3, valve components, pump rotors, gears and some other components subjected to high wear and thermal load are shown. All these components are produced from ceramics. In some cases, only the surfaces of certain components are coated with ceramic materials.

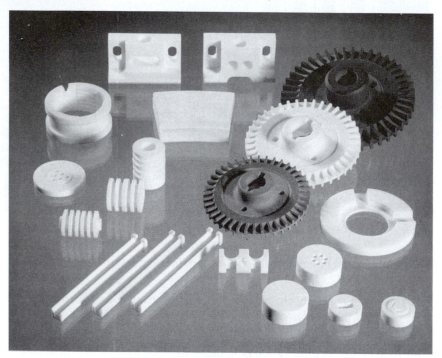

Figure 1.3
Valve components, pump rotors, gears and other components
made of ceramic materials

Ceramics are the most favored materials in many "high-tech" applications. It would have been impossible to develop machining processes with submicron accuracy without ceramic machine tools. The role of ceramic materials is particularly important in the aircraft industry and astronautics: supersonic airplanes, spacecraft, space shuttles have essential ceramic components, which can occasionally bear several thousand degrees of temperature. These ceramic materials can protect aircraft when exceeding the sonic barrier or space shuttles upon re-entering the earth's atmosphere.

1.3.4. Composite materials

Composite materials are mixtures of two or more materials: therefore, they are regarded as compound materials. This is the reason that they cannot be considered as a basic type in the classification of materials. However, it is reasonable to treat them in a separate section because they are applied to a rapidly growing extent. In addition, they have special production processes and application fields.

Composites can be of many types. Some of these composites are fibrous (composed of fibers in a matrix), whilst others are particulate (composed of particles in a matrix). In practical engineering, the most commonly applied composites are glass-fiber reinforced (i.e. glass-fiber in a polyester or epoxy-based matrix) and carbon fiber reinforced composites (carbon fibers mainly in an epoxy-matrix). For the particulate type of composites, the so-called hard metals are good engineering examples. These materials contain wear-resisting grains (e.g. *WC*-tungsten carbide, *TiC*-titanium-carbide, etc.) in a *Co*-cobalt matrix. Developments in space technologies and in the aircraft industry play an essential role in the development of composite materials.

In Figure 1.4, we can see examples of the application of composite materials as an aircraft structural material. In this figure, the parts of an airplane body made of composite materials are shaded with a different pattern.

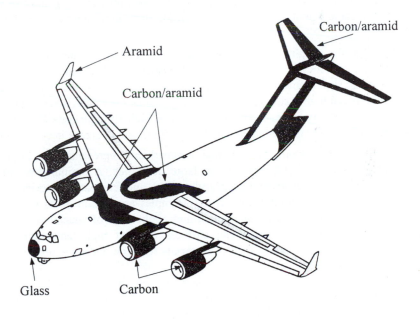

Figure 1.4
Application of fiber-reinforced composite materials in airplane construction
(Source: Journal of Advanced Composites, 1998)

1.4. Recent trends and application tendencies

Nowadays – like in any sector of the economy – there is a global competition among materials for present and future markets. The significance of different materials is changing from time to time according to several factors. Among these factors, besides the functional and technological aspects, costs play a crucial role. Obviously, there are some exceptions, such as space technologies where the requirement for certain special properties establishes priorities in spite of huge expenses.

The car industry – which is one of the leading industrial sectors in the economy of industrially developed countries – can serve as a realistically representative example of material application.

The greatest car producer in the world is the United States of America. In 1975, an average American car weighed 1800 kg. More than 60% of it consisted of iron-based alloys (various steels, cast irons). At that time, only 10-20% of the materials were plastics and 3-5% aluminum. The rest, approximately 15%, was made of other materials (glass, textile, etc.).

In the 70s, because of the oil crisis sweeping across the world, the car industry made a strong effort to decrease the weight of cars. This can be clearly seen from the fact that an average American car weighed only 1400 kg in 1985. The rate of steel – with higher rigidity than the earlier steels – was only 50-55%, plastics were used to the same extent (10-20%) but the use of aluminum increased, reaching 8-10%.

By 1995, the weight of an average American car had become only 1130 kg. Not only the tendency of the growing use of plastics (20-25%) and aluminum (10-15%) but the developments of material sciences, resulting in the evolvement of steels with much higher strength and stiffness (such as the HSLA steels) played an important role in this change. The application of high-strength low-alloyed steels ensures equivalent or greater safety beside the significant decrease in car weight.

The car industry is only one – though a remarkably representative – branch of industrial sectors applying materials in large quantities. A similarly prominent industrial sector is the power industry where the applied materials occasionally have to meet special requirements. Good examples are turbine parts working at very high temperature. The development and application of special super-alloys having the appropriate strength and stiffness even at high temperatures will prospectively be an important field of materials research and development in the near future.

2. ATOMIC STRUCTURE OF MATERIALS

2.1. Structure and properties of materials

In nature, every material exists in discrete and constant forms. Particles are discrete in appearance and amongst these particles, there are various interactions and relationships, which may be characterized by continuous fields. These two types of existence of substances determine together the properties of materials. Therefore, from the point of view of material properties, it is very important to know what kind of particles (atoms, molecules) they are composed of and what kind of order and relationship (interaction) exists among them. This knowledge is presumed to have been acquired from earlier studies of physics and chemistry; hence, in this chapter we summarize only the most important aspects.

2.1.1. Atomic structure

All materials in nature consist of atoms. Atoms can be further divided into subatomic particles: the nucleus – composed of protons and neutrons – which is surrounded by the electron shell. Atoms are extremely small in terms of the everyday concept of mass: the diameter of the nucleus is in the order of 10^{-12} cm, while the size of the electron shell can be found in the 10^{-8} cm domain. The mass of the proton is 1.673×10^{-24} g and it is electrically charged. The charge of the proton is equal to 1.602×10^{-19} C (Coulomb) and it has a positive charge. Protons and neutrons have approximately the same mass (1.675×10^{-24} g). Neutrons are electrically neutral, i.e. they do not carry any charge. The mass of an electron is significantly smaller than that of the proton (9.109×10^{-28} g), but it has the same charge (1.602×10^{-19} C) with a negative sign. These facts clearly show that almost the whole of the mass of an atom (99.97%) can be found in the nucleus, which is surrounded by an electron shell with a very low density. At the same time, the size of the atom can be taken to be equal to the size of the outermost electron shell. Electrons (mainly the valence electrons occupying the outermost electron shell) determine the most important electrical, mechanical and chemical properties of materials. The most significant properties related to atomic structure can be found in the periodic table (Mendeleev, 1869 - Table 2.1). Here, we briefly describe the characteristic parameters that will be used in subsequent chapters.

2.1.1.1. The most important properties related to the atomic structure

Each chemical element is characterized by its *atomic number,* which indicates its position in the Periodic Table. The atomic number is equal to the number of protons in the nucleus (and obviously, it equals the number of electrons of the neutral atom).

This atomic number ranges in integral units from 1 for Hydrogen (*H*) to 103 for Lawrencium (*Lw*). (Today, there are also elements with higher atomic numbers: *Meitnerium* (109) currently has the highest atomic number. However, we should mention that, under artificial circumstances, elements with even higher atomic numbers have been produced, though only for a short period of time.)

In the previous section, we could see that the masses of particles are extremely small (e.g. the mass of protons is 1.673×10^{-24}g); this is why the so-called *atomic mass* (n) is used in practice. Its unit is a "mole" (abbreviated as mol). It is known from chemistry that in one "mole" of a substance there are 6.023×10^{23} atoms (or molecules in the case of a molecular system). The value $N_A = 6.023 \times 10^{23}$ molecules per mole is called *Avogadro's number*. According to international agreements, a scale has been established where one *atomic mass unit* is defined as 1/12 of the atomic mass of carbon (*C*). Carbon contains 6 protons and 6 neutrons and it can be found in the 6th place of the periodic table. It shows that the atomic mass of carbon is 12, which is equal to the atomic mass unit of 6.023×10^{23} *carbon* atoms. This quantity is called *the molar mass* (*M*) and its unit is defined as g/mole.

The electronic structure of atoms is of essential importance with respect to several physical, chemical and mechanical properties. The hydrogen atom has the simplest atomic structure, with only one electron in its electron structure. This electron revolves in an orbit around the nucleus. The position and motion of electrons are described by the rules of *Quantum Mechanics*. According to these laws, electrons can occupy only definite electron shells with restricted energy levels. Electrons are in a stable state at their lowest possible energy level. When excited by external means, electrons occupy an orbital of a higher energy level. During this process, atoms absorb discrete amounts of energy (quanta) which can be precisely defined as being equal to the difference between the energy levels of the two orbitals.

Table 2.1 The Periodic Table of Elements

Notations:

Elements → Li (← Atomic Number: 3)
Atomic mass → 6.94
Crystal structure → A2 / A3 (← Lattice parameter: 3.5 / 3.1 / 5.1)

Notes to the crystal structure

A1	Face centered cubic
A2	Body centered cubic
A3	Hexagonal close packed
A4	Cubic diamond
A5	Body centered tetragonal
A6	Face centered tetragonal
A7	As-type structure
A8	Se-type structure
C	Complex cubic
H	Hexagonal
M	Monoclinic
O	Orthorhombic
R	Rhombohedral
T	Tetragonal

IA	IIA	IIIB	IVB	VB	VIB	VIIB	VIII	VIII	VIII	IB	IIB	IIIA	IVA	VA	VIA	VIIA	0
H 1 / 1.01 / A3 3.8, 6.1																	He 2 / 4 / A3 3.6, 5.8 / A2 -
Li 3 / 6.94 / A2 3.5, A3 3.1, 5.1	Be 4 / 9.01 / A2 2.6, A3 2.3, 3.6											B 5 / 10.8 / H, T, R, 9.5	C 6 / 12 / A4 3.6, H 2.5, 6.7	N 7 / 14 / H 3.6	O 8 / 16 / C 4, R 6.6, 5.7	F 9 / 19	Ne 10 / 20.2 / A1 4.5
Na 11 / 23 / A2 4.3, A3	Mg 12 / 24.3 / A3 3.2, 5.2											Al 13 / 27 / A1 4.1	Si 14 / 28.1 / A4	P 15 / 31 / C, O, C 5.4	S 16 / 32.1 / O 10, M 3.3, R 19	Cl 17 / 35.5 / A1 10, A3 11, A2 6.5	Ar 18 / 39.9 / A1 5.5, 5.3
K 19 / 39.1 / A2 5.3	Ca 20 / 40.1 / A1 5.6, A2 4.5, A1	Sc 21 / 45 / A3 3.3, 5.3	Ti 22 / 47.9 / A2, A3 3.3, 4.7, 5.3	V 23 / 50.9 / A2 3.3	Cr 24 / 52 / A2 3, A1 2.9, C	Mn 25 / 54.9 / A1 3.7, A2 2.9, C 6.3, 8.9	Fe 26 / 55.8 / A2 2.9, A1 3.7, A3 2.9	Co 27 / 58.9 / A3 2.5, A1 3.5, 4.1	Ni 28 / 58.7 / A1 3.5	Cu 29 / 63.5 / A1 3.6	Zn 30 / 65.4 / A3 2.67, 4.95	Ga 31 / 69.7 / O 4.5	Ge 32 / 72.6 / A4	As 33 / 74.9 / A7 5.7, 4.1	Se 34 / 79 / A8 4.4, M 9.1, 13	Br 35 / 79.9 / 4.4, 9.1, 13	Kr 36 / 83.8 / A1 5.6
Rb 37 / 85.5 / A2 5.6	Sr 38 / 87.6 / A1 6.1, A3 4.3, 7.1	Y 39 / 88.9 / A3 3.6, 5.7	Zr 40 / 91.2 / A1 4.1, A3 3.7, 5.2, 5.7	Nb 41 / 92.9 / A2 3.3	Mo 42 / 95.9 / A2 3.3	Tc 43 / 97 / A3	Ru 44 / 101 / A2 2.7, A3 4.3, 4.4	Rh 45 / 103 / A1 3.8	Pd 46 / 106 / A1 3.9	Ag 47 / 108 / A1	Cd 48 / 112 / A3 4.1	In 49 / 115 / A6 4.6	Sn 50 / 119 / A5 6.5, A4 5.8	Sb 51 / 122 / A7 4.5	Te 52 / 128 / A8 57.1°, 4.5	I 53 / 127	Xe 54 / 131 / A1 6.1
Cs 55 / 133 / A2 6.1	Ba 56 / 137 / A2 5.0	La* 57 / 138.9 / A1	Hf 72 / 178 / A2 3.5, A3 3.2, 5.1	Ta 73 / 181 / A2 3.3	W 74 / 184 / A2 3.2, A3	Re 75 / 186 / A3 2.8, 4.5	Os 76 / 190 / A3 2.7, 4.3	Ir 77 / 192 / A1 3.8	Pt 78 / 195 / A1 3.9	Au 79 / 197 / A1 4.1	Hg 80 / 201 / R, T 3.01, 4	Tl 81 / 204 / A2 3.9, A3 3.5, 5.5	Pb 82 / 207 / A1	Bi 83 / 209 / A7 5.8, 6.5	Po 84 / 209 / R, C 4.5	At 85 / 210	Rn 86 / 222
Fr 87 / 223	Ra 88 / 226	Ac 89 / 227 / A1 5.3	Th 90 / 232 / A1 5.1	Pa 91 / 231 / T 3.9, 5.1	U 92 / 238 / A2, T 4.5, 11, O 2.9	Np 93 / 237 / O, M 3.5, 4.9, 6.2 / A7	Pu 94 / 244 / A3, H 3.5, 9.3, 6.2	Am 95 / 243 / A1 3.6, 12	Cm 96 / 247	Bk 97 / 247	Cf 98 / 251	Es 99 / 254	Fm 100 / 257	Md 101 / 256	No 102 / 254	Lw 103 / 257	

*Lantanide series (Rare earth metals)

57 La 138.9	58 Ce 140.1	59 Pr 140.9	60 Nd 144.3	61 Pm 145	62 Sm 150.4	63 Eu 152.0	64 Gd 157.3	65 Tb 158.9	66 Dy 162.5	67 Ho 164.9	68 Er 167.3	69 Tm 168.9	70 Yb 173.0	71 Lu 175.0

This phenomenon is illustrated in Figure 2.1, which shows the Bohr atomic model of a hydrogen atom. In Figure 2.1.a) the unstable state after energy emission and in Figure 2.1.b) the stable state after energy absorption can be seen.

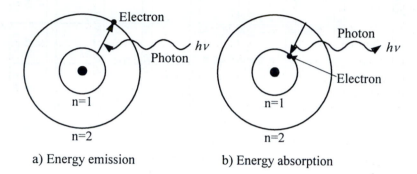

a) Energy emission   b) Energy absorption

Figure 2.1
Electron structure of hydrogen atom
a) Unstable state b) Following return to the stable state with photon emission

When returning to a lower energy level from a higher one, electrons release a discrete amount of energy – (energy quantum) in the form of electromagnetic radiation (photon emission). Conversely, when an electron enters a higher energy level, the same discrete energy quantum is absorbed. The energy difference between the energy levels can be expressed by the Planck equation:

$$\Delta E = h\nu,$$ (2.1)

where $h = 6.625 \times 10^{-34}$ (Js) is the Planck constant and ν is the frequency of the emitted photon. For electromagnetic waves, the expression $c = \lambda\nu$ is valid (where $c = 3 \times 10^8$ m/s is the velocity of light, λ is its wavelength), and therefore the energy change associated with photon emission can be calculated as

$$\Delta E = \frac{hc}{\lambda}.$$ (2.2)

Around 1913, *Bohr* elaborated an atomic model (he received the Nobel Prize in Physics in 1922, mainly for this work). In this atomic model, electrons are permitted to have only a specific energy level, which can be calculated from the following expression:

$$E = -\frac{2\pi^2 m e^4}{n^2 h^2},$$ (2.3)

where

e	the electron charge,
m	the electron mass,
n	the so-called principal quantum number (only integer numbers).

From the previous equation the possible energy levels of an electron in an excited atom can be calculated (e.g. for the excited hydrogen atom, where $n = 2$, $E = -3.4$ eV). From this equation, it can be also determined that 13.6 eV energy is required for the complete removal of the electron: this is called the *ionization energy* of the atom.

According to modern atomic physics, the motion of electrons is much more complex than the one described by the Bohr atomic model. Furthermore, every electron in an atom is characterized by four parameters called quantum numbers: the first (denoted by n) represents the principal quantum number, l *is* the second quantum number, the third is the magnetic quantum number (denoted by m), and s is the spin quantum moment. The further development of the Bohr atomic model is attributed to Sommerfeld. In contrast to Bohr, he assumed that electrons moved in both circular and elliptical orbits.

The *principal quantum number* (n) characterizes the main energy levels of electrons. This quantum number is an integer number between 1 and 7. Higher principal quantum numbers denote greater distances between the electrons and the nucleus, and consequently higher energy levels.

The *second or subsidiary quantum number (l)* specifies possible energy levels within the main energy levels and the subshells related to them. Providing that energy levels have been characterized by the second quantum number, the probability of finding electrons within these subshells is high. In Atomic Physics, letters denote subshells, as follows:

the subshell where $l = 0$ is denoted by s (sharp),
the subshell where $l = 1$ is denoted by p (principal),
the subshell where $l = 2$ is denoted by d (diffuse),
the subshell where $l = 3$ is denoted by f (fundamental).

The value of the second quantum number describes the shape and eccentricity of the orbital. The value must always be lower than that of the principal quantum number, otherwise the elliptical orbit would become circular. This is why the value of l changes between 0 and n-1.

The magnetic *quantum number* specifies the magnetic spatial orientation of the orbital and has no significant effect on the energy level of the electron. The possible values of the magnetic quantum number range from $-l$ to $+l$ (where l is the second quantum number). Therefore, when $l=2$, the magnetic quantum number can have a value between -2 and +2.

The *spin quantum number* (or in other terms the *spin moment*) specifies two possible electron orientations. Depending on the type of orientation, the value of the spin moment can be $s = +1/2$ or in the case of the opposite orientation, $s = -1/2$. The spin moment has an extremely small effect on the energy level of electrons. (It has been proved that when two electrons are located in a common orbital, their spin moments have opposite signs.)

The structure of electron shells can be defined on the basis and the laws of *Quantum Mechanics*. The principal quantum number (n) determines the number of main electron shells, while the possible number of electrons on each main shell is determined by the Pauli principle. According to this principle, the maximum number of electrons is equal to $2n^2$, where n is the principal quantum number. In *Table 2.2,* the maximum possible number of electrons and the electron configuration are shown.

Table 2.2 Electron configuration and possible number of electrons

Principal quantum number	The maximum number of electrons on the main shells $2n^2$	Electron configuration
1	2	s^2
2	8	s^2p^6
3	18	$s^2p^6d^{10}$
4	32	$s^2p^6d^{10}f^{14}$
5	50	$s^2p^6d^{10}f^{14}...$
6	72	$s^2p^6...$
7	98	$s^2...$

Each atom can be considered as a regular sphere with a given radius. As a first approximation, this radius can be considered equal to the radius of the outermost electron shell determined by the principal quantum number of the given element. It clearly shows that the greater the principal quantum number, the greater the size of the atom. However, there are certain exceptions (see Periodic Table - Table 2.1)

The electron structure – especially the structure of the outermost shell – is the basis for the most important physical, mechanical and chemical properties of materials from the viewpoint of engineering. The outermost electron shell structure determines the chemical properties of elements. The valence electrons occupying the outermost shell are extremely important as they determine the bonding among atoms and many physical and mechanical properties are based on them. The outermost electron shell determines the size of atoms, the electrical and thermal conductivity, and has a great effect on optical characteristics. For these reasons, we should pay great attention to the electron structure of atoms.

Consequently, we can state that the electron structure - and emphatically the structure of the outermost electron shell - has an extremely great effect on the chemical reactivity of electrons. The most stable and chemically least reactive elements are the so-called noble gases (*He, Ne, Ar, Kr, Xe* and *Rn*). The number of electrons on the outermost electron shell of these elements is equal to that determined by the principal quantum number. This kind of electron shell is regarded as "closed". They can be characterized by the s^2p^6 electron configuration: one exception is helium, which has the $(1s)^2$ electron configuration.

The closed outermost electron shell of noble gases is neutral from the chemical point of view and this results in an extremely stable structure. Contrary to noble gases, electropositive and electronegative elements can ensure a stable, closed outermost electron shell by giving up electrons (electropositive elements) or accepting electrons (electronegative elements).

Electropositive elements are those metallic materials, which give up electrons during chemical reactions. In this way, the electronically neutral atom becomes a positive ion, called a cation. Electrons given up by electropositive elements are called valence electrons. They are marked with positive oxidation numbers. Elements situated on the left-hand side of the periodic table (columns I-III) are included in this group.

Electronegative elements are mostly non-metallic in nature. They accept electrons during chemical reactions. Thus, an electronically neutral atom becomes a negative ion, called an anion. Electrons accepted by electronegative elements are valence electrons and are marked with negative oxidation numbers as well. This group includes elements of the right-hand side of the periodic table (columns VI-VII)

Certain elements can behave as both electronegative and electropositive elements. This dual behavior is typical of the elements occupying the columns IV-V of the periodic table, such as carbon, silicon, germanium, arsenic, and antimony. These are also known as transition elements.

2.1.2. The relationship between atomic order and states of matter

There are different interactions between atoms - as discrete elementary particles - due to the gravitational and the electrical fields existing among the particles. These interactions can be characterized by the potential energy. At the same time, particles are continuously moving, and therefore they also have kinetic energy.

The equilibrium conditions for every thermodynamic system can be fulfilled at the minimum of the potential and kinetic energy. Any state of matter can be characterized by the so-called state factors (i.e. temperature, pressure, specific volume and concentration: the latter one for multi-component systems). Accordingly, stable bonding between elementary particles can be realized if the resultant energy of the system is less than the summarized energy of the individual particles. The energy difference between the two states is called the bonding energy.

Before discussing the most important bonding – i.e. bonding in solid bodies – from the viewpoint of *Physical Metallurgy*, we should briefly summarize the most important characteristics of different states, with special regard to the relationships between particles and atomic order.

As regards atomic order, the different states of matter can be classified into three main groups; gases, liquids and solids. (At extremely high temperatures – several thousands of degrees – gaseous substances can change into a plasma state, which can be characterized by the appearance of ionization processes that finally lead to the complete destruction of electron shells of the atom. The Plasma State is often regarded as the "fourth" state of matter.)

According to the previously mentioned atomic order, the three basic states of matter can be characterized as follows:

- *In gases,* there is no atomic (or molecular) order, since atoms (molecules) are moving randomly. Their chaotic, disordered movement can be described by the rules of kinetic (statistical) theory of gaseous media. They always tend to spread over the whole volume available. The volume changes arise from their compressible nature. Bonding forces between particles are extremely weak, and the bonding energy is in the order of a few kJ/mole.

- *In liquids,* the position of particles can be characterized by the so-called short-range order. It can be explained by the fact that in the neighborhood of any particles - composing liquids - at a well-defined distance and order, it is most likely to find neighboring particles. However, there is little likelihood of the appearance of this relative atomic order at great distances - concerning atomic dimensions.

  This type of short-range order can be explained as follows: neither the potential nor the kinetic energy can be regarded as dominant in liquids, though the value of both types of energy can be considered significant. According to the relatively significant potential energy, the distance between atoms is determined by state factors, and this is the reason why a kind of short-range order can be observed. At the same time, due to a considerable kinetic energy, atoms can easily move relative to one another. Long-range order cannot therefore exist and liquids can take the form of any container. However, due to the relatively high potential energy, their volume can change only to a very small extent even in the case of high pressure, and therefore they are practically incompressible.

- *In solids,* the potential energy of particles plays a determining role. The ability of particles to change their location is minimal. Their kinetic energy is practically limited to the vibration around their atomic center. According to the atomic order, we can distinguish amorphous and crystalline solids:

  * *In amorphous solids,* the short-range order of particles is characteristic as in liquids. A good example is glass, which as an *amorphous* solid is often mentioned as an under-cooled liquid.

  * *In crystalline solids,* there is a long-range order of atoms. Atoms are located in well-defined points of geometrical configurations and this is regularly repeated in the three directions of space according to a rule determined by the crystallographic structure. In ideal cases, the

probability of finding further atoms at well-defined points in space is almost 100% due to the regular repetition of particles.

- As mentioned before, besides the three basic states of matter – i.e. solids, liquids and gases – a fourth state of matter, i.e. the plasma state, can exist. At extremely high temperatures – several thousands of degrees – gaseous substances can change into a plasma state, which can be characterized by ionization processes that finally lead to the complete destruction of electron shells of the atom. Therefore, the plasma state can be characterized by the presence of independent nuclei and individual, independently moving electrons. This kind of plasma state can be observed, for example, in a welding arc or during a thermonuclear reaction.

2.1.3. Atomic (molecular) bonding

As we considered earlier, the appearance of atomic (molecular) bonding is the result of the fact that the resultant (potential and kinetic) energy of the system originating from the bonding is smaller than the total energy of separate particles. The energy difference between the two states is called the bonding energy. According to the previously mentioned facts, atoms – in the case of certain conditions – are energetically more stable when they are bound, than in the free state.

Chemical bonding between atoms (molecules) can be classified into two basic groups: primary (strong) interatomic bonding and secondary (weak) bonding. In the following section, we are going to analyze interatomic bonds based on these two groups.

2.1.3.1. Primary (strong) interatomic bonds

Primary interatomic bonds can be characterized by the significant bonding forces between atoms, and they are therefore regarded as strong bonds. Primary bonds can be divided into three further types: metallic, ionic and covalent bonding.

Metallic bonding

Metallic bonding as a primary, strong bonding can be found in metals. The atoms of metallic materials are situated in certain points of regular geometrical forms spreading over the whole space (see details in the topic of crystalline structures).

Metallic materials have one, two, or three valence electrons. In metallic bonding, because of the close packing of atoms, the nuclei of neighboring atoms affect valence electrons. Therefore, the bonding of valence electrons to certain nuclei can be regarded relatively weak. They can be easily "separated" and they can form a so-called "electron cloud" or, in other terms, an "electron gas" having a low density and they can move more or less freely among the nuclei.

Metallic bonding in solid materials can be characterized by *metallic ions* located in special points of the crystal lattice (lattice sites). Metallic ions are positively charged and have no valence electrons. Valence electrons move freely among ions in the crystal lattice, so they are called *free electrons*. These free electrons provide an explanation to several metallic characteristics. Such materials are good electrical and heat conductors, due to the free valence electrons. A schematic illustration of metallic bonding can be seen in Figure 2.2.

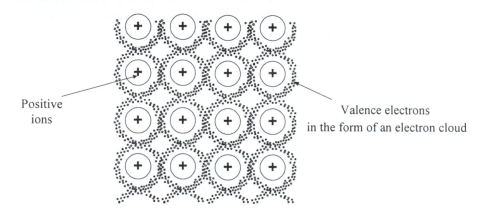

Positive ions

Valence electrons
in the form of an electron cloud

Figure 2.2
Schematic illustration of metallic bonding

The generally good formability of metals can be directly connected to the previously described characteristics of metallic bonding: atoms spread strictly over the whole space contacting each other by means of their outermost electron shells. Among all kinds of bonding atoms, the highest atomic packing is characteristic of metallic bonding. Consequently, on application of an external force crystalline planes can slide on each other relatively easily – along certain planes and directions – without interrupting the interatomic bonding. "Free electrons", which compose the electron gas, may be associated with a number of neighboring atoms. They compose a relatively freely moving, negatively charged electron field that continuously ensures metallic bonding during the movement of atoms.

Obviously, the resultant energy of atoms in metallic bonding is smaller than the total energy of the individual atoms. This is illustrated in Figure 2.3, where the changes in the energy of the two atoms are shown as a function of the distance between them for metallic bonding.

This figure clearly demonstrates that the energy minimum belongs to that distance of the atoms (a_o) at which they are in contact along their outermost electron shell. This is the equilibrium state and the related energy minimum (E_{min}) means the bonding energy.

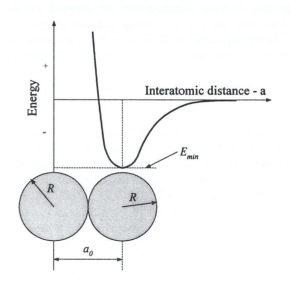

Figure 2.3
Energy vs. atomic distance
for metallic bonding

In Figure 2.3, the change in the bonding energy is shown for two neighboring atoms. Obviously, the variation of the bonding energy is somewhat different, if we consider all the atoms participating in the bonding. Naturally, this is still valid in this case as well, that is the total bonding energy of atoms participating in the bonding should be less than that of the individual atoms.

The bonding energy of different materials (and in close relation with this, their melting point) changes in a wide interval, as can be seen in Table 2.3. In general, we can state that metallic bonding is more characteristic of those materials having a lower number of valence electrons.

Table 2.3 Bonding energy and melting point of selected metallic materials

Chemical symbol of the element	Electron configuration	Bonding energy E (kJ/mol)	Melting point T (^oC)
K	$(4s)^1$	90	64
Zn	$(4s)^2$	131	419
Ca	$(4s)^2$	177	851
Ge	$(4s)^2(4p)^2$	377	960
Sc	$(3d)^1(4s)^2$	342	1397
Ti	$(3d)^2(4s)^2$	473	1812
V	$(3d)^3(4s)^2$	515	1730
Cr	$(3d)^5(4s)^1$	398	1903
Fe	$(3d)^6(4s)^2$	418	1536

Consequently, metallic bonding is most characteristic for alkaline metals having one valence electron. Therefore, the bonding energy and melting point is relatively low for these metals. At the same time, with the growth of the number of valence electrons in metallic bonding, both the bonding energy and the melting point increase, as shown in Table 2.3.

Ionic bonding

Ionic bonding is established between strong electropositive elements (mostly metals) and strong electronegative elements (mostly non-metals). In this bonding, the electropositive element (metal) gives up one or more electrons, while the electronegative element (non-metal) accepts them. Thus, both participant elements of this bonding have closed outermost electron-shell structures. In other words, they are in a stable state. At the same time, the electropositive element (by electron deposition) changes into a positively charged cation and the electronegative element (by accepting electrons) changes into a negatively charged anion.

Ionic bonding forces, that are the consequence of opposite charges, are due to the so-called electrostatic Coulomb attractive forces. This type of bonding force leads to a high bonding energy and significant, non-directional bonding forces.

Sodium chloride (*NaCl*) is a classical example of ionic bonding. During the ionization process, the sodium atom (*Na*) transfers its single valence electron to a chloride atom (*Cl*). This process is illustrated in Figure 2.4.

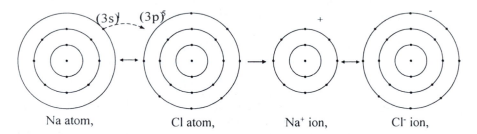

| Na atom, | Cl atom, | Na⁺ ion, | Cl⁻ ion, |

Figure 2.4
Schematic representation of ionic bonding in sodium chloride

After the establishment of ionic bonding, the participating atoms do not possess free electrons. Electrons are strictly localized on the closed electron shells; consequently, both their electrical and thermal conductivity is significantly less than that of materials with metallic bonding. The electron structure developing during the bonding process explains why materials having ionic bonding are quite brittle: they break along well-defined planes practically without any deformation. This is a consequence of the opposite electronic charge of ions.

Analyzing the interaction forces between oppositely charged pairs of ions, we can state that there are both attractive and repulsive forces between them (see Figure 2.5). When the two ions approach each other from a relatively large distance, an attractive force arises between the positively charged ion and the negatively charged electrons of the other ion. This means that they will mutually attract each other. The attractive force may be expressed by Coulomb's law. This is marked $F_{attractive}$ in Figure 2.5. At the same time, as the distance between the ions decreases, their electrons having the same kind of electric charge exert an increasing repulsion force ($F_{repulsive}$). The minimum energy level – i.e. the equilibrium state – corresponds to that distance between the two ions where the attractive and repulsive forces are equal.

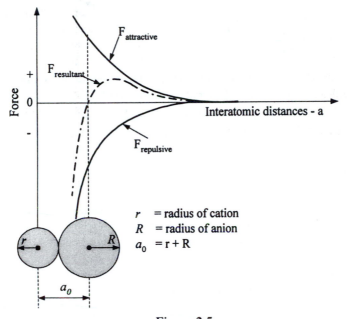

Figure 2.5
Attractive and repulsive forces vs. ionic distances for ionic bonding

The attractive force between the two ions can be calculated from the following formula according to Coulomb's law:

$$F_{attractive} = -\frac{(Z_1 e)(Z_2 e)}{4\pi \in_o a^2} = -\frac{Z_1 Z_2 e^2}{4\pi \in_o a^2}, \qquad (2.4)$$

where Z_1, Z_2 the number of electrons given up and accepted by the atoms when establishing ions (the sign of Z is positive for deposited and negative for accepted electrons)

e the charge of electrons,

a the distance between ions,

$\in_o$ the permittivity of free space, or in other words the dielectric constant of vacuum (its value is 8.85×10^{-12} C^2/Nm^2).

The repulsive force between ions is inversely proportional to the distance between them and can be calculated from the following formula:

$$F_{repulsive} = -\frac{nb}{a^{n+1}},$$
(2.5)

where a the distance between the ions,
         b and n constants (the value of n is generally between 7 and 9).

Summing up the attractive and repulsive forces between ions, we obtain for the resultant force:

$$F_{resultant} = -\frac{Z_1 Z_2 e^2}{4\pi \in_o a^2} - \frac{nb}{a^{n+1}}.$$
(2.6)

Using this equation, the equilibrium distance (corresponding to the minimum energy level) can be determined from the condition that $F_{resultant} = 0$.

The function of the potential energy of ionic bonding can be written as the sum of energies arising from the attractive and repulsive effects, i.e.

$$E_{resultant} = \frac{Z_1 Z_2 e^2}{4\pi \in_o a} + \frac{nb}{a^n}.$$
(2.7)

The first term in Equation (2.7) relates to the energy which is released when the ions get closer to each other (since Z_1 and Z_2 have opposite signs, their product of multiplication and consequently the released energy have negative signs.) The second term in Equation (2.7) represents the energy absorbed when the ions participating in the bonding move closer to each other: it has a positive sign. The resultant energy gives the pure potential energy of the bonding. Its minimum can be found at the equilibrium distance of the ions (a_o), as can be seen in Figure 2.6.

Previously, in the analysis of ionic bonding, we have taken into consideration only one pair of ions. However, ionic bonding is much more complex in a real, three-dimensional solid material with ionic bonding. Therefore, we can give only a brief summary of this.

Practically, ions can be considered as spheres with a radius determined by the outermost electron shell. In these spheres, the electric charge distribution can be characterized by spherical symmetry. When ions are joined in a solid material, according to the symmetrical charge distribution, there is no significant orientation originating from electrostatic attraction. At the same time, geometrical laws also influence the position of ions while ensuring the electrically neutral behavior. Consequently, solid materials with spatial ionic bonding can result in extremely complex structures.

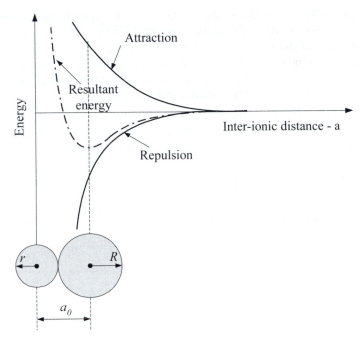

Figure 2.6
Potential energy vs. ionic distances for ionic bonding

To represent the previously mentioned facts, consider two relatively simple solid materials with ionic bonding (Figure 2.7): cesium-chloride (*CsCl*) and sodium chloride (*NaCl*). In Cesium-chloride, there are eight Cl^- ($r_{Cl} = 0.181\ nm$) ions positioned at the corners of a regular cube around a Cs^+ ion ($r_{Cs}=0.169\ nm$) located in the body-center of the cube. In sodium chloride, there are six Cl^- ions positioned in the corner of the cube and at the face-centers of the cube around the Na^+ ion ($r_{Na} = 0.095\ nm$) located in the edge-centers of the cube.

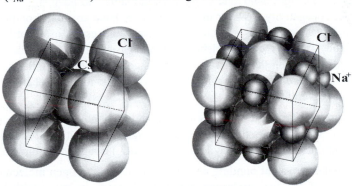

Figure 2.7
Ionic bond order for *CsCl* and for *NaCl* compounds

If we wish to analyze the ratio of cation/anion radii, (which is 0.169/0.181 = 0.934 for *CsCl*, while it is 0.095/0.181 = 0.525 for *NaCl*), the conclusion is that the smaller the ratio of cation/anion radii, the less anion can surround the centrally located cation. Beside the previous, geometrical law, the earlier mentioned requirement – stating that a solid material with ionic bonding must be electrostatically neutral – should still be valid.

The bonding force and melting point of materials with ionic bonding are relatively high, as can be seen from Table 2.4. The table shows the bonding forces and melting temperatures of selected materials with ionic bonding.

Table 2.4 Bonding forces and melting points for selected materials with ionic bonding

Compounds with ionic bonding	Bonding force E (kJ/mol)	Melting point T (^{o}C)
CsCl	649	646
KCl	686	776
NaCl	766	801
BaO	3127	1923
CaO	3583	2580
MgO	3932	2800

Covalent bonding

Covalent bonding is the strongest type of primary bonding. In contrast to ionic bonding (where the bonding exists between strong electronegative and electropositive elements), covalent bonding is typical of elements that are located next to each other on the right-hand side of the periodic table and have small differences in their electronegativity. Two or more atoms may have covalent bonding if they "share" their valence s and p electrons in a way that results in a stable electron configuration which is characteristic of noble gases. If bonding electrons are equally shared between the atoms (symmetrical covalent bonding), the result is an extremely high hardness. This symmetric covalent bonding is characteristic of diamond and several ceramics.

However, the sharing of electrons in covalent bonding is not always completely symmetrical. The hardness in this case is still significant, but compared with diamond, the hardness is significantly less. Quartz (SiO_2) is a good example of this.

The simplest covalent bonding can be found in the hydrogen molecule, where one electron of the two hydrogen atoms participates in the establishment of covalent bonding. Similar covalent bonding is characteristic of several molecules composed of two atoms, as can be seen in Figure 2.8 for F_2, O_2, or N_2 molecules.

In chemistry, the establishment of covalent bonding is denoted with dots around the chemical symbols of connected atoms, as shown in Figure 2.8. From the schematic illustration of Figure 2.8, it is obvious that electrons belong to both atoms participating in the bonding. This figure clearly demonstrates that neutral molecules (i.e. particles with noble gas configuration) are established in such a way that bonding electrons do not separate from their original nuclei and belong to another atom at the same time.

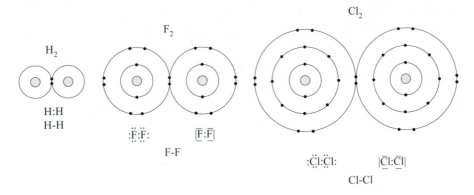

Figure 2.8
Schematic illustration and notation of covalent bonding

From the engineering point of view, those covalent bonds having carbon as one of the participant elements have prominent significance, since carbon is the basic element in many polymers. The carbon atom has the electron configuration $1s^2 2s^2 2p^2$ in its basic state. Due to electron hybridization, carbon can connect with single, double, triple and even fourfold bonds. Four equally stable covalent bonds arranged according to a regular tetrahedron can characterize the diamond version of carbon (see Figure 2.9). The symmetric covalent bonding of carbon results in an extremely high hardness and a high bonding energy ($E = 711$ kJ/mol) and quite a high melting point ($T_{melting} = 3550$ °C) for diamond.

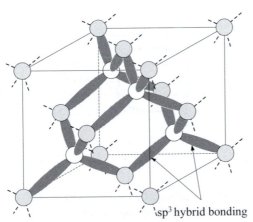

sp³ hybrid bonding

Figure 2.9
The covalent bonding of diamond as a special version of Carbon

Among covalently bonded materials, hydrocarbon molecules are of great significance. They consist of the carbon and hydrogen elements. The simplest hydrocarbon compound is methane (CH_4), in which carbon composes a regular tetrahedron covalent bonding with four hydrogen atoms (Figure 2.10). Though the bonding energy within molecules is quite high in methane ($E = 1650$ kJ/mol), the bonding energy between molecules is very low ($E = 8$ kJ/mol). This is the reason why an extremely weak bonding exists between the molecules and the result is a low melting point ($T_{melting} = -183$ °C).

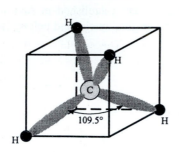

Figure 2.10
The covalent bonding of the
methane molecule

From both the theoretical and the practical point of view, benzene is one of the most significant hydrocarbons with covalent bonding of carbon and hydrogen. Benzene is the basis for several polymers (plastics). When considering its chemical symbol, benzene C_6H_6 should be classified as an unsaturated hydrocarbon; however, these compounds are usually extremely reactive which is not very characteristic of benzene. The explanation can be found in its ring-like structure as can be seen in Figure 2.11. According to this, the six carbon atoms are connected to one another forming a ring. A hydrogen atom belongs to every carbon atom. As carbon atoms have four valences, they are connected over single and double bonds in succession. Thus, it is possible to explain the benzene molecule structure exactly if we introduce the electron theory of chemical bonding.

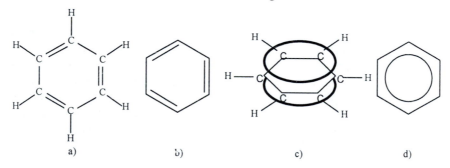

a)   b)   c)   d)

Figure 2.11
Structural formulae for the covalently bonded benzene molecule
a)   Straight-line bonding notation   b) Simplified notation
c) Bonding arrangement indicating the delocalization of the carbon-carbon
bonding electrons   d) Simplified benzene-ring notation

Based on this structure, it has been proved that the six carbon atoms of benzene are in the same plane and compose a regular hexagon. Two of the four valence electrons of the carbon atom establish a so-called σ-bonding with the neighboring carbon atoms and one with a hydrogen atom. This hexagonal-shaped bonding is often termed the benzene ring. The bonding angles are 120°. This arrangement forms the σ-skeleton of the benzene molecule. The fourth electron - which was left out of the σ-bonding (altogether there are six electrons belonging to the six carbon atoms) - composes three delocalized, π-bonds. There is a negatively charged cloud originating from these bonding. This cloud is equally distributed in the force fields of the six carbon atoms, composing a π-electron sextet. The stability of this is similar to the electron octet of noble gases. This explains the lack of reactivity in benzene compared to other hydrocarbons. It also explains its resistance to oxidation and several other properties of benzene.

In part a) of Figure 2.11, the so-called straight-line bonding notation of the benzene molecule is shown, while part b) illustrates its simplified notation. Part c) of Figure 2.11 represents the bonding arrangement indicating the delocalization of carbon-carbon bonding electrons in the benzene ring and part d) is the schematic representation of the benzene ring itself.

2.1.3.2. Secondary (weak) bonding - van der Waals bonding

In the case of primary (strong) bonding, we could see that the common characteristic is the significant role of valence electrons participating in the bonding. It is of equal importance that the energy reduction of the system motivates the establishment of bonding.

Secondary bonds are much weaker than primary ones. (The bonding energy ranges from 4 kJ/mol to 40 kJ/mol) and the bonding is motivated by the attractive force between electronic dipoles in atoms and molecules. Secondary bonding established with the help of electronic dipoles has two main types: the so-called *temporary (fluctuating)* and *permanent dipoles.*

An electronic dipole moment is created when two charges opposite in sign and equal in magnitude are separated, as illustrated in part a) of Figure 2.12. Furthermore, electric dipoles can arise in atoms or molecules if positive and negative charge centers exist. The permanent dipole bond has a special type, the so-called *hydrogen bonding*, where hydrogen makes up a dipole moment together with electronegative elements (*O, N, F, Cl*). A good example is the water molecule (H_2O), represented by part b) of Figure 2.12.

The dipole moment can be defined as the product of the multiplication of the charge volume and the distance between charges. It is represented by the

$$\mu = Q\,d \tag{2.8}$$

relationship, where μ is the dipole moment, Q is the magnitude of electric charge, d is the distance between the charge centers.

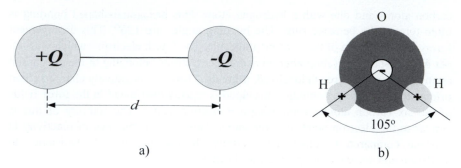

Figure 2.12
a) Electric dipoles   b) Water molecule as a permanent dipole

The unit of dipole moment is the Coulomb meter (*Cm*). Electrostatic (Coulomb type) attractive forces occur between electric dipoles. Compared to primary bonds, they are weak but important, especially in those cases, when they compose the only bonding forces between atoms or molecules.

Temporary and permanent dipole bonds are called *van der Waals bonds*, and bonding forces in them are the van der Waals bonding forces. Van der Waals bonding forces are present in every material to a certain extent, but their role is especially important in polymers where they often represent the only bonding force.

3. FUNDAMENTALS OF CRYSTAL STRUCTURES

3.1. The ideal crystal structures

It is obvious from the previous section that in the engineering practice, the basic characteristics of solid materials are determined by the atoms (molecules) that compose them and the atomic order and bonding existing among the atoms. Materials possessing regular crystal order in the solid state are regarded as one of the most important groups of materials. This group includes metals (pure metals and their alloys), most ceramics, and even certain polymers. This is why we deal in detail with the structure of crystals and their most important fundamentals in this section.

3.1.1. Crystal systems, crystal geometry

In crystals, atoms are positioned at distinct points of geometrical configurations (*lattice points*), determined by the laws of crystallography. In crystalline solids, there is a regular, *long-range order in the positions of atoms*. The atoms are located in a repetitive three-dimensional pattern. Therefore, it is often convenient to subdivide the crystal structure into small characteristic entities called *unit cells*. The *unit cell* is the smallest unit of the crystal structure possessing all the characteristic features of the given crystal system. Thus, by describing the unit cell, the crystal system itself is unambiguously defined.

Let us denote the three coordinate axes by *x, y* and *z,* respectively (Figure 3.1). Due to the regularly repeating order of atoms, three distances measured in the three directions of the crystalline axes (denoted by *a, b,* and *c*) and the three axial angles (denoted by α, β and γ) are required and sufficient to describe any crystal structure. These parameters are characteristic of the crystal structure, and they are therefore called *lattice parameters* or *lattice constants*. These parameters provide the unambiguous description of a crystal system; they are characteristic of any material and their value is constant for a chosen material.

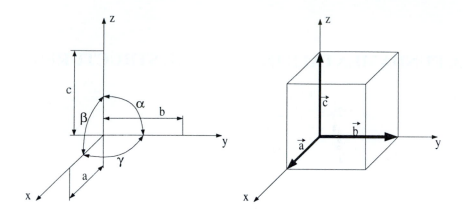

Figure 3.1
Coordinate system for the description of crystal systems with the applied symbols

It also follows from the previously introduced facts that any crystal system can be described by three crystal vectors, too. Denoting these vectors by $\vec{a}, \vec{b}$ and $\vec{c}$ vectors, they can be characterized by the distances measured in the directions of the x, y and z axes and the unit vectors in the axial directions $(\vec{e}_1, \vec{e}_2, \vec{e}_3)$ by the following equations:

$$\vec{a} = a\,\vec{e}_1,$$
$$\vec{b} = b\,\vec{e}_2, \qquad\qquad\qquad (3.1)$$
$$\vec{c} = c\,\vec{e}_3.$$

Atoms are equivalent in the crystal lattice. Consequently, in an ideal crystal, the neighborhood of any lattice point is completely the same. This is illustrated in Figure 3.2, where the unit cell of a *NaCl* crystal and the regular periodicity of atoms can be seen.

Furthermore, this regular order is repeated in the three directions of space and this regular atomic arrangement – in an ideal crystal – exists over large distances. This is called *long-range atomic order*, which is the most characteristic feature of solid crystalline materials.

All crystal systems existing in nature can be grouped into one of the 7 basic Bravais crystal systems. The unit cells of the basic crystal systems are shown in Figure 3.3 and their characteristic parameters are summarized in Table 3.1.

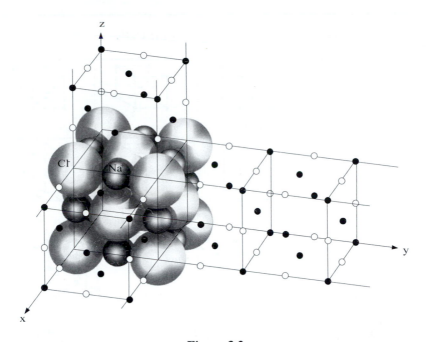

Figure 3.2
Illustration of long-range atomic order of a crystal system by the example of the
ionic bonded *NaCl* crystal structure

Table 3.1 Lattice parameters of the seven basic crystal systems

Crystal systems	Lattice parameters	
	Axial distances (a,b,c)	Axial angles (α, β, γ)
Cubic	$a = b = c$	$\alpha = \beta = \gamma = 90^o$
Tetragonal	$a = b \neq c$	$\alpha = \beta = \gamma = 90^o$
Hexagonal	$a = b \neq c$	$\alpha = \beta = 90^o \neq \gamma = 120^o$
Orthorhombic	$a \neq b \neq c$	$\alpha = \beta = \gamma = 90^o$
Rhombohedral	$a = b = c$	$\alpha = \beta = \gamma \neq 90^o$
Monoclinic	$a \neq b \neq c$	$\alpha = \gamma = 90^o \neq \beta$
Triclinic	$a \neq b \neq c$	$\alpha \neq \beta \neq \gamma \neq 90^o$

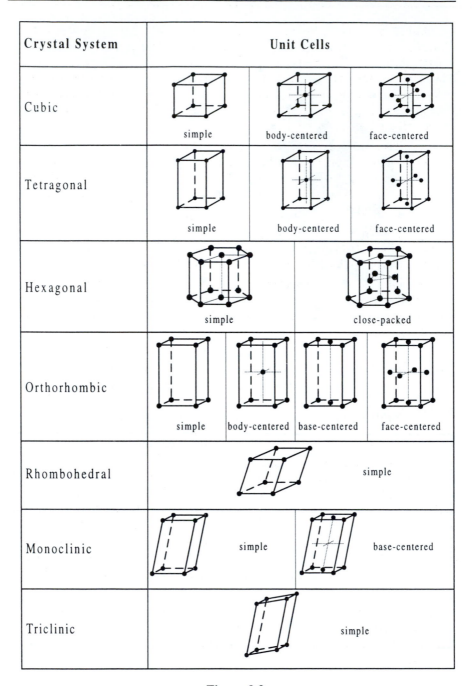

Crystal System	Unit Cells			
Cubic	simple	body-centered	face-centered	
Tetragonal	simple	body-centered	face-centered	
Hexagonal	simple		close-packed	
Orthorhombic	simple	body-centered	base-centered	face-centered
Rhombohedral	simple			
Monoclinic	simple	base-centered		
Triclinic	simple			

Figure 3.3
The Bravais lattices of crystal structures

Most of the crystalline solid materials used in industry (mainly metals and their alloys) crystallize according to the different modifications of three basic types of crystal systems, i.e. the *cubic, tetragonal* and *hexagonal* crystal systems. Therefore, we shall deal in detail only with these three crystal systems. However, before going into details, we shall summarize the previously mentioned basic concepts of crystals and their definitions considered extremely important for further understanding will be given.

The *lattice* is the regular order of atoms in space. This is repeated in the three directions of space, and the neighborhood of every lattice point is equivalent. The regular repetition of atoms in space is a characteristic feature of the crystal system and can be defined with regular geometrical configurations (e.g. regular cube, tetragon, and hexagon).

The *lattice point*s are those distinct points of the lattice (distinct place of geometrical configuration, such as the corner point, the body center, face centers, edge centers) where the atoms of the given crystal system are located.

The ***unit cell*** is the smallest unit of the crystal structure possessing all characteristic features of the given crystal system by virtue of its geometry. (In other words, the geometrical configuration of a given crystal system and the atoms located at the lattice points together form the unit cell).

3.1.1.1. The cubic crystal system

The geometrical configuration of the cubic crystal system is a regular cube with the lattice parameter a. Since, in cubic crystals $a = b = c$ and $\alpha = \beta = \gamma = 90^{\circ}$ are valid, only one lattice parameter, the a lattice constant, is sufficient to describe the cubic lattice.

Within the cubic crystal system, there are further variations depending on the distinct points of the geometrical configuration where atoms of the given material are positioned. In every crystal system, it is a basic requirement to have atoms at the corner points of the lattice. If there are atoms only at the corner points, the crystal structure is called a *simple* or *primitive* lattice. If we also find atoms at the body center, the crystal structure is called *body-centered*, while the crystal structure with atoms at the face centers are called *face-centered*. In the cubic crystal system, all three modifications – i.e. simple, body-centered and face-centered – can be found.

It is important to mention that about 90 % of industrially important metals crystallize according to one of the modifications of the cubic crystal system. In the next section, we introduce the most important properties of the three crystal modifications of the cubic crystal system.

Simple cubic crystal structure

In Figure 3.4, two types of graphical illustration – generally applied in crystallography – can be seen for the simple cubic structure. In part *a)*, the so-called *point model*, while in part *b)* the so-called *hard ball model* can be seen. The hard ball model is more realistic than the former one, since atoms are positioned in distinct lattice points as contacting spheres along their outermost electron shells – for the simple cubic crystal, actually at the corner points. For more complex crystal structures, the point model gives a clearer overview of the crystal, and this is why both types of graphical illustration are widespread in crystallography.

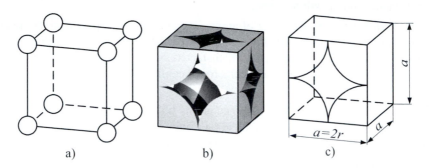

a)   b)   c)

Figure 3.4

Simple cubic crystal
a)   Point model b) Hard ball model
c) Relation between the atomic radius and the lattice parameter

Since atoms can be regarded as spheres contacting along their outermost electron shells, the relationship between the atomic radius (r) and the lattice parameter (a) can be easily calculated according to part c) of Figure 3.4, i.e.

$$a = 2r.$$ (3.2)

In crystallographic calculations, another important value is the number of atoms belonging to a unit cell (N). For its determination, part b) of Figure 3.4 provides a graphical aid. It clearly shows that each atom at the lattice points of the cube belongs to eight unit cells; therefore, only 1/8 of each atom can be considered to belong to one unit cell.

Naturally, this also follows from the periodicity of the lattice structure illustrated in Figure 3.5. From this figure, it is obvious that each atom at the corner points of the cube belongs to eight neighboring cells. Therefore, each atom at the corner points can be taken into account with the value of 1/8.

Considering that each cube has eight atoms at the corner points, the number of atoms belonging to a unit cell can be calculated from the following equation:

$$N = 8 \times \frac{1}{8} = 1.$$ (3.3)

Thus, in simple cubic crystal structures, the number of atoms belonging to a unit cell is $N = 1$.

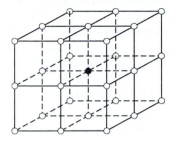

Figure 3.5
Schematic illustration for determination of the value of an atom belonging to a unit cell at the corner point of the lattice

A characteristic feature of metals is that atoms tend to occupy positions as close as possible to one another and to form a lattice structure as dense as possible in this way. For the characterization of different crystal structures, the atomic packing factor is generally used, which indicates the part of the volume of a unit cell occupied by atoms. According to this, the atomic packing factor (*APF*) can be determined by the following definition:

$$APF = \frac{V_a}{V_c} = \frac{The\ volume\ of\ atoms\ belonging\ to\ the\ unit\ cell}{The\ volume\ of\ the\ unit\ cell} \tag{3.4}$$

The volume of atoms belonging to a unit cell can be calculated from the following equation:

$$V_a = N\frac{4r^3\pi}{3}, \tag{3.5}$$

where N is the number of atoms belonging to a unit cell (the other part of the product is the volume of the atom with a radius r. The atom is assumed to be a regular sphere).

The volume of the unit cell – as a regular cubic lattice – can be calculated from the equation $V_c = a^3$, which results in the expression

$$V_c = a^3 = 8r^3 \tag{3.6}$$

considering the relationship between the lattice parameter *(a)* and the atomic radius *(r)* given by Equation (3.2). The atomic packing factor, after simple substitutions of Equations (3.5) and (3.6) into Equation (3.4), leads to:

$$APF = \frac{N\frac{4r^3\pi}{3}}{8r^3} = \frac{N\pi}{6}. \tag{3.7}$$

Since, for a simple cubic lattice, $N = 1$, the value of the atomic packing factor for the simple cubic crystal is equal to

$$APF = \frac{\pi}{6} = 0.52. \tag{3.8}$$

This means that in the simple cubic crystal structure, slightly more than half of the lattice (exactly 52 %) is occupied by atoms. As mentioned earlier, in metals atoms tend to occupy positions as close as possible to one another and to form a lattice structure as dense as possible. This is the explanation for the fact that we can hardly find any metal crystallizing according to the simple cubic structure. Among elements having great practical importance, phosphorus (P) is one of the elements that has an allotropic modification crystallizing according to this crystal structure.

An important parameter in crystallography is the so-called *coordination number (C)*, which refers to the number of nearest, neighboring atoms being at an equal distance from a given atom. (In other terms, the coordination number gives the number of neighboring atoms contacting the chosen one.)

To determine the coordination number, we again refer to part b) of Figure 3.4. From this figure, we can see that any atom – considering the regular spatial repetition – is in contact with six neighboring atoms, thus the value of the coordination number is six, i.e.$C=6$. To determine the value of the coordination number, select the atom located at the right hand corner point of the upper face of the cube in Figure 3.4. With the help of this figure, we can easily understand that this atom is contacting four other atoms in the upper face plane of the cube. There is one further atom at the corner points above and below this plane, exactly at one lattice parameter distance, i.e. altogether $4+2 = 6$ atoms are for the same distance in contact with the selected atom. Therefore, the coordination number is equal to six.

Body-centered cubic crystal structure – bcc crystals

Two kinds of illustrations for the body-centered cubic lattice can be seen in Figure 3.6. In contrast to simple cubic crystals, the body-centered cubic crystal has a further atom at the cube center beside the atoms located at all eight corners. The figure clearly illustrates that the atom in the body center is located inside the lattice. It means that it belongs completely to the examined unit cell. Therefore, the number of atoms belonging to a unit cell can be calculated from the following equation:

$$N = 8 \times \frac{1}{8} + 1 = 2 .$$
(3.9)

The relationship between the lattice parameter and the atomic radius can be determined by the principle that atoms are contacting spheres along their outermost electron shells. From parts *b)* and *c)* of Figure 3.6, it can be seen that in body-centered cubic crystals atoms contact one another along the cube diagonals. Therefore, the following relationship may be written between the lattice parameter and the atomic radius:

$$a = \frac{4r}{\sqrt{3}} .$$
(3.10)

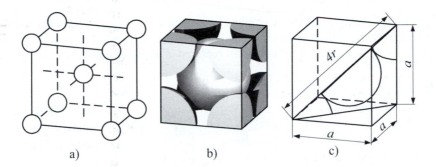

Figure 3.6
Body-centered cubic crystal
a) Point model b) Hard ball model c) Relationship between the atomic radius and
the lattice parameter

From the basic equation of the atomic packing factor (3.4), substituting data and relationships valid for the body-centered lattice, the following equation can be derived:

$$APF = \frac{N \dfrac{4r^3\pi}{3}}{a^3} = \frac{N \dfrac{4r^3\pi}{3}}{\dfrac{64\,r^3}{3\sqrt{3}}} \ . \tag{3.11}$$

Substituting the value $N = 2$ into equation (3.11) and performing the calculations, the following result can be obtained for the atomic packing factor:

$$APF = \frac{\pi\sqrt{3}}{8} = 0.68 \ . \tag{3.12}$$

From this equation, it is obvious that 68 % of the body-centered cubic lattice is occupied by the atoms. Consequently, the body-centered crystals possess a significantly higher density than the simple cubic ones. This crystal structure can be found in a large number of metals having great practical importance, such as α-Fe, *Cr, W, V, etc.*

Based on part *b)* of Figure 3.6, the coordination number can be easily determined. It can be seen, for example, that the atom at the body center is in contact with eight neighboring atoms at the corner points of the lattice; therefore, the coordination number is equal to 8, i.e. $C = 8$. Obviously, this is also valid for any arbitrarily chosen atom.

Face-centered cubic crystal structure – fcc crystals

In Figure 3.7, two types of graphical illustrations of the face-centered cubic crystal can be seen. In face-centered cubic lattices, in addition to the atoms located at the corners, there are six further atoms centered at all the faces of the cube.

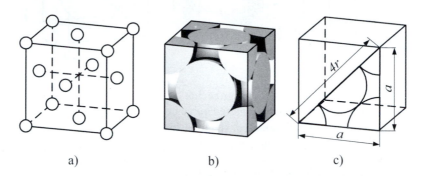

a)   b)   c)

Figure 3.7
Face-centered cubic crystal
a) Point model b) Hard ball model c) Sketch to illustrate the relationship between the atomic radius and the lattice parameter

Any atom located at the face center belongs to two unit cells, therefore half of these atoms may be assigned to a given unit cell. Since there is an atom at all six face centers, the number of atoms belonging to a unit cell can be calculated from the following equation:

$$N = 8 \times \frac{1}{8} + 6 \times \frac{1}{2} = 4 \ .$$

(3.13)

The relationship between the lattice parameter and the atomic radius can be determined based on the principle that atoms are contacting spheres along their outermost electron shells. In part *b)* and *c)* of Figure 3.7, it can be seen that, in face-centered crystals, atoms are in contact along the face diagonals. Therefore, the following equation can be written between the lattice parameter (*a*) and the atomic radius (*r*):

$$a = \frac{4r}{\sqrt{2}} \ .$$

(3.14)

From the basic equation of the atomic packing factor (3.4), substituting the data and relationships valid for the face-centered lattice, the following equation can be derived:

$$APF = \frac{N\frac{4r^3\pi}{3}}{a^3} = \frac{N\frac{4r^3\pi}{3}}{\frac{64r^3}{2\sqrt{2}}} \ . \tag{3.15}$$

Substituting the value $N = 2$ into equation (3.15) and performing the calculations, the following result can be obtained for the atomic packing factor:

$$APF = \frac{\pi\sqrt{2}}{6} = 0.74 \ . \tag{3.16}$$

This means that 74 % of the face-centered cubic lattice is occupied by atoms. This version of the cubic lattice has the densest structure. This structure can be found in many metals having great practical importance, such as γ-Fe, Al, Cu, Ni, etc.

From part *b)* of Figure 3.7, the value of the coordination number can be determined as follows: select, for example, the atom located at the center of the right-hand side face of the cube. This atom is in contact with the four corner atoms in this plane (the distance between them and the selected atom is $t = a\sqrt{2}/2$). The selected atom is also in contact with four further atoms in the planes to the right and left being at equal distance from the selected plane (half of the lattice parameter, i.e. $d=a/2$). All the atoms in these two planes can be found at the same distance ($t = a\sqrt{2}/2$) from the selected atom; therefore, the number of neighboring atoms being at an equal distance from the selected one is equal to 3x4 = 12. This means that the coordination number for the face-centered crystal is $C = 12$.

Diamond lattice, as a special crystal structure in the cubic system

Beside the simple, body-centered and face-centered cubic crystal structures, there is also a special cubic crystal structure. This is called the *diamond lattice*, since diamond, as an allotropic modification of carbon, crystallizes according to this crystal structure. The geometrical configuration of the diamond is also a regular cube.

The diamond lattice can be visualized in the simplest way as follows: in the face-centered cubic crystal, atoms are located at the corner points and at the face centers. In the diamond lattice, there are four further atoms located inside the lattice. Their centers can be found on the body diagonals at a height of 1/4 and 3/4 respectively, measured from the base plane.

In part *a)* of Figure 3.8, the spatial representation of the diamond lattice can be seen. In this illustration, the covalent bonding structure between the atoms can be easily recognized. In part *b)* of Figure 3.8, a so-called *projection view* of the diamond lattice is shown, where the projection of its spatial structure on the basic plane is illustrated. In this figure, the symmetry of atoms located on the body diagonals can be recognized. (Numbers beside the atoms at the lattice points refer to the position of the lattice point measured from the base-plane of the cube. For the sake of simplicity, the value of the lattice parameter is taken as a unit value. As we can find atoms at the corners of the base and the top plane of the cube, their relative positions are 0 and 1. Atoms on the side faces are located at a height of 1/2 of the lattice parameter. The positions of atoms alternating inside the lattice at a height of 1/4 and 3/4 of the body diagonals are denoted by the numbers 1/4 and 3/4, respectively.)

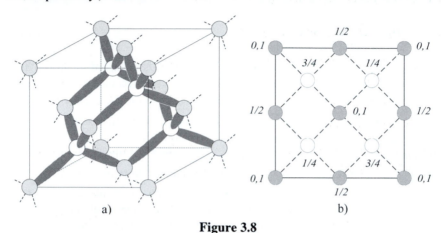

Figure 3.8
The diamond crystal lattice
a) Spatial illustration with covalent bonding   b) Projection view

The number of atoms belonging to a unit cell in the diamond crystal can be determined easily based on the face-centered structure. As known, the number of atoms is 4 in the face-centered cubic crystal. In the diamond lattice, there are 4 further atoms inside the lattice: they belong fully to that unit cell. Hence, the number of atoms belonging to the unit cell in the diamond lattice can be determined by the following equation:

$$N = 8 \times \frac{1}{8} + 6 \times \frac{1}{2} + 4 \times 1 = 8 .$$
(3.17)

The relationship between the atomic radius of carbon and the lattice parameter of the diamond lattice can be derived by considering that the body diagonal should be equal to four times the diameter of the carbon atom, i.e.:

$$a\sqrt{3} = 4d_c ,$$
(3.18)

where $d_C = 0.154\ nm$ is the value of the carbon atom diameter. Substituting this value into Equation (3.18), the following expression can be obtained for the lattice parameter of the diamond:

$$a = \frac{4 \times 0.154}{\sqrt{3}} = 0.356nm .$$
(3.19)

Beside the diamond modification of *carbon, silicon* and *germanium*, there are two further elements, which crystallize according to the diamond lattice. It is worth mentioning that all these elements (i.e. *C, Si* and *Ge*) crystallizing according to the diamond lattice, can be found in column IV of the Periodic Table and have 4 valence electrons. Among metals, *α-Sn* at a low temperature (i.e. when $T < 13.2\ ^{o}C$), as well as *GaAs,* which are applied widely in the semi-conductor industry, both have diamond crystal lattices.

3.1.1.2. The tetragonal crystal system

The geometrical form of tetragonal crystals is a regular tetrahedron that may be characterized by two different lattice parameters:

$$a = b \neq c \quad \text{and}$$

$$\alpha = \beta = \gamma = 90^{\circ}.$$

(3.20)

Consequently, it is sufficient to know the *a* and *c* lattice constants to describe the tetragonal crystal lattice.

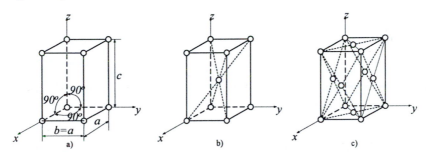

Figure 3.9
Modifications of tetragonal crystals
a) simple *b)* body-centered *c)* face-centered tetragonal crystals

In Figure 3.9, the schematic illustration of the three tetragonal crystal modifications can be seen. In part *a)* the simple tetragonal crystal is shown. In this crystal, atoms are located only at the corner points of the lattice. Metals with practical significance do not crystallize in this way.

In part b), the body-centered modification of the tetragonal system can be seen. In this crystal lattice, beside the atoms at the corner points, one further atom can be found at the intersection point of the body diagonals (i.e. in the geometrical body-center of the lattice). An allotropic modification of tin (so-called *β–Sn*) crystallizes according to the body-centered tetragonal crystal system.

In part c) the face-centered modification is shown. In this crystal, atoms are located at the corner points and at the intersection point of the face diagonals, (i.e. at the face centers of the lattice). As an example, indium (*In*) which crystallizes according to the face-centered tetragonal lattice should be mentioned .

3.1.1.3. The hexagonal crystal system

The geometrical form of the hexagonal crystal lattice is a regular hexahedron (see Figure 3.10).

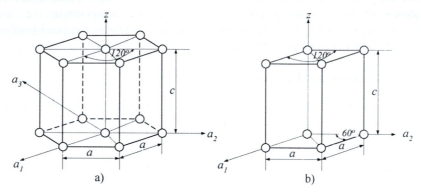

Figure 3.10
Hexagonal crystals – simple hexagonal lattice
a) four axes hexagonal representation b) three axes rhombic representation

As noted in the general introduction to crystallography, three directions and three distances in the direction of the coordinate axes are sufficient for the unambiguous description of any crystal lattice. This is also valid for hexagonal crystals. However in crystallography, a four-axis coordinate system is applied for the description of hexagonal crystals, since it reveals more clearly the hexagonal symmetry of this system. This is illustrated in part a) of Figure 3.10. In the four-axis coordinate system, the lattice parameters are:

$$a_1 = a_2 = a_3 = a \neq c,$$
$$\alpha_1 = \alpha_2 = \alpha_3 = 120^o \, ; \gamma = 90^o. \tag{3.21}$$

This means that the angles between the three axes in the so-called base-plane are 120^0 and the lattice constants in these directions are also equal. The fourth direction is perpendicular to the base-plane determined by the other three directions. The lattice constant in the fourth direction is different from the lattice constant in the other three directions. Therefore, it is sufficient to give the a and c lattice parameters for the description of the hexagonal lattice.

Actually, the three-axis representation is the fundamental one, since the hexagonal representation is composed of three rhombohedrons as elementary units, shown in part b) of Figure 3.10. This explains the unusual fact, that in the hexa-gonal representation of simple hexagonal crystals, atoms can be found beside the corner points at the face centers, too. From the previous facts, it naturally follows that the hexagonal description contains redundant information, as we will see later in the section dealing with the crystallographic planes and directions. It should be noted that no metals of practical importance crystallize according to the simple hexagonal system due to the low value of the atomic packing factor. However, it should also be mentioned that graphite has a very similar crystal structure.)

A modification of the hexagonal crystal system can be found in a great number of metals. This is termed the *hexagonal close-packed crystal*, and it is illustrated in Figure 3.11. All parameters examined in the cubic system can also be determined for hexagonal crystals. However, it requires significantly more complex calculations and further geometrical considerations.

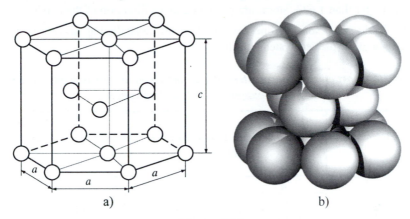

a)   b)

Figure 3.11
Hexagonal close-packed crystal: a) Point model b) Hard ball model

In order to determine the numbers of atoms belonging to a unit cell, we have to consider the following facts: atoms at the corner points belong to six neighboring lattices, therefore they contribute with one-sixth of an atom to a unit cell. Atoms at the face centers belong to two adjacent lattices, therefore they contribute with one-half of an atom to a unit cell, while atoms within the lattice should be considered with their total values. Thus, the number of atoms belonging to a unit cell can be calculated from the following equation:

$$N = 12 \times \frac{1}{6} + 2 \times \frac{1}{2} + 3 \times 1 = 6,$$
(3.22)

i.e. the number of atoms belonging to a unit cell in the hexagonal close-packed crystal is equal to six. Using relatively simple geometrical calculations, it can be easily proved that for an ideal hexagonal close-packed lattice – where atoms are as close as possible to one another – the following relationship should be valid for the lattice constants: c/a=1.633. In this case, in the layer between the bottom and top planes of the lattice (i.e. inside the lattice) there are three atoms located close to the atoms of the top and bottom faces. From this, we can easily understand that any atom in the lattice is in contact with altogether 12 neighboring atoms; therefore, the coordination number for the hexagonal close-packed lattice is C=12.

To prove this let us select the atom in the center of the top face (see Figure 3.11). This atom is in contact with six atoms located at the corner points of the same plane and with three atoms in the upper and three further atoms in the lower layers at a distance of $c/2$. This means that altogether 12 atoms are contacting one another; therefore, the number of nearest neighboring atoms, i.e. the coordination number, is twelve.

If the ratio of lattice constants c/a is different from the ideal 1.633 value, the atomic arrangement will also be different from the ideal one. If the c/a ratio is smaller than 1.633, atoms are compressed in the direction of the c axis, while if the ratio is larger than 1.633, atoms are elongated in the direction of the c axis.

In Table 3.2, some properties of elements crystallizing according to the hexagonal close-packed system are shown.

Table 3.2 Characteristic parameters of selected elements crystallizing according to the hexagonal close-packed system

Element	Value of lattice parameters (nm)		Atomic radius	Relation between lattice parameters
	a	*c*	*r*	*c/a*
Cadmium	0.2973	0.5618	0.149	1.890
Zinc	0.2665	0.4947	0.133	1.856
Magnesium	0.3209	0.5209	0.160	1.623
Cobalt	0.2507	0.4069	0.125	1.623
Zirconium	0.3231	0.5148	0.160	1.593
Titanium	0.2950	0.4683	0.147	1.587
Beryllium	0.2286	0.3584	0.113	1.568

From this table, it can be seen that *Cd* and *Zn* have larger and *Mg, Co, Ti* and *Be* have smaller c/a ratios than the ideal 1.633 value. The relationship between the lattice parameters (a and c) and the atomic radius (r) for the ideal hexagonal close-packed crystal (characterized by the $c/a=1.633$ ratio) can be determined as follows:

$$a = 2r \text{ and}$$
$$c = 1.633 \times 2r = 3.266r. \tag{3.23}$$

For the calculation of the atomic packing factor, first determine the volume of the hexagonal lattice. Using the notations shown in Figure 3.12, the area of the base hexagon can be calculated from the following equation:

$$A = 3a^2 \sin 60^o = \frac{3\sqrt{3}}{2} a^2, \tag{3.24}$$

while the volume of the lattice can be determined from the expression:

$$V_c = A \, c = \frac{3\sqrt{3}}{2} a^2 c \,. \tag{3.25}$$

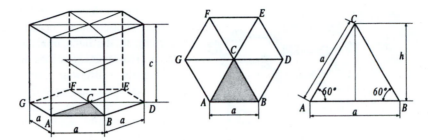

Figure 3.12
Schematic illustration for calculation of the volume of a hexagonal lattice

Considering that $c/a=1.633$, the volume of the lattice is equal to

$$V_c = \frac{3\sqrt{3}}{2} 1.633 a^3 \,. \tag{3.26}$$

The atomic packing factor can be determined using the previous results:

$$APF = \frac{V_a}{V_c} = \frac{N \dfrac{4r^3\pi}{3}}{\dfrac{3\sqrt{3}}{2} 1.633 a^3} \,. \tag{3.27}$$

Substituting the relationship $a=2r$ into the Equation (3.27), we obtain after elementary transformations:

$$APF = \frac{N\pi}{9\sqrt{3} \times 1.633} \,. \tag{3.28}$$

Substituting the number of atoms belonging to a unit cell in a hexagonal close-packed crystal ($N = 6$), we get the value for the atomic packing factor:

$$APF = 0.74 \,. \tag{3.29}$$

It can be seen that it is equal to the value of the atomic packing factor of the face-centered cubic crystal.

Previously, we could see that the coordination number of the hexagonal close-packed crystal ($C = 12$) is also equal to the coordination number of the face-centered cubic crystal. From this, two significant conclusions can be drawn. The first is a general law of crystallography stating that if the coordination numbers of different crystal systems are equal to each other, their atomic packing factors must be equal, too. The second conclusion formulates another important relationship between face-centered cubic and hexagonal close-packed crystals: these two crystal systems are composed of crystallographic planes having the same atomic arrangement, but the order of the planes is different. This rule will be later illustrated by Figure 5.13 in Chapter 5.

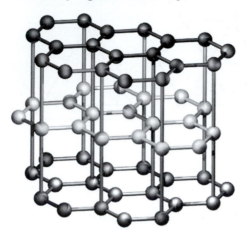

Figure 3.13

Scheme of graphite layer structure of hexagonal crystal structure

Graphite, which is of great practical importance, crystallizes according to the hexagonal crystal system. However, graphite has a significantly different structure from other hexagonal crystals. This explains several special characteristics of graphite. Among carbon atoms located in the graphite layers with hexagonal crystal structure there is a strong σ-bonding, which has already been introduced when we mentioned the covalently bonded benzene (see section 2.1.3.1). Due to this atomic bonding, the delocalized valence electrons are able to move relatively easily in a two-dimensional structure similarly as in benzene. At the same time, there is only relatively weak, secondary bonding between neighboring graphite layers. These layers can be regarded as special "*planar macro-molecules*" and this structure results in several unusual properties. As an example, the significant electrical and thermal conductivity of graphite in the plane (*two-dimensional conductivity*) should be mentioned, while it shows an extremely poor conductivity in the third direction. This means that graphite has significant *anisotropy*, i.e. the properties of graphite are different depending on the direction in which they are measured.

Further special characteristics of graphite may also be connected to its special crystal structure. The extremely strong (two-dimensional) σ-bonding retains the solid crystal structure in the layer up to more than 2200 °C, whilst the weak (secondary) bonding between the planes permits graphite layers to slide over each other applying relatively small shear stresses. This is the reason why graphite, as a solid material, has good lubricating characteristics.

3.1.2. Crystallographic planes and directions

To perform crystallographic calculations, it is often necessary to specify crystallographic planes and crystallographic directions. Crystallographic planes are denoted by the so-called Miller indices (in hexagonal crystal systems, the Miller-Bravais indices) while crystallographic directions are denoted by direction vectors (direction vector components).

3.1.2.1. Crystallographic planes – Miller indices

For determination of Miller indices of crystallographic planes, consider a general plane in the x, y, z coordinate system shown in Figure 3.14.

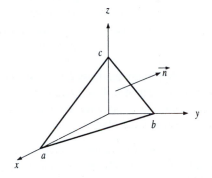

Figure 3.14

Schematic illustration to derive crystallographic plane indices

The vector equation of the plane, known from vector algebra, can be written as the following equation:

$$(\vec{r} - \vec{r}_o)\,\vec{n} = 0 \ , \tag{3.30}$$

where

$\vec{r}$ is the position vector of an arbitrary point,

$\vec{r}_o$ is the position vector of a fixed point in the plane examined,

$\vec{n}$ is the normal vector of the plane.

For description of a plane, the scalar equation using the coordinate intercepts is also widely used which leads to the following equation:

$$\frac{x}{a} + \frac{y}{b} + \frac{z}{c} = 1 \ , \tag{3.31}$$

where

x, y and z are the running coordinates,

a, b and c are the coordinate intercepts with the x, y and z coordinate axes, respectively.

All the above variables are suitable for the unambiguous description of crystallographic planes. In 1839, William H. Miller proposed a special notation for crystallographic planes based on the coordinate intercepts given in equation (3.31). This method is widely used in crystallography to describe crystallographic planes and it is called the Miller notation system. The Miller notation system applies the so-called Miller indices. The procedure of determining the Miller indices is as follows:

1. First, the plane should be displaced with a parallel displacement (if necessary) to a position, where it does not pass through the origin of the coordinate system.

2. The a, b and c intercepts of the plane with the x, y and z coordinate axes should be determined.

3. The reciprocals of these numbers should be determined; these are denoted with the letters h, k and l, respectively, i.e.:

$$\frac{1}{a} \to h; \quad \frac{1}{b} \to k; \quad \frac{1}{c} \to l .$$
(3.32)

4. The reciprocals usually result in fractions. Applying mathematical transformations (multiplication or division), the smallest set of h, k and l values as the simplest integer numbers should be determined. These three numbers in round brackets $(h\,k\,l)$ give the Miller indices of the examined plane.

It follows from the mathematical transformations mentioned above (multiplication, division) that the real size of the lattice has no influence on the Miller indices. Therefore, in crystallographic calculations, the lattice constant is taken as unit length.

This procedure is generally valid for the derivation of Miller indices of crystallographic planes. However, due to the differences between the three- and four-axis coordinate description, it is advisable to perform the analysis for the cubic and the hexagonal crystal system separately.

Miller indices in cubic crystal structures

For the analysis, consider the cubic lattices shown in Figure 3.15. In part a) an arbitrarily selected side face of the cube is indicated with shading. According to the previously described procedure for the derivation of the Miller indices, first we have to determine the intersection points formed by the plane and the coordinate axes. The intersection point between the plane and the x-axis is $a=1$. Since the plane is parallel to the y and z-axes, these intercepts may be regarded as an infinite value, therefore $b=\infty$ and $c=\infty$. Taking the reciprocals of the above values, we get $h=1$, $k=0$ and $l=0$; therefore, the Miller indices for the selected plane are (1 0 0).

The Miller indices can be determined similarly for any crystallographic planes. For example, the Miller indices of the right hand side face of the cube are (0 1 0), those of the top face are (0 0 1). It is also obvious that the side faces of the cube have the same atomic arrangement. In other terms, they can be regarded as *crystallographically equivalent* planes. It also follows from the previous analysis that the Miller indices of these planes consist of the same three numbers, i.e. (1 0 0). Crystallographically equivalent planes are called *family planes* and are denoted by the same three numbers in braces. Thus, the notation {1 0 0} represents all the side faces of the cube collectively, as a family plane.

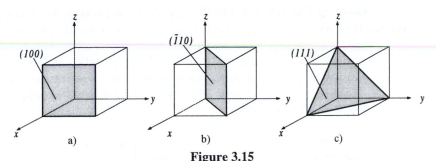

Figure 3.15
Miller indices of cubic crystal planes

In part b) of Figure 3.15, a so-called face diagonal plane is indicated. Its Miller indices are ($\bar{1}10$). As can be seen from the figure, this plane passes through the origin of the coordinate system; therefore, the intercepts may be determined only after a parallel displacement. After the parallel displacement of the plane, it intersects the x-axis at $a = -1$ value: negative intercepts are indicated in crystallography by an overbar above the appropriate index. In part c), a so-called octahedral plane is shown. This plane forms equal angles with the coordinate axes. Its Miller indices are (111).

Miller-Bravais indices in hexagonal crystal systems

Although to designate hexagonal crystal planes, the three numbers of Miller indices would also be sufficient, they are indicated by four numbers of Miller-Bravais indices since by using this four-axis hexagonal representation, the crystal symmetry is revealed more clearly. Beside the h, k, l indices applied in cubic systems, there is a fourth index (i) in hexagonal systems.

The procedure for determination of Miller-Bravais indices in hexagonal systems is the same as for the cubic systems. The differences follow from the four-axis representation. The three axes in the basal plane are denoted by a_1, a_2 and a_3. The reciprocals of the intercepts formed with these axes are denoted by h, k and i indices, while l is used to denote the intercept with the z-axis. Consequently, an arbitrary plane in hexagonal systems can be described by the general ($h\,k\,i\,l$) indices.

A further significant difference from the cubic systems is that in hexagonal systems there are two different lattice parameters. The a lattice parameter is measured along the three axes in the so-called basal plane, while the c lattice parameter is measured in the direction of the z-axis perpendicular to the basal plane. In crystallographic calculations, both are taken as unit length. This convention leads to the following equation: $a_1 = a_2 = a_3 = a \neq c$.

The basal and top planes in hexagonal crystal systems are of utmost importance (see the planes indicated in part a) of Figure 3.16). The Miller-Bravais indices of these planes can be determined using the same procedure as described which results in the following four numbers: $(0\,0\,0\,1)$. In hexagonal systems, the concept of family planes is also used for the collective notation of all crystallographically equivalent planes. It is obvious, for example, that prism planes of the hexagonal lattice compose *family planes* and they can be denoted by the $\{1\,\bar{1}\,00\}$ indices. It is important to mention that, in hexagonal systems, indices can be permutated only according to the indices of a_1, a_2, a_3 axes. Therefore, the index in the direction of the z-axis must be 0 for all prism planes.

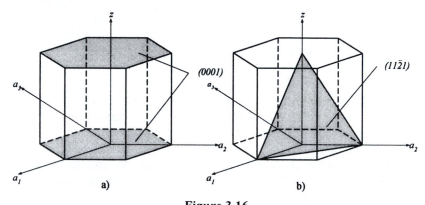

Figure 3.16
Miller-Bravais indices for various planes in the hexagonal system

Planes of the hexagonal crystal can be denoted unequivocally by three Miller indices in the three-axis representation. This follows from the fact that Miller-Bravais indices used in the four-axis representation contain redundancy, since the equation:

$$h + k = -i \qquad\qquad (3.33)$$

is always valid for the indices, that is the third index in the basal plane can always be calculated from the other two using Equation (3.33). This relationship can be clearly recognized among the Miller-Bravais indices of the plane shown in part b) of Figure 3.16. For this plane, the four-coordinate Miller-Bravais indices are $(11\bar{2}1)$. The convention that in the four-coordinate description two different unit lengths are used (one in the basal plane and another in the direction perpendicular to it) can also be seen from these indices.

3.1.2.2. Crystallographic directions – crystal vectors

In crystallography, crystal vectors are used to denote crystallographic directions. The components of vectors determining the directions are regarded as direction indices. The following procedure can be followed to determine the crystal vector components, i.e. the direction indices (see Figure 3.17):

1. The vector indicating the crystallographic direction is displaced parallel to itself in such a way that its origin passes through the origin of the coordinate system.

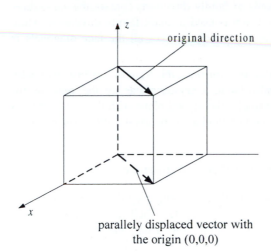

parallely displaced vector with
the origin (0,0,0)

Figure 3.17
Scheme for the determination of
crystallographic vector components

2. The end point of the direction projected perpendicularly to the coordinate axes. Denote the coordinates of the end point by u, v and w, respectively, along the x, y and z axes.

3. These three numbers are multiplied or divided by a common factor to express them as the combination of the smallest integer values. The resultant three numbers enclosed in square brackets $[u \, v \, w]$, without a separating comma, form the components of the crystal vector describing crystallographic direction, i.e. the u, v and w are the indices of the crystallographic direction.

It also follows from the above-described procedure, that the real size of the lattice is indifferent from the viewpoint of determination of the vector components. That is why the lattice parameter is taken as unit length when we derive the vector components.

Crystallographic directions in cubic systems

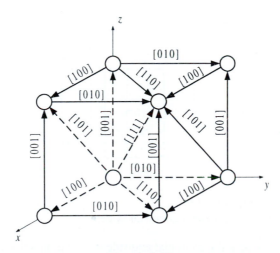

Figure 3.18
Characteristic crystal vector components
in the cubic crystal system

Some typical crystallographic directions are shown for a cubic lattice in Figure 3.18 applying the usual crystallographic notations.

Similarly to the Miller indices of crystalline planes, negative vector components are also indicated by an overbar above the appropriate index (e.g. $[1 \, 0 \, \bar{1}]$).

Crystallographic directions characterized by the same atomic order can be regarded as *equivalent in crystallographic context*. These directions are termed *family directions.* Vector components of family directions contain the same three numbers. The general notation $< u\ v\ w >$ is used to mark family directions. Thus, for example, the notation $< 1\ 1\ 0 >$ collectively denotes all the face diagonals in the cubic lattice.

An important relationship – valid only for the cubic system – is that the vector components for directions perpendicular to a crystal plane are the same as the Miller indices of that plane. For example, the crystal plane with the (1 0 0) Miller indices (shown in Figure 3.15) is perpendicular to the crystal direction given by the same three numbers [1 0 0].

Crystallographic directions in hexagonal crystal structures

The determination of vector components describing crystallographic directions can be performed by the same procedure as that described for the cubic system. Obviously, all the additional features following from the four coordinate representation should be considered.

Consequently, it is also true for the crystal vectors, as for crystal planes, that the four-coordinate description represents better the crystal symmetry of a hexagonal system. The same redundancy also follows for the vector components $[u\ v\ t\ w]$, which may be given by the following equation:

$$u + v = - t.$$ (3.34)

This means, that any of the vector components in the basal plane (or any plane parallel to it) can be calculated from the other two components using equation (3.34). In Figure 3.19, some typical crystal directions are shown indicating the appropriate vector components.

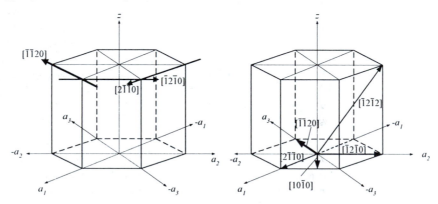

Figure 3.19
Crystal directions in the hexagonal crystal system

It is also true for the hexagonal crystal system that equivalent crystal directions compose *family directions.* They can be described by the combination of the same vector components considering the special features of a hexagonal crystal system.

(It should be kept in mind that the indices can be freely permuted only in the basal plane, i.e. for the indices along the a_1, a_2, and a_3 axes.) Family planes are collectively indicated by vector components between edge brackets. For a general crystal direction, this may be given in the following form $< u \, v \, t \, w >$.

3.1.3. Crystallographic calculations

In the previous sections, we have already determined some of the very important characteristics of various crystal structures. Thus, for example, we have analyzed the number of atoms belonging to a unit cell, the relationship between the lattice parameter and the atomic radius, the atomic packing factor, and the coordination number. Similarly, as regards the concept of atomic packing factor, we can define the meaning of planar and linear packing factors.

Beside the mentioned, very important crystallographic parameters, we have to know several further crystallographic characteristics for later metallographic analysis. Therefore, we briefly introduce the procedure for calculation of linear, planar and volume density, the determination of distances between crystallographic planes, the calculation of angles measured between different crystal directions.

3.1.3.1. Linear atomic density

For these calculations, first we have to determine the number of atoms belonging to the examined section. According to Figure 3.20, it is obvious that atoms within the section can be considered with their total value, and atoms at the ends of the section with half of their value. This is valid for both cubic and hexagonal crystal systems. For calculations, let us select one body diagonal of the body-centered α-Fe lattice, which can be characterized by the $[\bar{1}11]$ crystal vector components.

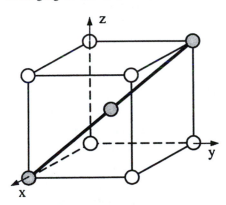

Figure 3.20
Schematic illustration for the determination of linear atomic density

According to the basic definition, the linear atomic density can be calculated as the ratio of the number of atoms belonging to the selected length of a line (N_{linear}) and the length of the selected section (l). The number of atoms belonging to the selected length of a line corresponds to the number of atomic diameters intersected by the examined section. Thus, the linear atomic density can be calculated with the following equation:

$$\rho_{linear} = \frac{N_{linear}}{l} \, , \qquad (3.35)$$

where ρ_{linear} the linear atomic density (atoms/mm),

N_{linear} the number of atoms belonging to the examined crystallographic direction,

l the length of the examined crystallographic direction (mm).

The value of the lattice parameter for α-Fe is equal to $a = 0.286\ nm$, the length of the body diagonal is $l = a\sqrt{3} = 0.495nm$. According to Figure 3.20, we can state that the number of atoms along the body diagonal is equal to:

$$N_{[\bar{1}11]} = 1 + 2 \times \frac{1}{2} = 2\ . \tag{3.36}$$

Substituting these values into Equation (3.35), we get for the linear atomic density in the direction of $[\bar{1}11]$ in α-Fe:

$$\rho_{[\bar{1}11]} = \frac{N_{[\bar{1}11]}}{l_{[\bar{1}11]}} = \frac{2}{0.495} = 4.037\ \frac{atoms}{nm} = 4.037 \times 10^6\ \frac{atoms}{mm}\ . \tag{3.37}$$

This means that there are more than four million atoms in a one-millimeter length in the direction of the body diagonal of the α-Fe.

3.1.3.2. Planar atomic density

According to the definition, the planar density can be calculated as the ratio of the number of atoms belonging to the selected area (N_{planar}) and the area of the selected plane (A). It can be expressed by the following equation:

$$\rho_{planar} = \frac{N_{planar}}{A}\ , \tag{3.38}$$

where ρ_{planar} the planar atomic density (atoms/mm^2),

N_{planar} the number of atoms in the selected crystallographic plane,

A the area of the examined crystallographic plane (mm^2).

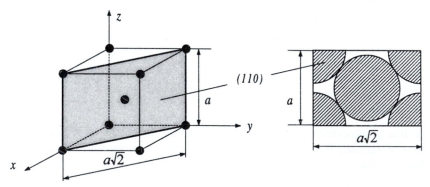

Figure 3.21
Scheme for the calculation of planar atomic density

For the calculation of the planar atomic density, we have to determine the number of atoms belonging to the examined plane. For this purpose, consider Figure 3.21. We can clearly see from this figure that atoms within the plane can be considered by their total value, and atoms at the corner points of the area, with a quarter of their value.

As an example, let us determine the planar atomic density for the (1 1 0) face diagonal plane of α-Fe. From Figure 3.21, we can see that the number of atoms in a face diagonal plane can be calculated as:

$$N_{(110)} = 1 + 4 \times \frac{1}{4} = 2 \; . \tag{3.39}$$

From the previous example, we know the value of the lattice parameter for α-Fe. It is equal to $a = 0.286 \; nm$. Therefore, the area of the face diagonal plane is equal to:

$$A = a^2 \sqrt{2} = 0.116 \; nm^2 \; . \tag{3.40}$$

Consequently, the planar atomic density of the (1 1 0) face diagonal plane in the α-Fe can be calculated from the following equation:

$$\rho_{(110)} = \frac{N_{(110)}}{A_{(110)}} = \frac{2}{0.116} = 17.3 \frac{atoms}{nm^2} = 1.73 \times 10^{13} \frac{atoms}{mm^2} \; , \tag{3.41}$$

This means, that there are about 1.73×10^{13} atoms in a unit area of the face diagonal plane in the α-Fe.

3.1.3.3. Volumetric atomic density

Volumetric atomic density can be calculated as the ratio of the number of atoms belonging to a unit volume. It can be expressed by the following equation:

$$\rho_v = \frac{N}{V} \; , \tag{3.42}$$

where ρ_v the volumetric atomic density (atoms/mm^3),

N the number of atoms belonging to a lattice element,

V   the volume of a lattice element (mm^3).

Since the number of atoms belonging to a unit cell was already introduced when we determined the atomic packing factor, it does not need any further explanation here. As an example, consider the lattice of α-Fe (which is a body-centered cubic crystal lattice). We know that the number of atoms belonging to a unit cell of the body-centered crystal is $N = 2$. The volume of the lattice is:

$$V = a^3 = 0,286^3 = 0,0234 \; nm^3 \; . \tag{3.43}$$

Consequently, the volumetric atomic density of α-Fe is:

$$\rho_{\alpha-Fe} = \frac{N_{\alpha-Fe}}{V_{\alpha-Fe}} = \frac{2}{0.0234} = 85.5 \frac{atoms}{nm^3} = 8.55 \times 10^{19} \frac{atoms}{mm^3}. \qquad (3.44)$$

This means that there are approximately 10^{20} atoms in one cubic millimeter of the α-Fe lattice.

3.1.3.4. Determination of the volume density of materials

In everyday practice, the volume density of a material (or, simply, the density) is defined by the mass of a unit volume, which can be calculated as

$$\rho = \frac{m_{unit\ cell}}{V_{unit\ cell}}. \qquad (3.45)$$

In this equation, $m_{unit\ cell}$ means the mass of atoms belonging to a unit cell, and $V_{unit\ cell}$ refers to the volume of the unit cell. The mass of atoms belonging to a unit cell can be calculated from the mole quantity of the examined material.

As an example, let us determine the density of pure iron (Fe). The one mole quantity of iron is equal to $m_{Fe} = 55.847\,g$ (see the Periodic Table – Table 2.1), which is equal to the mass of 6.023×10^{23} atoms. Consequently, the mass of one iron atom is:

$$m_{Fe-atom} = \frac{m_{Fe}}{6.023 \times 10^{23}} = 9.27 \times 10^{-23}\,g. \qquad (3.46)$$

Substituting this into Equation (3.45), we obtain for the density of pure iron:

$$\rho_{Fe} = \frac{N \times m_{Fe-atom}}{V_{unit\ cell}} = \frac{2 \times 9.27 \times 10^{-23}}{2.34 \times 10^{-20}} = 7.92 \times 10^{-3} \frac{g}{mm^3}. \qquad (3.47)$$

According to the ASM Metals Handbook, the measured density of pure iron is equal to 7.85 g/cm^3 (equivalent to 7.85x10^{-3} g/mm^3). This value is 0.8 % lower than the theoretically calculated one. This small difference may be reasoned as follows. In ideal (theoretically perfect) crystals, atoms can be found at all the characteristic lattice points (for pure iron, this means that atoms are located at all the corner points and in the body center of the lattices). However, in real crystals there are various imperfections (e.g. vacancies). This means that atoms are missing at certain lattice points. These imperfections can cause the small difference between the calculated and measured density.

3.1.3.5. Further crystallographic calculations

Beside the previously detailed crystallographic calculations, there is a large number of further crystallographic calculations to determine important parameters for various kinds of crystal structures. Here, we can just briefly review some of these calculations. For further more detailed crystallographic calculations, we refer to the technical literature.

Distances of crystallographic planes

The distance between the nearest neighboring parallel planes can be determined based on pure geometrical considerations. However, in this case we have to consider very carefully the concept of the neighboring plane to correctly select the nearest neighboring plane. For this purpose, a fundamental crystallographic rule should be considered, which states that planes parallel to a selected crystallographic plane can be put through any atoms having the same atomic order as the selected plane. This follows from the regular spatial periodicity of crystals.

In cubic crystal structures, we can derive a simple equation for the determination of distances between the nearest crystallographic planes. This expression is based on the lattice parameter (a) and the Miller indices of the planes considered, as follows:

$$d_{hkl} = \frac{a}{\sqrt{h^2 + k^2 + l^2}}, \tag{3.48}$$

where d_{hkl} the perpendicular distance between the nearest neighboring parallel planes with the $(h\,k\,l)$ Miller indices,

     a the lattice parameter of the crystal,

     $(h\,k\,l)$ the Miller indices of that parallel plane which is nearest to the origin of the coordinate system.

As an example, consider Figure 3.22. In part a) of the figure, we can see the projection view of a body-centered cubic crystal, where the $(\overline{1}\,10)$ crystallographic planes are marked with shading. Substituting the Miller indices of the $(\overline{1}\,10)$ crystallographic plane - as the nearest parallel plane to the origin of the coordinate system - into equation (3.48), we obtain the following formula:

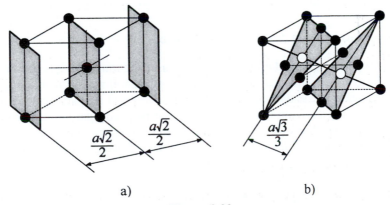

a)   b)

Figure 3.22

Scheme for determination of the distances between parallel crystallographic planes
a) body-centered cubic, b) face-centered cubic crystal

$$d_{\bar{1}10} = \frac{a}{\sqrt{2}} = \frac{a\sqrt{2}}{2} \ , \tag{3.49}$$

This equation is in total harmony with the calculations based on geometrical calculations illustrated in part *a*) of Figure 3.22.

In part *b*) of Figure 3.22, we can see a face-centered cubic crystal, where the $(\bar{1}1\bar{1})$ crystallographic planes are shaded. Substituting the Miller indices into equation (3.48), we get for the distances of the nearest parallel planes

$$d_{\bar{1}1\bar{1}} = \frac{a}{\sqrt{3}} = \frac{a\sqrt{3}}{3} \ , \tag{3.50}$$

which is also the same as the value resulting from pure geometrical calculations based on the figure.

Determination of interplanar angles

The determination of the interplanar angles between planes intersecting one another can be deduced for the determination of angles between normal vectors of the planes. Using the rule of cubic structures, which states that indices of the plane and direction, perpendicular to each other, are equal, the interplanar angles of the two planes can be determined by the following equation:

$$\cos\varphi = \frac{\vec{n}_1\vec{n}_2}{|\vec{n}_1||\vec{n}_2|} = \frac{u_1u_2 + v_1v_2 + w_1w_2}{\sqrt{u_1^2 + v_1^2 + w_1^2}\sqrt{u_2^2 + v_2^2 + w_2^2}} \tag{3.51}$$

where

$\vec{n}_1, \vec{n}_2$ are the normal vectors of the two planes,

u_i, v_i, w_i are the components of the two normal vectors (i = 1,2).

The relations between directions and the planes

In many cases (particularly in the case of plastic deformation), it is important to know if a given direction can be found in a given plane. If a direction can be found in a given plane, the scalar product of the normal vector of the plane and the vector of the examined direction is zero (since they are perpendicular to each other). In other words, the direction characterized by [u v w] vector components will be found in the plane with *(h k l)* Miller indices, if the following equation is fulfilled:

$$\frac{hu + lv + kw}{\sqrt{u^2 + v^2 + w^2}\sqrt{h^2 + k^2 + l^2}} = 0. \tag{3.52}$$

3.2. Methods of studying crystal structures

The crystal structures analyzed in the previous sections had only been a matter of theory as far as an appropriate method was developed for studying the lattice structure. Those techniques can only be applied which have a resolution power in the order of crystal lattices. (From the previous chapters, it is known that the lattice parameters are generally in the order of 10^{-7} mm.)

Optical (light) microscopes generally applied in metallography usually have magnifications up to 3×10^3. Obviously, this is not sufficient for studying of lattice structures, since the resolution power of the human eyes for visual observation is in the order of 0.1 mm. Therefore, the minimum distances to be distinguished should be enlarged at least to this value. The appropriate method may be found in the domain of electromagnetic waves, where the wavelength is equal to the lattice parameter or even smaller. These requirements are fulfilled by the so-called X-ray beam, which has a wavelength in the interval of 10^{-7} - 10^{-10} mm.

3.2.1. The theoretical basis of X-ray examinations

The production of an X-ray beam is illustrated in Figure 3.23. An X-ray beam is generated in a special device called X-ray tube. It is generally made of glass, with a rotational symmetrical shape. There are two electrodes, a cathode and an anode in the tube, which is under a high vacuum (in the order of 10^{-7} Hgmm).

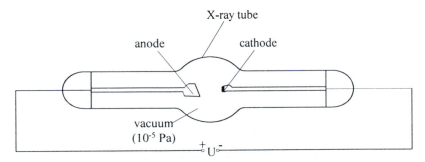

Figure 3.23
X-ray tube for producing X-ray radiation

The cathode is heated by a low-voltage electrical source. Due to thermal electron emission, electrons are emitted from the heated cathode. These electrons are accelerated by a high-voltage ($U = 10 \div 400\,kV$) electric field between the negative and positive electrodes. The velocity of electrons can be determined from the following formula:

$$eU = \frac{mv^2}{2}, \qquad (3.53)$$

where

e	the electric charge of electrons,
m	the mass of an electron,
U	the high voltage between the anode and cathode,
v	the velocity of electrons.

The high-speed, electrically charged electrons collide with the anode and lose part of their velocity. Consequently, a significant part of their kinetic energy is transformed into heat and a smaller part of it is transformed into electromagnetic radiation. The wavelength of the electromagnetic radiation can be determined from the Einstein equation:

$$\frac{mv^2}{2} = \frac{hc}{\lambda}.\tag{3.54}$$

where

h	Planck's constant (6.625×10^{-34} Js),
c	the velocity of light (3×10^8 m/s),
λ	the wavelength of the electromagnetic radiation.

The decrease in the velocity of the electrons is a multi-step procedure. The consequence of this fact is that not only one definite X-ray radiation (a so-called monochromatic X-ray) is produced, but a mixture of them with different wavelengths. The change in intensity with the wavelength depends on the tube voltage. This is illustrated in Figure 3.24.

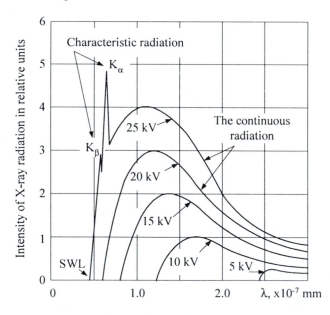

Figure 3.24
X-ray spectrum of molybdenum as a function of applied voltage

Equation (3.54) is applied to the determination of the shortest wavelength, indicated with *SWL* in the figure, which can be calculated from the equation

$$\lambda_{SWL} = \frac{hc}{eU} \, .$$
(3.55)

Combining all the constants into one expression in Equation (3.55), we obtain the following expression:

$$\lambda_{SWL} = \frac{12\,400}{U} \, .$$
(3.56)

From Equation (3.56), it can be seen clearly that the short wavelength limit is only a function of the high voltage (U) between the two electrodes. The total X-ray energy radiated in unit time is proportional to the area under the curves of the continuous spectrum and can be calculated as follows:

$$E_{unit} = AI_t ZU^m \, ,$$
(3.57)

where

A	a proportionality constant,
I_t	the so-called tube current, which is the measure of the number of electrons striking the anode per second,
Z	the atomic number of the anode material,
U	the high voltage between the two electrodes in the tube,
m	is a constant, with a value of about 2 for most materials.

From this expression, it is obvious why heavy metals (elements with a high atomic number, such as tungsten, or cobalt) are generally used as anode materials in X-ray tubes. Further, from Equation (3.57), we can also clearly see why the voltage is increased to achieve the highest possible intensity of radiation.

As the voltage reaches a critical value, the energy of the electrons – accelerated by the electric field strength – is high enough to knock an electron out of an inner electron shell. In this case, one of the outer electrons – with a higher energy level – falls into the vacancy, emitting the energy difference between the two shells in the form of electromagnetic radiation. This is the so-called characteristic radiation, which depends only on the material of the anode.

From a practical point of view, the most important case is when an electron is knocked out of the K-shell (the principal quantum number is equal to 1 for the K-shell) and an electron from the L-shell (with the principal quantum number 2) falls into the vacancy. This is called K_α characteristic radiation, which has the following frequency:

$$h\nu = E_L - E_K \, .$$
(3.58)

(As the emitted radiation is obviously a characteristic feature of the anode material, this phenomenon can be used for studying material composition: this is the basis of X-ray spectroscopy.)

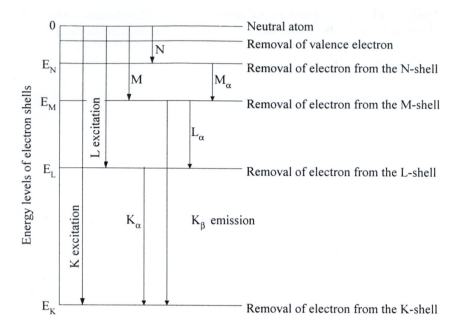

Figure 3.25
Energy levels of different electron shells

The *K*-shell vacancy may be filled by an electron from any of the outer shells. If the electron of an electron shell with a higher quantum number falls into the *K*-shell vacancy, the radiation is indicated by a series of *K*-radiation, e.g. K_α, K_β, K_γ. A series of *L*-radiation arises when the high-speed, colliding electrons, knock an electron out of the *L*-shell. It also follows from the analysis of the energy levels of electron shells (Figure 3.25) that to remove an electron from an electron shell with a higher quantum number requires less energy than that for a lower quantum number. From the illustration, it is also evident that the *L*-series characteristic radiation requires a lower tube voltage than the *K*-series characteristic radiation. However, this means that the *K*-series characteristic radiation is necessarily followed by the *L*-series characteristic radiation. The energy spectrum of characteristic radiation shows a great difference from the continuous spectrum of radiation (illustrated in Figure 3.24). In Figure 3.26 the spectrum of the characteristic radiation is shown for a copper anode.

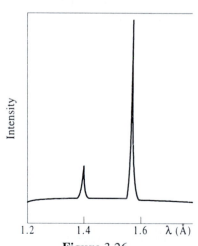

Figure 3.26
Intensity distribution of
characteristic radiation
for a copper anode

If the energy required for removing an electron from the K-shell is marked with E_K, the critical tube voltage (U_{crit}) can be determined from Equation (3.53). This critical tube voltage is necessary for the emission of K_α characteristic radiation. This value can be calculated from the following equation:

$$e\,U = \frac{mv^2}{2} = E_K .$$
(3.59)

The critical tube voltage from the above expression is:

$$U_{crit} = \frac{E_K}{e} .$$
(3.60)

The U_{crit} tube voltage is also called the *critical excitation voltage*. The increase in voltage above the critical value increases the intensity of characteristic radiation, but the wavelength does not change.

The intensity of the characteristic radiation (measured by a value which is above the continuous spectrum) is directly proportional to the tube current (I_t) and the amount by which the applied voltage exceeds the critical (excitation) voltage:

$$I_K = BI_t (U - U_{crit})^n ,$$
(3.61)

where B is the proportionality constant, and $n \approx 1.5$ is a constant value.

The intensity of the characteristic radiation – depending on the tube voltage – can be much higher than that of the continuous spectrum immediately adjacent to it. Besides being very intense, the characteristic radiation appears as a very narrow, sharp (i.e. characteristic) line in the radiation spectrum. This characteristic makes X-ray diffraction techniques suitable for examining lattice structures, since many examinations can be performed only by monochromatic (or approximately monochromatic) radiation. For a better understanding, consider the following analysis as an example. X-rays – regardless of material – are partly absorbed, partly reflected and partly pass through when entering the examined material. The change in intensity of the beam as it enters and exits can be expressed as follows:

$$I_1 = I_o\, e^{-\mu d} ,$$
(3.62)

where

I_o	the intensity of incident beams,
I_1	the intensity of transmitted beams passing,
μ	the so-called linear absorption coefficient,
d	the materials thickness.

The μ linear absorption coefficient depends on the materials density (ρ), the wave-length of the radiation (λ) and the atomic number of the material (Z), which may be characterized by the following formula:

$$\mu = k\,\rho\,\lambda^3 Z^3 .$$
(3.63)

This formula is often given in the following form, where (μ/ρ) is the so-called mass absorption coefficient:

$$\frac{\mu}{\rho} = k\,\lambda^3 Z^3 \; . \tag{3.64}$$

In Figure 3.27, the variation of the mass absorption coefficient is shown as a function of the wavelength (of an arbitrary material). From this figure, it can be seen that the curve has a sharp discontinuity at a certain wavelength. The wavelength, where this discontinuity appears, is called the absorption edge. We can also see from the illustration that the change in the mass absorption coefficient follows the same function on both sides of the absorption edge.

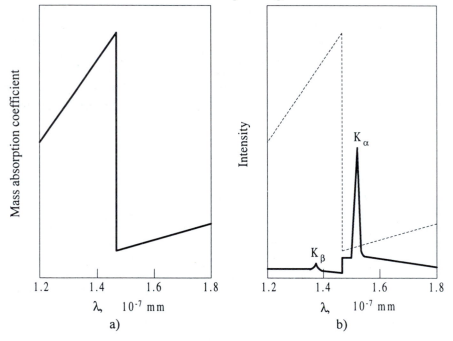

Figure 3.27
a) The change in the mass absorption coefficient as a function of the wavelength
b) The principle of production of monochromatic X-ray radiation

This characteristic change in the mass absorption coefficient plays an important role in the production of monochromatic X-ray radiation. From Figure 3.26 illustrating the distribution of the intensity of characteristic radiation, we can see that this radiation consists of a sharp, K_α radiation at high intensity and a weaker K_β characteristic radiation (besides the continuous spectrum). The intensity of undesirable components can be significantly reduced compared to the K_α radiation by passing the beam through a filter made of a material having a sharp absorption edge between the K_α and K_β wavelengths of the anode, as can be seen in part b) of Figure 3.27.

The principle shown in Figure 3.27 can be applied with excellent results to produce monochromatic X-rays (containing only K_α characteristic radiation). For any anode material, we can find a filter material that has an absorption edge between the K_α and K_β wavelengths of the anode. This filter material can generally be found near the anode material in the periodic table, its atomic number being only greater or smaller by 1 or 2 than that of the anode material. For example, *Ni* filter – with the atomic number 28 – is applied for *Co*-anode with the atomic number 27.

3.2.2. X-ray diffraction analysis

The theoretical basis of X-ray diffraction can be summarized as follows. X-ray beams penetrate the examined material and excite its atoms. These excited atoms are sources of electromagnetic waves emitting so-called scattered rays. Scattered waves are of the same wavelength as the incident X-ray beam. The scattered waves propagate in all directions. Actually, this is the phenomenon of X-ray diffraction, which can be regarded as a reflection phenomenon. This is shown in Figure 3.28, where X-ray beams entering two parallel crystalline planes at an angle of θ and the scattered rays generated by the incident beams are illustrated as reflected beams.

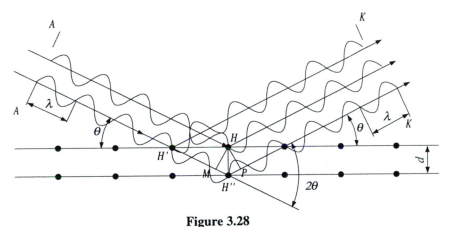

Figure 3.28
Theoretical scheme for examination of crystal structures
by the X-ray diffraction technique

Since the wavelength of X-rays is in the order of the distance between crystallographic planes, the scattered electromagnetic beams, generated by X-rays entering the regularly arranged crystal planes, interact with each other. If the reflected beams meet in the same phase, they strengthen one another and give a perceptible reflected sign on the X-ray film placed in the way of beams, while beams encountered in the anti-phase – according to the rules of wave physics – extinguish each other.

This condition is expressed by *Bragg's law,* which gives the following equation for the path difference requirement between the reflected beams:

$$n\,\lambda = 2d\,\sin\theta\,, \tag{3.65}$$

where

λ	the wavelength of the X-rays (mm),
d	the distance between parallel crystallographic planes (mm),
θ	the incident/diffraction angle of X-rays
n	is an integer number (its value is n = 1 for first order diffraction).

Naturally, X-ray diffraction occurs not only on two neighboring atomic planes shown in Figure 3.28, but other atomic planes also reflect. The condition that the reflections from further crystalline planes remain in phase can be derived from Bragg's equation (3.65). The path difference between the reflected beams must be equal to one or more integer wavelengths of the X-ray (i.e. in equation (3.65) n = 1, 2, 3... should be substituted).

3.2.2.1. The Laue method

The phenomenon of X-ray diffraction was discovered and first applied for studying crystal structures by Laue in 1912. The theoretical scheme of the Laue method using incident X-ray beams can be seen in Figure 3.29.

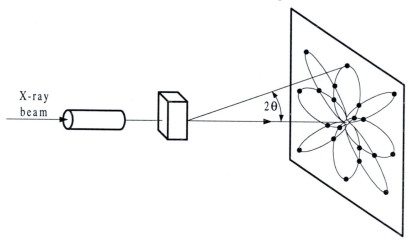

Figure 3.29
Theoretical scheme of Laue X-ray diffraction method

Laue, in his early experiments, used polychromatic (multi-colored, multi-wavelength) X-ray radiation. For this reason, it was not suitable for the determination of lattice parameters, since in this case there are two unknown parameters in Bragg's law: the distance between lattice planes (d), and the wavelength λ (since, we do not know which wavelength of the polychromatic X-rays gave a perceptible reflection). However, this method can be successfully applied to analyze the crystal symmetry and it is still used for the determination of the crystallographic orientation of single crystals.

For the quantitative analysis of crystal structures (e.g. for the determination of the distance between crystallographic planes and the value of lattice parameters), the so-called Debye-Scherrer method is used. This is still the most widely applied experimental technique for the examination of crystal structures by X-rays. In the following section, this method will be summarized briefly.

3.2.2.2. Debye-Scherrer X-ray diffraction analysis

The Debye-Scherrer X-ray diffraction analysis is based on the previously described X-ray diffraction phenomenon (see Figure 3.28). A schematic view of this method can be seen in Figure 3.30.

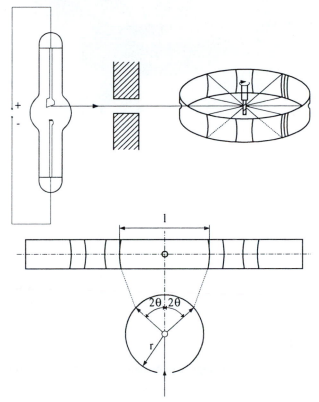

Figure 3.30
Theoretical scheme of the Debye-Scherrer X-ray diffraction method

In this process – in contrast to the Laue method – monochromatic X-ray beams are applied which can be characterized by one characteristic wavelength. A fine powder of the material to be examined is formed into a very thin filament and is placed at the center of a cylindrical filmstrip with the radius r.

A collimated beam of X-rays – passed through appropriate filters and lens systems – is directed at the powder. Since there is a large number of powder particles with many crystalline orientations, the diffracted beam emerges as a cone of radiation at an angle 2Θ from the incident beam.

Different reflection cones emerge from the differently oriented crystallographic planes. Beams reflected as cones expose reflection rings to the filmstrip as shown in Figure 3.31.

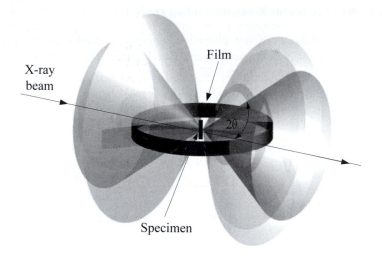

Figure 3.31
Reflected cones in the Debye-Scherrer diffraction method

Following the development of the film and rectifying it into the plane, the 2θ reflection angle can be calculated from the distance of reflection rings. First, the following ratio can be stated:

$$\frac{l}{2r\pi} = \frac{4\theta}{360} \tag{3.66}$$

and from this, the 2θ diffraction angle can be determined from the expression

$$2\theta = \frac{360}{4r\pi}l . \tag{3.67}$$

Considering Bragg's law – Equation (3.65) – the distance between the neighboring reflecting atomic planes can be calculated as follows:

$$d = \frac{n\lambda}{2\sin\theta} . \tag{3.68}$$

If we know the distance between the atomic planes, the lattice parameter can also be determined. In cubic crystal systems, a relationship exists between the d

distance of reflecting parallel crystallographic planes, the lattice parameter and the *(hkl)* Miller indices of the plane examined, as follows:

$$d_{hkl} = \frac{a}{\sqrt{h^2 + k^2 + l^2}} \ . \tag{3.69}$$

During the evaluation of the results of the diffraction method, further geometrical and crystallographic relationships should also be considered. It requires a thorough crystallographic knowledge. Below, some general rules for face-centered and body-centered cubic crystal structures have been summarized:

1. In body-centered cubic crystals, reflections can be obtained from those crystallographic planes for which the sum of the Miller indices (i.e. *h+k+l*) is equal to an even number.

2. In face-centered cubic crystals, reflections can be obtained from those crystallographic planes for which all the Miller indices are either odd or even numbers (the "0" Miller index is considered as an even number).

For example, the (110) plane reflects only in body-centered, the (111) plane in face-centered cubic crystals, while the (200) plane reflects both in body-centered and in face-centered cubic crystals.

4. BASIC RULES OF CRYSTALLIZATION

Crystallization is a transformation process of a substance from the liquid to the solid (crystalline) state. The process of crystallization is governed by the laws of thermodynamics. For this reason it is necessary to briefly overview the thermodynamic principles of crystallization.

4.1. Thermodynamic principles of crystallization

Before going into the details of thermodynamic principles of crystallization, we have to define some basic concepts of thermodynamics, which are often referred to in the following section.

The term *thermodynamic system* refers to a real or imaginary part of the space separated for experimental purposes. Such a thermodynamic system is, for example, a liquid metal or alloy in a melting pot for which we want to analyze the crystallization process. The system may comprise a number of components. If there is only one component in the examined system (e.g. a pure metal), it is called a *single-component thermodynamic system*, while a system consisting of two or more elements is called a *multi-component system*.

The (metallurgical, thermodynamic) *phase* may be defined as a part of a system that is separated from the others by an interface. Thus, the phase is a part of the system, which has uniform chemical (e.g. composition) and physical (e.g. state of matter and crystal structure) characteristics. If there is only one phase within the examined system, it is called a *homogeneous system*. The liquid metal in the melting pot can be regarded as a homogeneous system as far as crystallization (the appearance of solid phase) begins. When the solid phase appears beside the liquid phase, the earlier homogeneous system becomes *heterogeneous*.

The state of thermodynamic systems can be characterized by *state factors*. The state factors of metallurgical thermodynamic systems are: the *temperature (T)*, the *pressure (p)*, the *volume (V)* and for multi-component systems the *concentration (c)*. (Since, in metallurgical systems we deal with liquid and solid phases, the volume may be taken as constant. Thus, in metallurgical systems, the basic state factors are: temperature, pressure and concentration.)

In thermodynamic systems the *degree of freedom* characterizes the number of state factors that can be freely changed without any alteration in the equilibrium of the system (in other words, without decreasing or increasing the number of phases within the system).

The relationship between components, phases and the degree of freedom can be described by the *Gibbs phase rule:*

$$P + F = C + 2 ,$$
(4.1)

where P is the number of phases present in a system,
        F is the number of degrees of freedom in a system,
        C is the number of components in a system.

For most metallurgical systems, the pressure is constant during phase transformation processes. Therefore, Gibbs' phase rule can be written in the following form:

$$P + F = C + 1 ,$$
(4.2)

since due to the $p = constant$ condition, the number of degrees of freedom is one unit lower on the left hand side of Equation (4.1), and consequently to keep the equation true, we also have to decrease the value on the right hand side of the equation by one.

During crystallization – and later in other phase transformation processes – the following basic question arises: "What is the reason behind the phase transformation processes"? The answer is given by the rules of thermodynamics. It is well known from the basic studies of thermodynamics that all spontaneous transformations in nature occur only if the new state, under the new conditions, is energetically more stable, i.e. it possesses a lower energy level. This is also true for any metallurgical system. For these systems, we should apply the second main rule of thermodynamics which may be written in the following form for the free energy (G) of the system as a function of the total enthalpy (H) and the product of absolute temperature (T) and entropy (S), as follows:

$$G = H - TS ,$$
(4.3)

where

      G is the so-called Gibbs free energy of the system,
      H is the enthalpy (the thermal energy at constant pressure),
      T is the absolute temperature and
      S is the entropy.

From the viewpoint of changes occurring in the system, the direction of the changes of the free energy plays a decisive role, not the absolute value. To analyze this, let us derive the total differential of Equation (4.3):

$$dG = dH - TdS - SdT .\tag{4.4}$$

Since, we are analyzing the crystallization processes at constant temperature ($T = constant$), Equation (4.4) can be reduced to the following formula due to the condition $dT = 0$:

$$dG = dH - TdS .\tag{4.5}$$

This equation may be written in the form of finite differentials, as follows:

$$\Delta G = \Delta H - T\Delta S .\tag{4.6}$$

This form reflects the following. The amount of heat energy introduced in the system at constant pressure to achieve the required changes is identified by the enthalpy and denoted by ΔH. A part of it is accumulated in the material (i.e. the so-called bonded energy characterized by the $T\Delta S$ amount of entropy), while the *change in the free energy (ΔG)* means the *driving force of the process.*

The equilibrium of a thermodynamic system is stable if the total energy of the system has its smallest value. This means the minimum of Equation (4.3), which can be determined from the following equation based on extreme calculus:

$$dG = 0 .\tag{4.7}$$

It also follows from this equation, that only those changes that can occur spontaneously result in a decrease in the free energy of the system. For these changes, the following equation should be valid:

$$dG < 0\tag{4.8}$$

Actually, these equations form the basis of the phase transformation occurring during crystallization. Based on these equations, we can understand those driving forces that result in the solidification of liquid phases. Let us first consider the crystallization process of a pure metal as an example of application of the thermodynamic principles described in this section.

4.1.1. Analysis of crystallization processes in pure metals

A thermodynamic system consisting of only pure metals can be regarded as a single-component system; therefore, the number of components in the system is equal to $K = 1$. The changes in the free energy of liquid and solid phases of pure metals as a function of temperature can be seen in Figure 4.1.

From this figure, it is obvious that as far as $T > T_o$ is true for the temperature, the free energy of the liquid phase is smaller, and consequently this phase is stable. In this temperature interval (i.e. in the $T > T_o$ domain), using Gibbs' phase rule, we can easily justify that the temperature can be changed without any changes occurring in the equilibrium state of the system. For this purpose, let us express the degree of freedom from Equation (4.2) by substituting the appropriate values of parameters into it, i.e.:

$$F = C + 1 - P = 1 + 1 - 1 = 1, \tag{4.9}$$

where $C = 1$ since we analyze a single-component, pure metallic system,
         $P = 1$ since there is only a liquid phase present in the system.

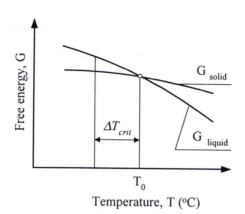

Figure 4.1

Free enthalpy for liquid and solid phases
as a function of temperature

A similar analysis can be performed for the temperature interval $T < T_o$, where the free energy of the solid phase is smaller, therefore, this is the stable phase. In this temperature domain, the only free state factor that can be changed is the temperature, which can be justified as we did above using Equation (4.9).

Using the basic equations of thermodynamics, we can also prove that in the two previous temperature domains (i.e. in the intervals $T > T_o$ and $T < T_o$), the change in temperature follows an exponential function as shown in Figure 4.2.

The $T = T_o$ temperature requires special analysis (this is the so-called *equilibrium temperature* corresponding to the melting and the solidification point of a metal). At this temperature, the free energies of the liquid and solid phases are equal ($G_f = G_s$). Consequently, changes do not exert driving forces at this temperature. To initiate crystallization, a ΔT undercooling is necessary. As a result of this, the free energies of the solid and liquid phases will change, i.e. $\Delta G = G_s - G_f < 0$ (see Figure 4.1). Consequently, the solid phase becomes more stable and crystallization can start. When the first solid crystal appears in the system, the originally homogeneous liquid phase becomes heterogeneous (i.e. liquid and solid phases are present). Therefore, in Gibbs' phase rule, the number of phases is equal to $P = 2$.

Considering this, the degree of freedom for the system can be calculated as follows:

$$F = C + 1 - P = 1 + 1 - 2 = 0. \tag{4.10}$$

In other words, this means that none of the state factors in the system can be changed without changing the equilibrium of the system. Consequently, during the crystallization of pure metals, the temperature should be constant ($T = T_o = constant$) until one of the phases disappears from the system. Naturally, this should be the liquid phase during crystallization. Therefore, we can conclude that the crystallization of pure metals is a non-variant process (non-variant means that the number of degrees of freedom is equal to zero).

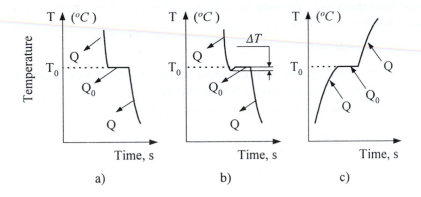

Figure 4.2
Temperature-time diagrams of a pure metal
a) theoretical cooling curve   b) real cooling curve   c) heating curve

Theoretically, the crystallization occurs exactly at the $T = T_o = constant$ temperature, as shown in part a) of Figure 4.2. This diagram can be regarded as the
theoretical cooling curve of pure metals.

In reality, crystallization can start only after a certain amount of ΔT undercooling. The necessity of this was proved by the previous analysis concerning the
changes in free energy and the necessary driving forces of crystallization. Therefore, the real cooling curve of pure metals corresponds to that shown in part b) of
Figure 4.2. Additional phenomena occurring during the actual crystallization processes of pure metals require further thermodynamic analysis and consideration.

The free energy function for the liquid and solid phase at an arbitrary
temperature can be given by the following expressions:

$$G_l = H_l - TS_l \qquad \text{for the liquid phase,}$$
$$G_s = H_s - TS_s \qquad \text{for the solid phase.}$$

(4.11)

The difference between the free energies of the liquid and solid phase at the
temperature $T = T_o$ can be expressed by the following equation:

$$\Delta G = (H_l - H_s) - T_o(S_l - S_s) = 0 .$$

(4.12)

Introducing the following notations:

$$\Delta H = H_l - H_s \quad \text{and} \quad \Delta S = S_l - S_s ,$$

(4.13)

the following matters should be taken into consideration. ΔH is actually the heat
quantity removed from the system during crystallization at a constant pressure and
at a constant temperature $T = T_o$. This amount of heat energy should be removed
from the system for the liquid to solid phase transformation to occur. In single-
component systems – such as the examined pure metallic system – this phase
transformation occurs at constant temperature, while heat should be continuously

removed from the system. This amount of heat is called the *latent heat (L)* and can be calculated from the following expression:

$$\Delta H = H_l - H_s = L .$$ (4.14)

(Obviously, the same amount of heat should be put into the system during the phase transformation from the solid to the liquid state.)

Substituting Equations (4.13) and (4.14) into Equation (4.12), we obtain the following expression:

$$\Delta G = \Delta H - T_o \Delta S = 0 ,$$ (4.15)

from which we can derive:

$$\Delta S = \frac{\Delta H}{T_o} = \frac{L}{T_o} .$$ (4.16)

In Equation (4.16), the quantity ΔS is known as the bonding energy (in other words, the entropy of fusion), which in harmony with experimental observations is constant for all metals. Its value, according to Richard's law, is equal to $R = 8.3$ *J mol⁻¹K⁻¹*.

For small values of ΔT undercooling, we can write that $c_l^p - c_s^p \approx 0$ for the difference between the specific heat of the liquid and the solid phase. In addition, ΔH and ΔS can be regarded as independent of temperature. Taking all this into consideration, the following formula can be obtained from Equation (4.12):

$$\Delta G \approx L - \frac{LT}{T_o} .$$ (4.17)

From this equation, the relation between the free energy change (ΔG) and the latent heat (L) can be expressed as follows:

$$\Delta G \approx \frac{L \Delta T}{T_o} .$$ (4.18)

This latter equation plays an important role in the analysis of elementary processes of crystallization.

4.1.2. Elementary processes of crystallization

The process of crystallization principally consists of two elementary processes: the first is the nucleation of small crystalline particles called crystallization nuclei or crystallization centers; the second is the growth of crystals from these nuclei or centers.

4.1.2.1. The formation of crystallization nuclei in the liquid phase

There are two main mechanisms describing the formation of crystallization nuclei: the so-called homogeneous and the heterogeneous mechanism of nucleation.

In *homogeneous nucleation*, the crystallizing material itself provides atoms as crystallization centers or nuclei. In this mechanism, the conditions of formation of crystallization nuclei are the same at all points of the material. In the *heterogeneous nucleation mechanism,* the crystallization nuclei or centers are so-called foreign atoms put into the liquid to support and accelerate the crystallization process. Atoms of alloying elements or insoluble impurities can be regarded as foreign crystallization nuclei, as well.

Homogeneous mechanism of crystallization nucleus formation

Consider a continuously cooling homogeneous, single-phase metal liquid. When the temperature becomes lower than the T_o equilibrium temperature by a ΔT undercooling – sufficient to form crystallization nuclei – a large number of randomly distributed homogeneous crystallization nuclei are formed within the liquid metal. This is due to atoms (moving relatively freely in the liquid) taking up positions that are favorable for bonding. The fact that the crystallization nuclei are able to grow or they just dissolve in the surrounding liquid is consistent with the laws of thermodynamics.

For this analysis, consider a system containing liquid metal and – due to ΔT undercooling – also solid crystallization nuclei. For simplicity, the crystallization nuclei are regarded as spheres with a radius r. During the nucleation of these tiny crystallization nuclei, the free energy of the system increases by the surface energy necessary for the formation of a sphere with radius r (denoted ΔG_f). At the same time, the free energy of the system decreases by the volume energy resulting from the solidification of the sphere-shaped crystallization nucleus with the radius r. This amount of energy decrease can be connected to the latent heat (denoted ΔG_v). The total change in the free energy of the system (ΔG_t) can be calculated as the sum of the two free energies, i.e.:

$$\Delta G_t = \Delta G_f + \Delta G_v .\tag{4.19}$$

The increase in free energy arising from the formation of the sphere with the radius r can be determined from the following equation:

$$\Delta G_f = 4r^2\pi\sigma ,\tag{4.20}$$

where σ is the so-called surface stress, which denotes the surface energy per unit area, and $4r^2\pi$ is the surface of the crystallization nucleus with the radius r.

The energy release during the formation of a crystallization nucleus with the radius r (considering Equation (4.18), which describes the energy release from a unit volume) can be determined from the following equation:

$$\Delta G_v = -\frac{4r^3\pi}{3} \frac{L\Delta T}{T_o}.$$

(4.21)

The total change in the free energy of the whole system can be determined by substituting Equations (4.20) and (4.21) into Equation (4.19). This will result in the following formula:

$$\Delta G_t = 4r^2\pi\sigma - \frac{4r^3\pi}{3} \frac{L\Delta T}{T_o}.$$

(4.22)

In Figure 4.3, the free energy changes are shown as a function of the size of the crystallization nucleus.

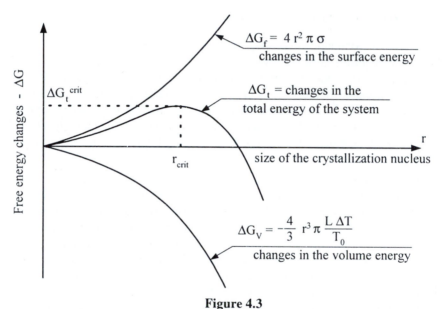

Figure 4.3
Free energy changes as a function of crystallization nuclei size

It is obvious from the figure that the free energy function described by Equation (4.22) has an extreme (maximum) value at $r = r_{crit}$. This maximum can be determined by differentiation of Equation (4.22) with respect to the r variable, and the maximum can be found where the following equation is fulfilled:

$$\frac{d}{dr}(\Delta G_t) = 8r\pi\sigma - 4r^2\pi\frac{L\Delta T}{T_o} = 0.$$

(4.23)

From this equation, the critical value of the crystallization nucleus (r_{crit}) can be determined by using the following expression:

$$r_{crit} = \frac{2\sigma T_o}{L\Delta T} \; . \tag{4.24}$$

It follows from the previous equations that the total free energy attains its maximum value at the critical size of the crystallization nucleus. If the size of the crystallization nuclei were smaller than the critical size of the nucleus determined by Equation (4.24), the free energy of the whole system would increase during the growth of the nucleus. This is impossible according to the basic rules of thermodynamics. Consequently, any crystallization nucleus in the range of $r < r_{crit}$ is unable to grow, it dissolves and disappears in the liquid metal. Therefore, these nuclei are regarded as unstable ones.

If the size of the crystallization nuclei were greater than the critical size of the nucleus determined from Equation (4.24), the free energy of the whole system would decrease. This means, that any nucleus in the range of $r > r_{crit}$ is able to grow and therefore it is regarded as a stable nucleus.

It is a very important feature of crystalline materials that, in their liquid, crystallization nuclei can nucleate which are then able to grow. This phenomenon is termed the *nucleation ability* and denoted by N. This is characterized by the number of crystallization nuclei formed in a unit volume in unit time. According to this, its unit is [1/(mm^3 s)]. This is the function of undercooling and the cooling rate (the unit of the cooling rate is $^\circ$C/s). This is illustrated in Figure 4.4

The N_1 curve refers to a slow, and the N_2 curve refers to a higher cooling rate ($v_2 > v_1$). From the figure, it can be seen that at the same value of undercooling (ΔT), higher nucleation ability is associated with a higher cooling rate.

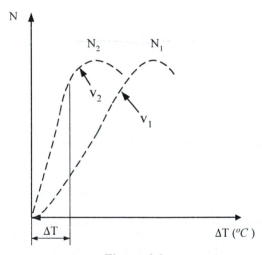

Figure 4.4
The change of crystallization ability in the function of undercooling

It follows from Equation (4.24), that the critical size of the nucleus depends on both the constants characterizing the examined material (σ, L, T_o) and the rate of undercooling (ΔT). This function is illustrated in Figure 4.5.

It is clear from Figure 4.5 that the critical size of the nucleus is hyperbolically decreasing with the increase in undercooling (i.e. with the decrease in temperature). In other words, the higher the rate of undercooling, the smaller the sizes of crystallization nuclei able to grow. This figure provides further significant

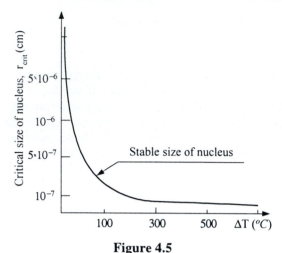

Figure 4.5
The change in the critical size of crystallization
nucleus as a function of ΔT undercooling

evidence for the necessity of undercooling. It was pointed out earlier that as far as the temperature is equal to the theoretical crystallization temperature of the examined material $(T = T_o)$, there is no driving force for the nucleation. At this temperature, the free energies of the liquid and solid phases are equal, i.e. $\Delta G = 0$.

Consequently, it is clear from Figure 4.5 that undercooling is necessary to initiate crystallization, since at $\Delta T = 0$ the critical size of the nucleus would become extremely large, (i.e. $r_{crit} \to \infty$). The nucleation of a nucleus with infinite size is perfectly impossible.

Conversely, Figure 4.5. suggests that the greater the rate of undercooling, the finer the grains. However, undercooling is limited since the decrease in temperature will also decrease the diffusion ability, which also limits the growth of crystals. The maximum undercooling (which is in closely related to the rate of undercooling) can be very significant for certain metals, and may reach hundreds of degrees. This is illustrated in Table 4.1, where some important parameters from the viewpoint of homogeneous nucleation (i.e. the melting temperature, the bonding energy, the surface energy and the maximum rate of undercooling) are shown for some metals widely applied in practice.

*Table 4.1 Characteristic parameters of homogeneous crystallization
for selected metals*

Symbol of the metal	Melting temperature (^{o}C)	Bonding energy (J/cm^2)	Surface energy (J/cm^2)	Maximum undercooling (^{o}C)
Pb	327	280	33×10^{-7}	80
Al	660	1066	93×10^{-7}	130
Cu	1083	1826	177×10^{-7}	236
Ni	1453	2660	255×10^{-7}	319
Fe	1536	2098	204×10^{-7}	295

Heterogeneous mechanism of crystallization nucleus formation

This mechanism can be characterized by the presence of foreign atoms that serve as the starting nuclei for the center of crystallization. These atoms are added to the liquid in order to facilitate or to accelerate the crystallization process. These nuclei play a significant role because the great undercooling necessary for the formation of homogeneous crystallization nuclei cannot always be ensured under industrial conditions. (Some examples of the amount of undercooling necessary in homogeneous crystallization are shown in Table 4.1).

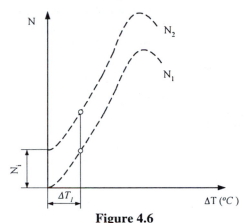

Figure 4.6

The effect of foreign crystallization nuclei on the crystallization ability

The effects of these foreign nuclei on the crystallization ability of metals are shown in Figure 4.6. It can be seen from this figure, that foreign crystallization nuclei increase the crystallization ability (N) by an additive value (N_i) compared to homogeneous crystallization.

From the viewpoint of crystallization ability, it is very important that the liquid of the material to be crystallized "*wets*" the surface of the nucleus. Another requirement is that the material should be easily "*bonded*" to the surface of the nucleus.

Liquid metal containing foreign crystallization nuclei has a lower surface energy compared to homogeneous crystallization. Consequently, the size of the nucleus is also smaller – see Equation (4.24) – therefore, significantly less undercooling is necessary for the formation of a stable nucleus, which then is able to grow.

4.1.2.2. Growth of crystals, crystallization mechanisms

In the previous section, we could see that crystallization nuclei formed in the liquid metal are able to grow under certain conditions. These stable nuclei can grow by different mechanisms. Crystallization – during continuous cooling – is continued until the whole liquid metal solidifies to a crystalline solid.

The growth of crystals always starts at the crystallization centers described in the previous section and occurs during the growth of nuclei by different mechanisms. The growth of nuclei can be characterized by the rate of crystallization growth. This is denoted by G and it means the linear growing rate of the crystal (therefore, its unit is mm/s). This rate of crystal growth is a direction dependent vector quantity and plays an important role in the different mechanisms of crystal growth.

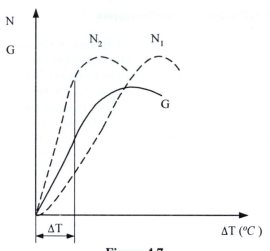

Figure 4.7

Nucleation ability (*N*) and rate of crystal growth
(*G*) vs. undercooling (*ΔT*)

In Figure 4.7, the rate of crystal growth is shown as a function of undercooling. In this figure, the variation of nucleation ability is also shown for various cooling rates. It is well known that the cooling rate has a significant effect on the grain structure (the grain sizes), and in this way, the mechanical properties of metals and alloys can be changed on a broad scale.

The grain structure is closely connected with the strength and deformation properties of metals. It is also well known that the yield limit (R_p) for most metals and alloys changes linearly as a function of $d^{1/2}$ (where *d* is the grain diameter). This relationship is described by the *Hall-Petch* rule:

$$R_p = R_o + kd^{-\frac{1}{2}}, \tag{4.25}$$

where R_p is the yield limit of the polycrystalline material, R_o is the yield limit of the monocrystalline material, *d* is the grain diameter and *k* is a material constant.

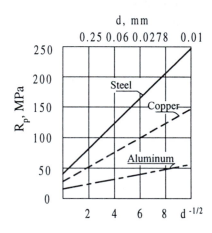

Figure 4.8

Yield limit vs. grain diameter for
various material qualities

The variation of the yield limit for various material qualities is illustrated in Figure 4.8 as a function of grain size.

Fine-grain structure plays a decisive role not only from the viewpoint of strength parameters, but it also affects the resistance of materials to dynamic loads. This is why a fine-grain structure is so important for many practical applications. The theoretical basis of producing fine-grain structure is illustrated in Figure 4.7.

It can be seen from this figure, that the rate of crystal growth is actually independent of the cooling rate, but the nucleation ability depends significantly on it. It is generally valid that fast cooling results in significant undercooling. Consequently, due to a greater nucleation ability, many nuclei formed in the liquid metal are able to grow. This results in a fine-grain crystal structure due to a lower rate of crystal growth.

In the next section, we will introduce the most typical mechanisms of polycrystalline crystallization.

Polyhedrous or grain type mechanism of crystallization

The polyhedrous or grain type crystallization is one of the most characteristic and general crystallization mechanisms of metals. This crystallization process is illustrated in Figure 4.9.

During crystallization, differently oriented crystallization centers are formed at certain points of the liquid metal as shown in part *a*) of Figure 4.9. Some of them are stable crystallization nuclei that are able to grow by bonding further and further atoms to these centers. Obviously, certain crystallization nuclei do not fulfil the conditions of forming stable crystallization nuclei described by Equation (4.24): these nuclei dissolve and disappear in the liquid.

Since nuclei formed in different parts of the liquid metal are oriented differently, the growing crystals will also be oriented differently, as indicated in part *b*) of Figure 4.9. During the process of crystal growth, differently oriented crystal parts meet one another and obstruct the regular growth and formation of ideal crystals. This is shown in parts *c*) and *d*) of Figure 4.9. Consequently, the final structure of crystallized material has irregularly shaped crystallites touching one another along these irregular grain boundaries, as can be seen in part *e*) of Figure 4.9.

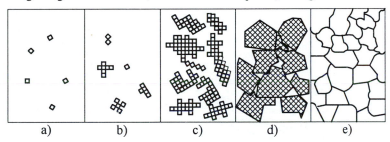

a)   b)   c)   d)   e)

Figure 4.9
Mechanism of polyhedrous (grain-type) crystallization

Mechanism of dendritic crystallization

In addition to polyhedrous (grain-type) crystallization, dendritic crystallization is another important group of polycrystalline crystallization. Dendritic crystallization is mainly due to vector characteristics of the rate of crystallization. There are two basic types of dendritic crystallization: *disordered* and *columnar dendritic crystallization*.

Figure 4.10
Disordered dendritic crystallization

In Figure 4.10, the result of the so-called *disordered dendritic crystallization* is shown. In this case, crystallization centers are formed at different points of the liquid metal. At first, differently oriented nuclei (resulting from the vector characteristics of the rate of crystallization) quickly grow like needles in the direction of one crystallographic axis. Because of this crystallization process, further atoms will be bonded to the tips of the needles. Due to the latent heat released during crystallization, local flows are generated in the liquid metal in the neighborhood of the needle tips that resulting in cross-linked atoms bonded to the formerly crystallized needles. These cross-linked atoms form sub-trees perpendicularly along the main axis of primary needles. In the crystallization process, along these sub-trees further smaller needles are formed. The result of all these processes is the characteristic "pine-tree" like structure illustrated in Figure 4.10. This is the reason why this crystallization mechanism is named after the Greek word for tree (*dendron*).

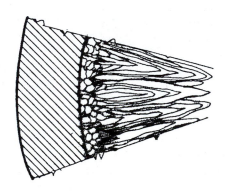

Figure 4.11
Mechanism of columnar (radial) dendritic crystallization

The other type of dendritic crystallization is the so-called *columnar dendritic* one (Figure 4.11). It is related to the direction of cooling and the vector characteristic of the rate of crystallization.

When the liquid metal is poured into the metallic mold, a large number of tiny nuclei are formed on the surface of the cold metal mold due to the rapid undercooling. Among these nuclei, only those favorably oriented from the viewpoint of direction of cooling can grow quickly: i.e. those having an orientation that coincides with the radial direction of the metal mold. These nuclei grow through the other less favorably oriented ones and form great columnar grains in the radial direction, as shown in Figure 4.11.

Spherulitic crystallization mechanism

The name of this type of crystallization originates from the Greek name of sphere (sfaira) and stone (lithos), since this crystal structure can be found in many minerals (stones). Among metals, spherulitic crystallization is very characteristic of the so-called nodular cast iron. To improve the mechanical properties of cast iron (to refine the shape, size and distribution of graphite lamellas), foreign nuclei – magnesium ($T_{melt} = 650\ ^{o}C$) or cerium ($T_{melt} = 775\ ^{o}C$) – are added to the liquid metal. Around these nuclei, as crystallization centers, graphite crystallizes in the form of spherical or nodular graphite, as shown in Figure 4.12.

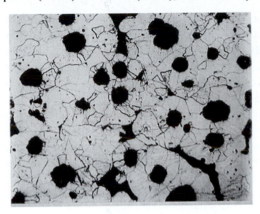

Figure 4.12
Nodular graphite cast iron, as an example of spherulitic crystallization

Production of single crystals

In the previous section, we summarized the different mechanisms of polycrystalline crystallization resulting in the so-called polycrystalline structure. In industrial practice, mainly polycrystalline materials are applied. However, there are some special application fields where functional requirements call for a material completely composed of a single crystal. From this point of view, certain electronic applications are regarded as the most important fields. Particularly, the production of transistors is the field where single crystals, composed of semi-conductors, are largely applied. The application of single crystals is an essential requirement in this field, since the grain boundary of polycrystalline materials would "destroy" the special electronic characteristics of semiconductors.

Obviously, apart from the semiconductor industry, single crystals are also applied in other areas. Single crystals are almost indispensable, for example, in the analysis of elementary processes of plastic deformation. In single crystals the elementary processes of deformation are disturbed neither by the presence of grain boundaries nor by the interaction between differently oriented grains.

The production of single crystals requires special technology. During single crystal production, such conditions should be ensured that only one nucleus formed in the liquid metal is able to grow and solidify into a single crystal in the entire volume. The growth procedure of single crystals – with the precisely controlled melting/cooling process – utilizes the characteristics of nucleation ability and crystallization rate. As a result, the nucleation ability is reduced and localized so that only one nucleus can be formed. The undercooling is insufficient to form further nuclei, while the cooling rate is sufficient for the growth of the single formed nucleus.

There are several methods for single crystal production – these utilize the principles introduced previously but they differ in their practical realization. One of the simplest procedures is usually applied to produce single crystals for studying the elementary processes of plastic deformation. In this case, a tube made of heat-resistant glass with a sharp end is used. The tube is filled with the material to be crystallized and is pulled through a vertical (generally an inductively heated) tube-furnace in such a manner that the metal melts before it leaves the furnace. The speed of pulling is kept at a low value to ensure a slightly lower pulling speed than the rate of crystallization of the examined metal. At the sharp end of the tube, which leaves the furnace first, only one nucleus can be formed. Due to the low pulling speed, the metal melted in the furnace continuously solidifies into a single crystal in its total volume, continuously bonding atoms to the nucleus formed at the tube end.

In the semiconductor industry, mainly the so-called *Czochralski-method* is applied to produce high-quality single crystals. The theoretical scheme of the process is shown in Figure 4.13.

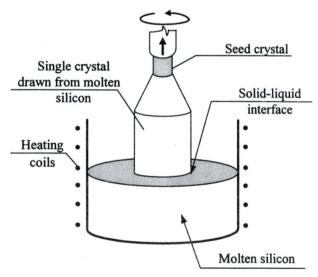

Figure 4.13
Production of a semiconductor single crystal by the Czochralski method

In this process, the polycrystalline silicon is first melted in a melting pot that does not react with the base material. The molten silicon is held at a constant temperature, only a few degrees above the melting point. A high-quality crystallization nucleus, i.e. a seed crystal of silicon with the desired orientation, is added to the molten liquid metal while it is continuously rotated. A part of the nucleus surface is molten to provide the necessary conditions for bonding further atoms to the seed crystal through this surface. The growing nucleus is still rotated and slowly moved upwards uniformly and continuously. As it is raised from the molten metal, silicon atoms bond to the seed crystal providing its growth both in the axial and diagonal direction producing a large-diameter single crystal of silicon.

After the single crystal has been produced, it is ground to a precise diameter and then sliced into very thin layers. These layers are chemically etched and polished with successively finer polishing abrasives until a defect-free, perfect mirror-like surface is achieved. The produced high quality, perfect semiconductor layer forms the basis for transistor production.

4.2. Diffusion in solid materials

Diffusion – in the general context – is the collective term for those processes in which matter is transported within matter. Diffusion can occur in all kinds (solid, liquid, gaseous) of materials. In this book, we generally consider the movement of atoms in solid materials by diffusion. In the next sections, we shall use this narrower interpretation of diffusion.

The movement of atoms in solid materials, i.e. diffusion, plays an extremely important role in metallurgical processes and reactions. Starting from crystallization, the different phases of recrystallization processes following plastic deformation, the various phase transformations – as the basis of heat treatment processes – are the most important evidence for the importance of diffusion.

There are two basic types of diffusion: so-called self-diffusion and concentration diffusion. *Self-diffusion* is the movement of atoms composing the solid material caused by the energy difference between the different parts of the material. *Concentration diffusion* is the movement of atoms caused by the concentration difference between various parts of the material.

Diffusion (either self-diffusion caused by an energy difference or concentration diffusion caused by a concentration difference) has certain common characteristics. *Diffusion* in any case *requires driving force*: in self-diffusion, the driving force is the energy difference, while in concentration diffusion, the driving force is the concentration difference in different parts of the material. Generally, the driving force itself is insufficient to initiate the diffusion processes. They also need *activation energy*. In most diffusion processes, the activation energy is the heat energy. A further common characteristic of both types of diffusion processes is that they are *irreversible*.

There are two basic mechanisms of concentration diffusion: the so-called substitutional and the interstitial diffusion mechanisms. In the first, *vacancies, i.e. the empty lattice points,* play a dominant role.

A theoretical illustration of the *substitutional diffusion* mechanism is shown in Figure 4.14. If any atoms, next to a vacancy in the lattice, possess the activation energy required to initiate the process, diffusion can start. In part a) of Figure 4.14, the neighborhood of a vacancy can be seen, while in part b) of Figure 4.14, the potential energy of the atom (denoted by a black spot) is shown as a function of position change. Consequently, the atom next to a vacancy can move to the position of the vacancy (thus creating a vacancy site at its own original location) if it is able to overcome the "energy barrier" denoted q_m. This is equal to the activation energy required for a vacancy movement. Naturally, the substitutional diffusion mechanism occurs not only in the neighborhood of the vacancy as illustrated in Figure 4.14.

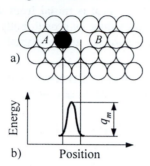

Figure 4.14
Theoretical illustration of the substitutional diffusion mechanism

If the required driving force and activating energy is available, diffusion passes through the material by the described vacancy mechanism, as can be seen in Figure 4.15.

Direction of the diffusion

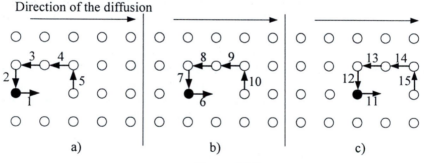

a)   b)   c)

Figure 4.15
The movement of atoms in solids with the vacancy mechanism

The previously described diffusion mechanism can be proved by several experimental facts, as it follows:

- Diffusion is much greater at the grain boundaries than within the grains. It is well known that lattice defects at the grain boundaries (e.g. vacancies, dislocation) have a much higher density than within the grains. This fact proves the role of vacancies in diffusion.

- It has also been shown by experiment that diffusion is greater in polycrystalline materials than in single crystals. This is obvious, since there are no grain boundaries in single crystals.

- It is also well known that the diffusion rate is higher in materials with fine grain structure. In fine-grained materials, the specific grain boundary surface and, consequently, the specific density of vacancies is much higher.

- The diffusion rate is also increased by plastic deformation. It is known that during plastic deformation, the number of dislocations (the density of dislocations) increases with plastic deformation. Thus, it provides more vacancies for the diffusion mechanism, as well.

- The diffusion rate increases exponentially with an increase in temperature. The increase in temperature can contribute to the increase in diffusion rate in several ways. On the one hand, lattice parameters are increased with increasing temperature providing a greater space for the movement of atoms. On the other hand, the temperature rise also provides so-called "thermal vacancies". Furthermore, the increase in temperature provides the necessary activation energy for the diffusion processes.

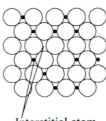

Interstitial atom
diffusing into
interstitial vacancy

Figure 4.16
Schematic illustration of
interstitial diffusion

The so-called *interstitial diffusion mechanism* occurs in a similar way. The greatest difference can be found in the following. In interstitial diffusion mechanisms, foreign atoms located in an *interstitial* position diffuse into neighboring (empty) interstitial positions, as illustrated in Figure 4.16.

In the previous analysis, phenomenological knowledge related to diffusion was summarized. The first mathematical description of diffusion was proposed by Fick. His basic equation refers to the diffusion rate. According to it, in quasi-stationary diffusion, the quantity of atoms passing through the cross section (*A*) perpendicular to the direction of diffusion, per unit time, is directly proportional to the concentration gradient, ie.:

$$J = -D\frac{dc}{dx} \, , \qquad\qquad (4.26)$$

where D the so-called diffusion coefficient,
       J the material quantity passing through a unit surface per unit time (its unit is the number of atoms $/m^2 \cdot s$)

$\dfrac{dc}{dx}$ the concentration gradient.

The negative sign in the equation refers to the fact that atoms diffuse from a point with higher concentration to a point with lower concentration.

Actually, from the viewpoint of engineering practice, non-stationary diffusion processes are more important, since concentration often changes as a function of time. It is described by Fick's second law. According to this rule, the temporal change in concentration is proportional to the changes in the concentration gradient:

$$\frac{dc}{dt} = \frac{dc}{dx}\left(D\frac{dc}{dx}\right). \tag{4.27}$$

If the diffusion coefficient (D) is regarded as a constant in the space, the previous equation can be rewritten as follows:

$$\frac{dc}{dt} = D\frac{d^2c}{dx^2} \tag{4.28}$$

The diffusion coefficient D appearing in both Fick laws is highly dependent on the temperature. This relation is described by the so-called *Arrhenius-type* equation. According to this equation:

$$D = D_o e^{-\frac{Q}{RT}}\,, \tag{4.29}$$

where D_o is a constant value depending on the material grade and the type of diffusion (its value can change between 10^{-6} - 10^2 mm^2/s for concentration diffusion and between 10^{-1} - 10 mm^2/s for self-diffusion)

R the universal gas constant, its value is equal to 8.314 J/(mol·K),

Q the activation energy (J/mol).

It follows from Equation (4.29) that the diffusion coefficient increases exponentially with increasing temperature. This explains why in metallurgical processes, where diffusion plays an important role, the time necessary for the process decreases exponentially with the increase in temperature.

From this well-known rule, it is also obvious that temperature and time in metallurgical processes – within certain limits – can be considered as parameters substituting each other. This suggests, that if the time required to perform a process is equal to t_1 at a lower temperature (T_1), at a higher temperature (T_2) the time necessary to complete the process t_2 decreases exponentially (i.e. $t_2 < t_1$).

4.3. Polymorphism and allotropy

Some crystalline materials can exist in more than one crystal structure in different temperature domains: this phenomenon is called *polymorphism* of crystalline materials. The polymorphism of metals is called *allotropy*. Phase transformations from one crystal structure to another are called *allotropic transformations*. Different crystal structures resulting from allotropic transformations are called *allotropic modifications*. They are usually marked with Greek letters in the order of their existence as a function of temperature.

Allotropic transformation is actually a phase transformation occurring in the solid state. This, similarly to the so-called primary crystallization, a liquid to solid transformation, involves a significant energy change. This is the reason why phase transformations in the solid state, similarly to primary crystallization, result in characteristic break points on the cooling curves of metals. (For example, on the cooling curves of pure metals, a horizontal straight line appears on the cooling curve, indicating a non-variant, constant temperature process.) The non-variant process of allotropic transformation of pure metals can be justified using the well known Gibbs' phase rule.

Allotropic transformations are of great significance from the viewpoint of ensuring certain properties. However, in certain cases they can have unfavorable consequences, as well. A good example of this is the allotropic transformation of tin (Sn). These unfavorable consequences are well known. However, at first people did not assign them to allotropic transformations. Pure tin crystallizes in a body-centered tetragonal crystal structure between 13.2 - 161 $^{\circ}$C (this is the so-called β-Sn), which under 13.2 $^{\circ}$C is transformed into the α-Sn modification having a lattice similar to the diamond structure. α-Sn – which is called gray tin due to its characteristic gray color – is an extremely brittle phase, while β-Sn, which has a white color, is a ductile and stable modification of Sn. Since β-Sn is extremely stable, the $\beta \rightarrow \alpha$ allotropic transformation spontaneously occurs at a temperature much lower than the equilibrium transformation temperature ($T_{\alpha \rightarrow \beta}$=13.2 $^{\circ}$C). This temperature is lower by approximately 50 - 60 $^{\circ}$C, i.e. the phase transformation usually occurs at T=(-30) - (- 40) $^{\circ}$C. At the same time, this modification of tin may collapse into gray powder as a result of a minor impact.

It is well known that organ pipes are made of tin. In countries located in a geographically cold climate, there were such cold winter periods that the previously described phase transformation occurred in unheated churches. Because of dynamic effects caused by sound waves, tin pipes collapsed: shining, bright pipes turned into gray powder. In medieval times, people thought it was an unknown "illness" of tin and called it tin plague. Obviously, today we already know that the explanation can be found in the allotropic transformation process of tin. Tin powder can regain a normal crystal structure, by melting and recrystallizing.

In the previously described phase transformation – which can be considered as a curiosity – the allotropic transformations of iron are of much greater importance in industrial practice. The allotropic transformations of pure iron, with the indication of transformation temperatures, are illustrated by Figure 4.17.

Under equilibrium cooling conditions, the crystallization of pure iron from the liquid to the solid state occurs at the temperature $T = 1536$ $^{\circ}$C (this is the melting/crystallization point of pure iron). The result of this primary crystallization process is a body-centered cubic crystal called δ_{Fe}. This δ_{Fe} modification of iron is stable until it reaches the temperature $T = 1392$ $^{\circ}$C. At this temperature, due to the $\delta_{Fe} \rightarrow \gamma_{Fe}$ phase transformation, the body-centered crystal is transformed into a face-centered cubic crystal (called γ_{Fe}). Under equilibrium cooling conditions, at the temperature $T = 911$ $^{\circ}$C the face-centered cubic crystal (γ_{Fe}) will change again to a body-centered modification called α_{Fe}, as a result of $\gamma_{Fe} \rightarrow \alpha_{Fe}$ allotropic transformation. (The $\gamma_{Fe} \rightarrow \alpha_{Fe}$ phase transformation is extremely important from

the viewpoint of heat treatment of practical iron-carbon alloys. This will be
analyzed in detail later.)

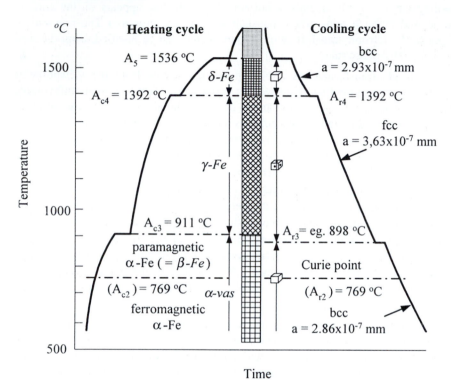

Figure 4.17
Allotropic transformation of pure iron (*Fe*)

At the same time, it is obvious from Figure 4.17 that $\gamma_{Fe} \to \alpha_{Fe}$ phase
transformation occurs at a temperature below the equilibrium temperature
($T = 911$ $^{\circ}$C). This phenomenon is called the *temperature hysteresis of allotropic
phase transformations*. The value of the temperature hysteresis increases with the
increase in cooling rate. It is also of great practical significance in heat treatment
processes.

Considering Figure 4.17, it is apparent that the modification of β_{Fe} is missing
from the series of allotropic modifications ($\alpha \to \gamma \to \delta$). This can be reasoned as
follows. Pure iron is not magnetic (i.e. paramagnetic) at temperatures above
$T = 769$ $^{\circ}$C, but it is magnetic (i.e. ferromagnetic) at lower temperatures. The
temperature, which separates para- and ferromagnetic states, is called the *Curie
point*. The two states are different only from the viewpoint of magnetization, but
up to the allotropic transformation temperature $T = 911$ $^{\circ}$C, the same body-
centered α_{Fe} can be found in both cases. (Earlier, ferromagnetic material was
regarded as a new allotropic alteration and it was indicated as α_{Fe}, while the

paramagnetic modification in the temperature range $T=769-911\ ^\circ C$ was indicated as β_{Fe}).

The previously described allotropic transformations occur similarly during heating. Transformational temperatures on heating are less dependent on the heating rate than on cooling. Therefore, during heating, the hysteresis of transformation temperatures is smaller compared to the equilibrium values. For the indication of transformation temperatures – following a suggestion of *Osmond,* a French metallurgist – the letters A_1 to A_5 are used. (*A* originates from the French word *arret = stop,* while the number indices indicate the increasing rank of transformation temperatures.) The two transformation temperatures – valid for heating and cooling, respectively – are differentiated by the *c* and *r* indices (*c*-means *chauffage = heating,* while *r*-means *refroidissement = cooling*).

Concerning the notations in Figure 4.17, there is an apparent contradiction. Allotropic transformation temperatures include A_2 to A_5 values during both heating and cooling. (The temperature A_2 does not mean an allotropic transformation. It corresponds to the *Curie* point separating ferro- and paramagnetic domains. Furthermore, A_1 is missing from the figure. This may be reasoned as follows: iron alloys, containing more carbon than $C = 0.025\ \%$, go through a further transformation at $T = 723\ ^\circ C$, which is denoted A_1. (In fact, this is not the earlier explained type of phase transformations. It is the temperature of the so-called pearlitic transformation of the γ-phase. It will also be detailed in the next sections.) Obviously, this transformation does not occur in pure iron; therefore, the A_1 temperature does not appear on the cooling curve of pure iron. However, at that time, when the cooling curve of "pure iron" was first determined, it was not possible to produce high quality pure iron containing less than C=0.025% carbon. Therefore, the A_1 temperature was experienced and it was included in the series of allotropic transformation temperatures of pure iron, as well.

5. IMPERFECTIONS IN CRYSTALLINE SOLIDS

Ideal crystals can be characterized by an ordered, regular repetition of the lattice in any direction of space. This also means that at each lattice point an atom composing the given crystalline material can be found. In other words, an ideal crystal can be regarded as a perfect crystalline structure.

However, crystalline materials are never perfect, they contain various types of crystallographic defects. Though, these imperfections are often minor, they can have very significant effects on certain properties. As an example, we can mention that the strength of crystalline metals having great practical importance might be less than 1% of the theoretical strength calculated on the basis of the ideal crystal structure. As another example, silicon can be mentioned. If silicon is contaminated with 10^{-8} mass percentage of boron, its conductivity is doubled compared to pure silicon (which is widely applied in different fields of the electronic industry). Naturally, several other examples could be listed here.

Consequently, in real crystals the perfect structure of ideal crystals is disturbed by various disorders and crystal defects. There are certain properties that can be perfectly defined on the basis of the perfect (ideal) lattice structure – these are the so-called *structure-insensitive properties*, such as thermal and electrical conductivity or specific heat, etc.

However, some other properties show significant differences in real crystals compared to ideal crystals, i.e. there are large differences in certain properties concerning the theoretically calculated and experimentally measured values. These differences are in the magnitude of several orders. This contradiction can be resolved only by the assumption of various imperfections in real crystals. Properties that can be defined in accordance with experimental results only, based on the real crystalline structure, are called *structure-sensitive properties*. For example, most of the mechanical properties of metals belong to this category, like different strength and strain properties, such as *yield limit, ultimate tensile strength* or *true strain* (these are very important from the viewpoint of plastic deformation).

Imperfections may have both advantageous and disadvantageous effects and they are always present in real materials. That is why their analysis and conscious utilization is of great significance from the viewpoint of practical engineering.

5.1. Main types of imperfections and their classification

Imperfections can be classified on the basis of their geometry and shape. According to this classification, imperfections belong to three main groups:
* point (zero-dimensional) defects,
* line (one-dimensional) defects, and
* surface (two-dimensional) defects.

5.1.1. Point defects – zero-dimensional imperfections

Point defects can be characterized by the fact that the effect of the imperfections extends over a relatively small distance, usually in the order of some lattice parameters.

Two main types of point defects are the so-called *vacancies* and *interstitialcies*. While a vacancy means an atom missing from the regular repetition, interstitialcy can be caused either by the atoms composing the lattice or by foreign atoms. The simplest point defect is a *vacancy* or, in other words, an *empty lattice site*. Vacancy means a missing atom at any lattice site as illustrated by Figure 5.1.

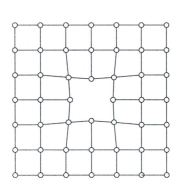

Figure 5.1
Vacancy, as a point defect

Vacancies can be formed during crystallization because of imperfect packing, but vacancies can also occur at elevated temperatures as *thermodynamic vacant lattice sites*. With an increase in thermal energy, there is an increased probability that certain atoms will jump out of their lattice sites that correspond to their lowest energy level. Vacancies play an extremely important role in diffusion processes. The number of thermodynamic vacancies increases exponentially with temperature, as can be seen from the following equation:

$$n_v = Ne^{-\frac{E_v}{kT}} ,$$ (5.1)

where n_v the number of vacancies at the temperature T,
      N the total number of lattice sites,
      k the Boltzmann's constant (its value is $13.8 \times 10^{-24} J/K$)
      E_v the energy necessary to form a vacancy,
      T the absolute temperature in Kelvin degrees.

Vacancies may arise in several ways. The so-called Frenkel-type defect (Figure 5.2) is formed when an atom is displaced from a lattice site into an interstitial position. Thus, the Frenkel-type point defect is an example of interstitialcy, also. However, to form a vacancy and an interstitialcy simultaneously requires such a high energy that this type of vacancy formation has a very low probability. This high-energy may be provided by neutron irradiation in atomic reactors. Therefore, in reactor materials, both vacant lattice sites and interstitial atoms originated by the Frenkel mechanism may be found.

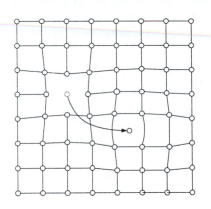

Figure 5.2
Frenkel mechanism of a simultaneous formation of a vacancy and an interstitial atom

A more realistic way to form vacancies was suggested by Schottky. According to this mechanism, vacancies first arise at crystal boundaries and move inside the lattice by diffusion, as can be seen in Figure 5.3.

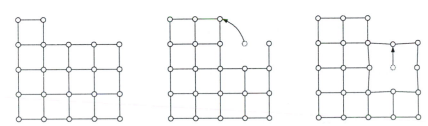

Figure 5.3
Vacancy formation by the Schottky mechanism

In the previously mentioned types of point defects, imperfections were caused by so-called own atoms (i.e. atoms composing the given crystal structure). In many cases, point defect type imperfections are caused by foreign atoms. There are two main types of point defects depending on the position of the foreign atoms. Foreign atoms may be located at lattice sites (substitutional point defects) and at lattice voids in intermediate positions (called interstitial point defects).

The occurrence of interstitial foreign atoms as point defects means that a foreign atom with small atomic diameter is positioned in the lattice voids (not at the lattice sites) as shown in Figure 5.4 a). Interstitialcy can be formed only by elements having a much smaller atomic radius than that of the own atom composing the lattice. This is due to the fact that only elements with a small atomic radius can fit into the voids or interstices among the original atoms. Consequently, only elements with a very small atomic radius (such as hydrogen, nitrogen, oxygen, boron or carbon) form interstitial foreign atoms in the lattice of crystalline materials with a greater atomic radius.

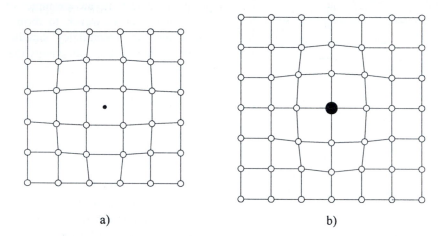

a)   b)

Figure 5.4
Point defects caused by foreign atoms
a) Interstitial foreign atom b) Substitutional foreign atom

Substitutional foreign atoms as point defects mean that foreign atoms substitute the host atoms at the lattice sites as shown in Figure 5.4 b). There are several features of the solute and solvent atoms that determine the main properties of substitutional solid solutions. This topic will be discussed in detail in a later chapter on the formation of solid solutions.

The previously introduced point defects result only in small distortions in the crystal lattice. However, their presence is very significant from several points of view (e.g. diffusion), these imperfections are insufficient to explain the differences existing between the various parameters of real and ideal (perfect) crystals. As is well known, these differences can extend to several orders of magnitude in case of certain properties.

5.1.2. Line defects - one dimensional imperfections

Among the different imperfections, line defects are of utmost importance. *Line defects* are various kinds of dislocations that can be classified into two basic types, i.e. *edge dislocation* and *screw dislocation*. With the combination of these two basic types, various *mixed dislocations* as line defects can be produced.

Dislocations are regarded as line defects since they extend over several hundreds or sometimes several thousands of lattice parameters, whilst, in the other two directions, they extend only to a few lattice distances. Thus, dislocations can cause lattice imperfections (disturbances) to a great extent compared to the lattice parameters.

As already mentioned when we discussed point defects, the presence of some point defects having a small lattice extension is insufficient to explain the great differences between theoretical and experimental values of structure-sensitive properties in crystal structures. Due to the scientific results of Taylor, Polányi and Orowan we can understand and explain crystal structure anomalies on the basis of *dislocation theory* which is very important from the viewpoint of plastic deformation.

A dislocation mechanism theoretically developed by Taylor and Orowan was improved and first proved experimentally by Frank. He was the first scientist who demonstrated dislocations appearing on crystal surfaces: these experiments were performed by electron microscopic examinations.

5.1.2.1. Edge dislocations

Edge dislocations may be described as an edge of an extra half-plane of atoms within a crystal structure. For further description of edge dislocations, we can use the following vector notations: $\vec{u}$ denotes the displacement vector between two parts of the crystal, the *axis of dislocation* is denoted $\vec{t}$ and the so-called *Burger's vector* is denoted $\vec{b}$. Burger's vector can be defined based on the so-called *Burger's circuit* shown in Figure 5.5.

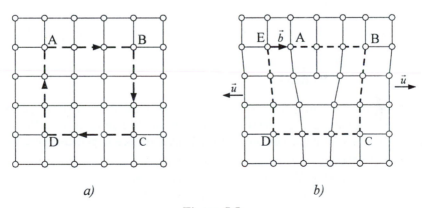

a)   b)

Figure 5.5
Derivation of Burger's vector and the Burger's circuit
a) Burger's circuit in a perfect crystal b) Burger's circuit around an edge dislocation

Burger's circuit can be drawn passing equal distances in the positive and negative translation directions of the crystal as shown in Figure 5.5. In perfect (ideal) crystals, Burger's circuit is closed (Figure 5.5. a), while Burger's circuit including an edge dislocation has a so-called closure failure (Figure 5.5. b).

Burger's vector can be defined as the vector spanning the gap between the starting and end points of Burger's circuit. Burger's vector characterizes dislocation both quantitatively and qualitatively.

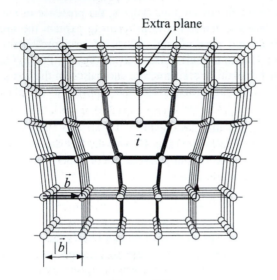

Figure 5.6
Edge dislocation as an extra half-plane

If the axis of dislocation and Burger's vector are perpendicular to each other, i.e. $\vec{b} \perp \vec{t}$ is true, it is regarded as an edge dislocation. Edge dislocations (as we will see later) are mostly formed during plastic deformation. Beside the previous ($\vec{b} \perp \vec{t}$) condition, we can also state that Burger's vector is parallel to the $\vec{u}$ displacement vector, which for edge dislocations leads to the condition that $\vec{u} \parallel \vec{b} \perp \vec{t}$.

An elastic distortion zone can be observed around the dislocations, which is illustrated by a so-called spring model in Figure 5.6. The crystal below the dislocation line is in a stretched state, while the part above it is in a compressed state. The energy level increased by the elastic distortion caused by the edge dislocation is proportional to the square of the absolute value of Burger's vector.

5.1.2.2. Screw dislocations

Screw dislocations constitute the second main type of line defects. A screw dislocation is like a spiral ramp with an imperfection line down its axis. The plane of atoms involved in the formation of the helical surface is perpendicular to the dislocation axis ($\vec{t}$).

Similarly to edge dislocations, we can define Burger's circuit and Burger's vector (Figure 5.7) for screw dislocations, as well. However, there is a basic difference: in the case of screw dislocations, Burger's vector is parallel to the axis of dislocation and to the slip plane as well, therefore $\vec{u} \parallel \vec{b} \parallel \vec{t}$ is valid.

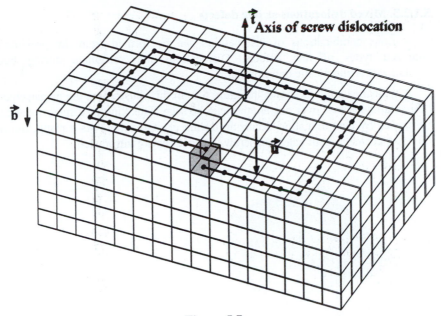

Figure 5.7
Schematic spatial illustration of a screw dislocation

The helical surface can be easily recognized on the two-dimensional illustration of the screw dislocation in Figure 5.8. In this figure, the slip direction ($\vec{u}$), the axis of dislocation ($\vec{t}$), and Burger's vector ($\vec{b}$) are also indicated.

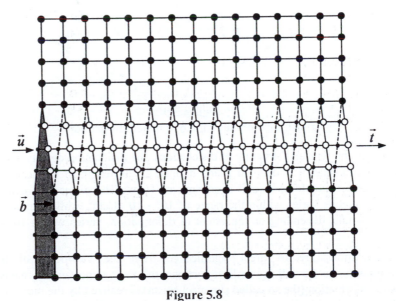

Figure 5.8
Two-dimensional illustration of a screw dislocation

5.1.2.3. Mixed dislocations as line defects

Various combinations of dislocations can be formed from the previously analyzed types of line defects. Dislocations may originate during both crystallization and plastic deformation.

A typical example is shown in Figure 5.9. In this illustration, the displacement of crystalline planes over one another can be seen. This kind of displacement can be caused by heat dilatation occurring during crystallization. As a result, during this displacement, both edge and screw dislocations are introduced.

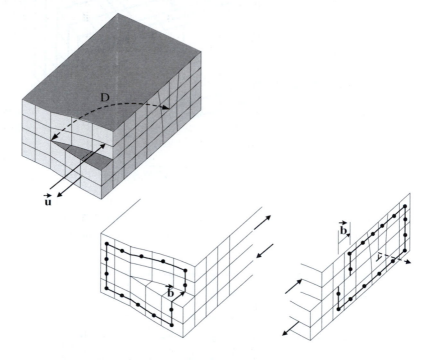

Figure 5.9
Illustration of the origin of a mixed dislocation

Beside the type and structure of dislocations, the dislocation density also is of great significance. It is denoted ρ and defined as the length of dislocations per unit volume. Therefore, its unit is cm/cm^3, or $1/cm^2$. The dislocation density of annealed metals varies within the range of 10^6 - 10^7 $1/cm^2$, while that for cold-formed metals, depending on the extent of deformation, varies within 10^8 - 10^{11} $1/cm^2$.

Dislocations are of great significance from the viewpoint of plastic deformation. As we will see in later sections, the most important mechanism of plastic deformation (the so-called slip mechanism) is realized by the movement of dislocations in the slip planes. Moving dislocations while can react with each other. During these reactions, certain dislocations can disappear. However, it is more characteristic for the slip mechanism that a great number of new dislocations are formed during plastic deformation. In other terms, dislocations are multiplied, their number and the dislocation density increase. Since an elastic distortion with higher energy can be observed around dislocations, it results in a continuously increasing resistance against plastic deformation. Because of this resistance, increasing plastic deformation can be performed only by an increasing level of stresses. This phenomenon is called *strain hardening* and it will be discussed in detail in a later chapter.

One-dimensional line defects result in crystal imperfections extending over a great area of atomic size. Concerning the so-called structure-sensitive properties, this is sufficient to explain the significant differences between the theoretically calculated and experimentally measured values.

5.1.3. Surface defects – two-dimensional imperfections

Surface defects can be characterized as follows: they extend in two directions over a relatively large surface in an atomic context, while their size in the third direction is equal only to the value of one or two lattice parameters.

This type of imperfections can be divided into further sub-categories. One of the subgroups is independent of crystal structures. It includes two different types: *grain boundaries* and *phase boundaries*.

Furthermore, certain surface defects are dependent on crystal structures, such as *coherent phase boundaries, twin boundaries* and *stacking faults.*

5.1.3.1. Surface defects independent of crystal structures

Grain boundaries as surface defects

Crystallites at the boundaries, originating from differently oriented crystalliza-tion nuclei, do not result in the formation of a regular crystal boundary as would be expected in ideal (perfect) crystals. Misorientations are gapped by extra half-planes (edge dislocations) on grain boundaries and they make crystallization processes complete. This process can also well illustrate the formation of edge dislocations during crystallization. Small-angle grain boundaries occur when the misorientation at the grain boundary is in the range of a few degrees (less than 5°). Regions surrounded by small-angle grain boundaries are called subgrains or subcrystals. These regions are basically free of dislocations. (These regions are also named *mosaic blocks.*)

If the misorientation is greater than the characteristic value for small-angle grain boundaries, it is called a *high-angle grain boundary*. Regions separated by high-angle boundaries are called grains. In Figure 5.10, a high-angle grain boundary can be seen which is formed during crystallization. Misorientation at grain boundaries results in a less ordered arrangement of atoms. Consequently, the

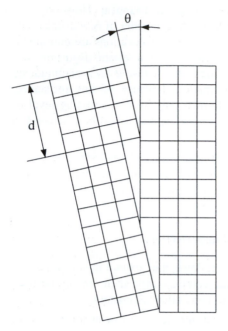

Figure 5.10
Grain boundary as a surface defect

atoms along the boundary have a higher energy than those within the grains. This higher energy level is determining from the viewpoint of several metallurgical processes: the higher energy level of grain boundaries has an advantageous effect on the processes of the formation of crystallization nuclei that are necessary for allotropic phase transformations. Furthermore, the higher energy level supports diffusion in solid bodies and has a significant influence on the movement of dislocations.

Phase boundary as a two-dimensional, surface defect

Phase boundaries occur at the boundary of two different phases, and are to a certain extent similar to grain boundaries. They are also two-dimensional surface defects. Certain types of phase boundaries are *independent of crystal structures* (such as *incoherent phase boundaries*). Some others can only occur under definite crystallographic conditions or in other words, they are *surface defects dependent on crystal structures* (such as *semi-coherent and coherent phase boundaries*).

Since incoherent and semi-coherent phase boundaries can be understood more easily based on the description of coherent phase boundaries, let us first describe this latter one. There are some special conditions for the formation of *coherent phase boundaries*: firstly, there must be the same atoms in the crystal lattices in the adjacent phases. Secondly, the two different phases must have a common crystal plane where the atomic order is the same. Finally, misorientation should be a definite value between the two phases along the phase boundaries.

If all the three conditions are fulfilled, a coherent phase boundary will be formed between the two phases. This surface defect has a low energy level, since atoms along the boundary are located at the lattice sites in both phases. An example of coherent phase boundaries is the one formed at the phase boundaries of α and γ iron under certain conditions. This is illustrated in Figure 5.11.

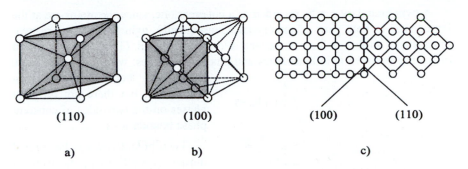

Figure 5.11
Coherent phase boundary in α and γ iron

If only the first two conditions are fulfilled (i.e. the two phases are composed of the same atoms and they have at least one crystal plane with the same atomic order), but misorientation at the boundary of the two phases is different from the value determined by crystallographic laws, a *semi-coherent phase boundary* can be formed. In this case, there will be a coherent phase boundary in certain regions. However, the misorientation is different from the value required to form coherent phase boundaries; therefore, a coherent phase boundary cannot be formed over a certain section, since it would require too high elastic stresses to gap the orientation differences. In these regions, misorientations are equalized by dislocations gapped into the phase boundaries and the formation of local coherent phase boundary is again possible. If the first or the second of the previously listed conditions is not completely fulfilled, only *incoherent phase boundary* can be formed. Naturally, an incoherent phase boundary will also be formed in those cases where none of the conditions is fulfilled. The incoherent phase boundary represents a two-dimensional surface defect with a high energy level.

5.1.3.2. Surface defects depending on crystal structures

This group of surface defects includes two-dimensional imperfections that are dependent on strict crystallographic laws.

Obviously, coherent phase boundaries are also included in this group. In this section, we will introduce further surface defects depending on crystal structures, namely twin boundaries and stacking faults.

Twin boundaries as surface defects

Twin boundaries are similar to phase boundaries from several points of view. A *twin boundary* separates contiguous volumes of material, which are mirror images of one another along the twin planes (Figure 5.12).

A twin boundary can be regarded as a coherent one, since the atoms are at the lattice sites in both parts of the crystal. Consequently, twin boundaries are two-dimensional surface defects possessing a low-energy level. (There is a significant difference between twin boundaries and phase boundaries: we can find the same phases on both sides of twin boundaries, but there are different phases on the two sides of coherent phase boundaries.)

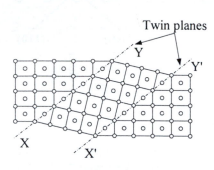

Figure 5.12

Schematic illustration of twin boundary formation

Twin boundaries can be formed during crystallization, as well as during plastic deformation. In α copper and austenite steels, crystallizing as face-centered cubic crystals, twin boundaries are characteristically formed during the crystallization process. Twin boundaries appear as parallel lines on the microscopic images. Twin boundary formation also plays a significant role in the formation of a favorable crystallographic orientation from the viewpoint of plastic deformation in metals and alloys crystallizing according to the hexagonal crystal system. This topic will be analyzed in more detail in the section dealing with the deformation mechanisms.

Stacking faults

A *stacking fault* is a very important type of surface defects depending on crystal structures. It can be regarded as a disturbance of the regular order of atomic planes composing the crystal.

A comparison of the face-centered cubic crystal structure with the hexagonal crystal may be considered a good example for studying of the stacking fault. Both crystal structures are composed by crystallographic planes possessing the same atomic arrangement, however, the order of the crystal planes is different in the two different crystal system. Face-centered cubic crystals with the Miller indices (111) have completely the same (equivalent) planes from crystallographic viewpoints as the (0001) base plane in the hexagonal crystal structure. Consider first an arbitrary plane with (111) Miller indices in the face-centered cubic system. Since there are two different positions of voids between the atoms, a second and a third plane can follow in two different manners until we get to a plane with the same position of atoms as the first one. This means that altogether three different positions in the order of the planes can be arranged. The order of crystalline planes may be indicated as follows: ABC ABC ABC (see Figure 5.13 a). In hexagonal crystal structures, there are only two different possibilities to locate the (0001) planes, since the third plane (i.e. the top plane) should be located exactly as the base plane. It results in the following order of crystal planes: AB AB AB (see Figure 5.13 b).

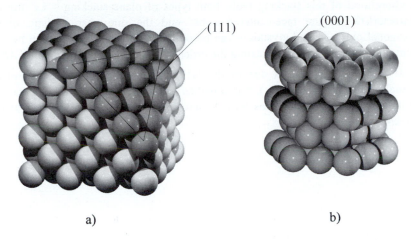

(111)   (0001)

a)   b)

Figure 5.13

Comparison of crystal structures with the same atomic arrangement
a) face-centered cubic crystal b) hexagonal close-packed crystal

If the regular arrangement of the face-centered cubic crystal structure is disturbed by any imperfection (e.g. the junction of vacancies, vacant lattice sites extending over a greater area), in these parts, the repetition of the planes becomes similar to that of the hexagonal crystals. This is illustrated in Figure 5.14.

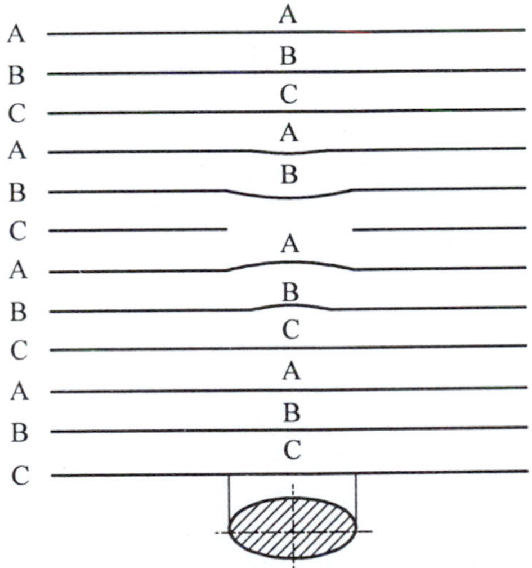

In this figure, the projection of (111) crystallographic planes of face-centered cubic crystals is shown. (This is why the planes are shown as lines in the figure.) In the perfect parts of the crystal, the regular plane repetition characteristic for the face-centered cubic crystal structures can be seen. The regular order of planes is the following: *ABC ABC ABC*.

In one of the planes, denoted by the letter *C* in the middle of the figure, an imperfection extending over a great area can be seen. Consequently, the order of planes at this region becomes the same as that characteristic of the hexagonal close-packed crystal structures, i.e.: *AB AB*. (Obviously, at

Figure 5.14

Schematic illustration of stacking fault formation

a certain distance from this two dimensional defect, the regular plane repetition characteristic of the face-centered cubic crystal will reappear. Therefore, in the neighborhood of this stacking fault, both types of plane stacking – i.e. the one characteristic of the face-centered cubic and the one characteristic of the hexagonal close-packed crystals – can be observed in the lattice, as can be seen from the following letters indicating the order of planes, i.e.: *ABC AB AB C ABC*).

If the stacking fault is due to the lack of certain parts of atomic planes (actually caused by vacancies), it is called as *intrinsic stacking fault*. If a stacking fault is caused by extra atoms wedged into the lattice, it is called an *extrinsic stacking fault*.

6. MECHANICAL PROPERTIES OF SINGLE-PHASE METALLIC MATERIALS

6.1. Brief overview of mechanical properties

All materials applied in practice are subjected to various loads originating from their functions. These loads may be physical, chemical or mechanical in nature or may be miscellaneous types. The properties related to various mechanical loads and the fundamental principles of these properties are significant for mechanical engineers designing and manufacturing different structures for defined purposes. In this chapter, first the mechanical properties will be overviewed, since they are most important for application in engineering practice. Then the theoretical and materials science background of these properties will be discussed.

6.1.1. Mechanical stresses and strains

In engineering, *mechanical stress* (or simply "stress") is defined as the amount of internal force on a unit surface. Consequently, the unit of stress is N/mm^2. In engineering practice MPa is also frequently used as the unit of stress (1 MPa = 1 MN/m^2, from which the equivalence of 1 MN/m^2 = 1 N/mm^2 also follows).

Depending on the characteristics of the applied load, normal and shear stresses can be distinguished. Normal stresses (denoted by σ) can be tensile or compressive – part *a*) and b) in Figure 6.1 – and shear stresses (denoted by τ) originate from shearing or torsional loads – part c) and d) in Figure 6.1. (Note that in Figure 6.1, the original form of the component is shown by a broken line, while the final shape is illustrated by a solid line.)

Obviously, beside these basic types of mechanical loads, various combinations of them may occur in components and structural elements in engineering practice.

As a result of applied loads, the shape and sizes of the components may change and become deformed. Normal deformations (denoted by ε) occur under the effect of normal stresses (σ). Under the effect of tensile load, the component elongates in the direction of the tensile force, and thus a positive normal strain occurs ($\varepsilon > 0$). Under the effect of a compressive load, the component becomes shorter, thus a negative normal strain occurs ($\varepsilon < 0$). Shear stresses (τ) initiate shear strains or angular distortions that are denoted by γ.

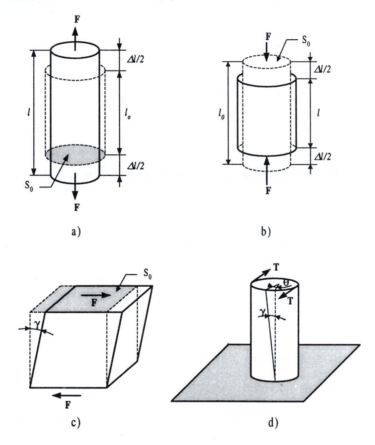

Figure 6.1
Schematic illustration of elementary mechanical loads
a) tensile, b) compressive, c) shear, d) torsional load

To further study of the terminology of mechanical stresses and strains, we can consider the uniaxial tensile test of a cylindrical specimen shown in Figure 6.2. The specimen may be characterized by a *gauge length*, l_o and an initial diameter, d_o. The uniaxial tensile test is a basic type of material testing technique that is applied for the determination of several important standardized mechanical characteristics. The standardized cylindrical test specimen has an initial diameter of $d_o = 10$ mm, and a gauge length of $L_o = 5d_o$, or $L_o = 10\,d_o$ is marked prior to the test. (It should be noted here that in accordance with the most recent standards, the gauge length is given in terms of the cross section of the specimen using the relationships $L_o = 5,65\sqrt{S_o}$, or $L_o = 11,3\sqrt{S_o}$. However, it should also be noted that it leads to the same results as the former expressions given for the determination of the measuring length.)

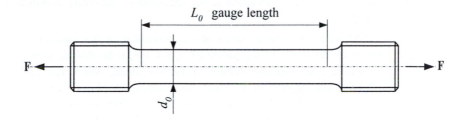

Figure 6.2
Schematic illustration of the uniaxial tensile test of a cylindrical test specimen

Using a universal material testing equipment or a simple tensile test instrument, the test specimen is loaded continuously at a constant speed until it is broken. During the test, the values of the loading force (F) and the elongation (ΔL) are recorded. The diagram plotted from these recorded values is called the tensile diagram. The tensile diagram of mild steel is shown in Figure 6.3. In this diagram, three characteristic sections can be distinguished. In the initial straight section, the loading force (F) is linearly proportional to the elongation (ΔL). The important feature of this section is that after unloading (that is removing the loading force) the deformation will cease and the plotter will return to its original position along the straight line. In this section, there are no changes concerning the dimensions of the test specimen. Therefore, this section is called the range of *elastic deformation*. Above a certain loading, the linear relationship between the loading force (F) and elongation (ΔL) terminates. Until the maximum loading (F_m)

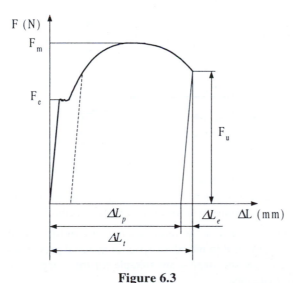

Figure 6.3
Characteristic tensile diagram of a mild steel

force is reached, an increasing degree of deformation can be achieved only by increasing the load. After unloading the specimen at any point in this section, the plotter moves back along a line parallel to the initial straight section, but does not return to the origin of the coordinate system. This means that a residual deformation occurs (see the dotted line in Figure 6.3). This section is called the region of *strain hardening* since, as noted, increasing deformation can only be achieved by increasing the load. This section is also often called *the region of uniform elongation* since the deformation of the specimen is uniformly distributed along its total length.

The third section of the diagram starts at the maximum of the loading force (F_m) and ends with breaking of the specimen (at $F = F_u$). In this section, the deformation is limited to a narrow region of the specimen, which has so far deformed uniformly. This appears in the form of a locally reduced diameter (called *necking* or *contraction*). Therefore, this section of the diagram is called the *contraction section*. This phenomenon is illustrated in Figure 6.4 showing the various deformation stages of the specimen. The figure shows the original state of the specimen (section a), an intermediate state at the end of uniform elongation (section b) and the state following the fracture of the specimen (section c. of Figure 6.4).

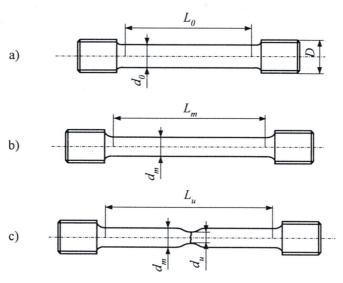

Figure 6.4
Shape and dimensions of a cylindrical specimen in various stages of the tensile test
a) original state   b) the end of uniform elongation   c) after fracture

It is obvious that the magnitude of the loading force is a function of the size of the test specimen. Therefore, it is practical to introduce specific material properties that – independently of the size of the specimen – are suitable for the comparison of the mechanical properties of materials.

Mechanical stress as a specific property perfectly fulfills this requirement. Stress by definition means the internal force on a unit surface. In general, mechanical stress can be calculated from the expression:

$$\sigma = \frac{F}{S}. \tag{6.1}$$

In material testing, the measured force (F) is divided by the initial cross section (S_o) of the specimen, i.e. the term *engineering stress* is used:

$$\sigma = \frac{F}{S_o}. \tag{6.2}$$

Similarly to stresses, the deformation of the specimen is expressed in terms of specific characteristics. In engineering practice, the generally applied deformation is called specific elongation (ε) and it is calculated as the ratio of the elongation of the specimen to its original length according to the expression:

$$\varepsilon = \frac{L - L_o}{L_o} = \frac{\Delta L}{L_o}. \tag{6.3}$$

Similarly to engineering stress, this specific elongation is called *engineering strain*, since in this case, too, the elongation is related to the original length of the test specimen.

It follows from the diagram shown in Figure 6.3, as well as from the previously introduced relationships that in the initial, elastic section of the diagram, a linear relationship exists between stresses and strains. This can be expressed by Hooke's Law, namely:

$$\sigma = E \varepsilon , \tag{6.4}$$

where the coefficient of proportionality, E is the modulus of elasticity of the tested material. This is usually called Young's modulus. (It is worth noting that the modulus of elasticity of steels approximately equals the value 2.1×10^5 MPa. The value of Young's modulus for ceramic materials is significantly higher, and that of polymers is much lower. The significant difference in the values of the modulus of elasticity for various types of materials can be reasoned by the essential differences in atomic bonding. However, it is also important to note that the value of the modulus of elasticity decreases with increasing temperature.)

Expressions similar to the Equation (6.4) can be given to describe the relationship between the shear stresses and shear strains, i.e.

$$\tau = G \gamma , \tag{6.5}$$

where G is the so-called shear modulus of elasticity. This can be related to Young's modulus with the following equation:

$$G = \frac{E}{2(1+v)} . \tag{6.6}$$

In the relationship (6.6), ν represents the so-called Poisson's ratio that expresses the ratio of transverse and longitudinal strain. In Table 6.1, the values of Young's modulus, the elastic shear modulus and Poisson's ratio of some materials of industrial importance are shown.

Table 6.1 Young's modulus, elastic shear modulus and Poisson coefficient of various materials

Material	Young's modulus E [N/mm^2]	Elastic shear modulus G [N/mm^2]	Poisson coefficient ν
Aluminum	69,600	26,400	0.33
Brass	101,000	37,000	0.35
Copper	124,800	47,090	0.35
Magnesium	44,800	16,670	0.29
Nickel	207,000	73,090	0.31
Steel	210,000	81,420	0.28
Titan	110,000	45,000	0.36
Tungsten	408,300	160,000	0.28

Using the characteristic values shown in Figure 6.3, important stress and strain properties can be calculated. Since these are very characteristic of the materials, they are called *characteristic material properties*. From the value of force F_e, indicating the limit of the elastic region of the diagram, we can calculate the *elastic limit stress* of the material using the relationship:

$$R_{eH} = \frac{F_e}{S_o} .$$
(6.7)

(Instead of the symbol of stress (σ) generally used in continuum mechanics, the symbol R with the appropriate subscripts are used in materials testing to denote the various types of stresses. Accordingly, in expression (6.7), R_{eH} is introduced to designate the *elastic limit stress*).

As noted before, in the second section of the diagram, residual deformation occurs together with elastic deformation. This means that the specimen undergoes plastic deformation. As we shall see later, the limit stress indicating the commencement of the plastic deformation is of special significance.

Some materials exhibit definite yielding phenomena. In this case, a characteristic "step" can be observed in the tensile diagram (for example, in the tensile diagram of mild steel as shown in Figure 6.3). This means that the straight section of elastic deformation and the strain-hardening section of residual (plastic) deformation can definitely be separated. This is typical for mild steels. For most materials, the elastic and plastic regions do not separate sharply (this is typical of aluminum or tempered steels – see part b) of Figure 6.5).

In fact, the transition between the elastic and the plastic region of deformation can only be determined with a degree of uncertainty for these materials. In order to avoid this, the yield point (i.e. the stress limit indicating the commencement of plastic deformation) is generally determined from that value of loading which results in 0.2 % residual plastic deformation. It can be determined using the following expression:

$$R_{p0,2} = \frac{F_{p0,2}}{S_o} \quad , \tag{6.8}$$

where:

$F_{p0.2}$ marks the force resulting in the 0.2 % of residual deformation (the value of this can be determined very precisely using special fine measurement techniques),

$R_{p0.2}$ the so-called *yield strength* measured in a loaded condition.

(In order to "visualize" the 0.2 % of residual elongation, this section is "*magnified*" compared to other parts of the diagram illustrated in part b) of Figure 6.5.)

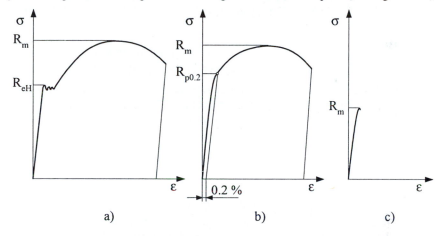

a)   b)   c)

Figure 6.5
Characteristic tensile-test diagrams for various materials
a) Ductile material with a definite yielding phenomenon (e.g. mild steel)
b) Ductile material without a definite yielding phenomenon (e.g. aluminum)
c) Brittle material (e.g. cast iron)

Among the stress characteristics, determined by material testing, the ultimate tensile strength corresponding to the maximum loading force plays a very important role. This is marked R_m. The ultimate tensile strength is calculated as the ratio of the maximum force measured during the tensile test (F_m) and the original cross section (S_o) of the specimen, namely:

$$R_m = \frac{F_m}{S_o} \quad . \tag{6.9}$$

Using the value of the force (F_u) measured at the moment of the specimen's fracture, the contraction stress or fracture stress can be calculated from an expression similar to the relationship (6.9) with the following expression:

$$R_u = \frac{F_u}{S_o} \quad . \tag{6.10}$$

Elongation, measured during the tensile test can be used to determine various specific deformation characteristics. The length of the test specimen (L_m) corresponding to the maximum loading force is used to calculate the so-called *uniform elongation* from the expression:

$$\varepsilon_m = \frac{L_m - L_o}{L_o} \quad . \tag{6.11}$$

Using the elongated length L_u measured after fracture, the *specific elongation at fracture* (or simply called *specific elongation)* is calculated from the relationship:

$$A = \frac{L_u - L_o}{L_o} \times 100 = \frac{\Delta L_u}{L_o} \times 100 \quad (\%). \tag{6.12}$$

For the specific elongation, the name *percentage elongation at fracture* is also used and is denoted by the symbol A as given in the Equation (6.12).

An important specific deformation characteristic can be derived from the diameter measured at the cross section of fracture. This material property is called *specific (percentage) reduction of cross section* (or in other words *contraction*). It can be determined from the following expression:

$$Z = \frac{S_o - S_u}{S_o} \times 100 = \frac{\Delta S_u}{S_o} \times 100 \quad (\%). \tag{6.13}$$

This is more characteristic of the deformation occurring in the vicinity of the fracture and thus more characteristic of the ductility of the material than the specific elongation calculated from the change of the total gauge length.

The original cross section, S_o, used for calculation of the elastic limit stress and the yield strength, hardly differs from the momentary cross section of the specimen (S) at these points. However, the cross section, S_m, taken into consideration when we calculate the ultimate tensile strength, and particularly the cross section, S_u, measured at the moment of fracture, is significantly different from the original cross section, S_o. Therefore, we introduce the term *true stress, which can be* determined as the ratio of forces and the momentary cross sectional areas at the appropriate points. (True stresses are denoted by σ' or in materials testing by R' for distinction from engineering stresses.) Consequently, significantly different values can be obtained when comparing the *true stresses* and the so-called engineering stresses. Thus, the true stress belonging to the maximum force (F_m) can be calculated from the expression:

$$R'_m = \frac{F_m}{S_m} \quad, \tag{6.14}$$

while the true stress corresponding to the moment of fracture is calculated from the expression:

$$R'_u = \frac{F_u}{S_u} \quad . \tag{6.15}$$

Similarly, we can define *true strains*. When calculating the true strains, elongation refers to the current (momentary) length (L) of the specimen rather than the original length (L_o). The true strain (denoted by the symbol φ) is distinguished from the engineering strain (denoted by ε). The incremental value of true strain is determined by using the general expression:

$$d\varphi = \frac{dL}{L} \quad . \tag{6.16}$$

After integrating equation (6.16), the total true strain may be calculated as

$$\varphi = \int d\varphi = \int \frac{dL}{L} \quad . \tag{6.17}$$

Thus, for example, the total true strain at fracture can be determined from the expression:

$$\varphi_u = \int_{L_o}^{L_u} \frac{dL}{L} = \ln \frac{L_u}{L_o} \quad . \tag{6.18}$$

Comparing the expressions (6.12) and (6.18), it is easily proved that the relationship between true and engineering strain can be given by the expression:

$$\varphi = \ln (1 + \varepsilon). \tag{6.19}$$

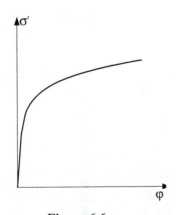

Figure 6.6

True stress vs. true strain

The $\sigma' - \varphi$ (true stress–true strain) curves clearly indicate what remains hidden in the case of the $\sigma - \varepsilon$ curves showing the relationship between engineering stress and engineering strain. As already stated concerning the plastic deformation, increasing plastic deformation can be achieved only by increasing true stress as illustrated in Figure 6.6. The true stress – true strain relationship can be described by Nádai's law:

$$\sigma' = K\varphi^n \quad, \tag{6.20}$$

where K is the so-called hardening coefficient and n is the exponent of hardening. The value of the hardening exponent is equal to the uniform true strain given by the following expression:

$$n = \varphi_m \tag{6.21}$$

Hardness is also a very important mechanical property of materials. This characterizes the resistance of materials against the penetration of an external object. The most widely applied hardness tests are the so-called *indentation hardness tests* that are performed by indenting hard tools of varying shapes into the surface of the material to be tested. The tool for the hardness test is a ball made of hardened steel for the *Brinell test*, a diamond pyramid for the *Vickers test* and a diamond cone for the *Rockwell-A and Rockwell-C tests*. The value of hardness is calculated from the applied force and the surface of the indentation in the Brinell and Vickers tests. In the Rockwell test, hardness is determined from the depth of the indentation. The conceptual sketches of these tests are shown in Figure 6.7.

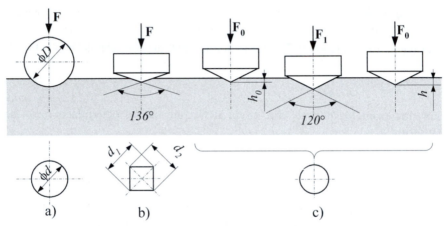

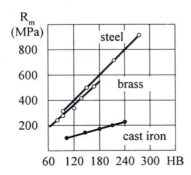

Figure 6.7
Various hardness tests
a) Brinell   b) Vickers   c) Rockwell-C hardness tests

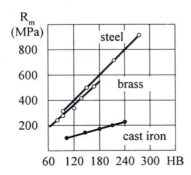

Figure 6.8
Ultimate tensile strength vs. hardness

For most metals, an experimental relationship exists between hardness and ultimate tensile strength. For example, the value of the tensile strength of mild steels is about three times the value of the Brinell hardness. A similar, practically linear correlation can be observed for other materials, as well. This correlation is demonstrated by Figure 6.8 for various material grades.

6.2. Theoretical background of mechanical properties

Until now, we have briefly summarized the fundamental knowledge of the most important mechanical properties. The theoretical background of these properties will be discussed in the forthcoming sections. From the analysis described in the previous sections, we could see that elastic and plastic deformations are of outstanding significance for mechanical properties. Therefore, we will start the overview of the theoretical background of mechanical properties with a brief review of the most important features of elastic and plastic deformations.

6.2.1. Elastic deformation

When applying external forces to crystalline solids, elastic deformation occurs first. The *elastic deformation* can be characterized by the following. During elastic deformation, atoms remain in the neighborhood of the same atoms (that is, they do not leave their original crystal lattices) and they return to the original lattice points when the external force is removed. This means that the elastic deformation ceases after unloading. Therefore, *elastic deformation is a reversible process.*

In order to illustrate the elastic deformation of a crystalline structure, consider Figure 6.9. In part a), a *lattice plane* is shown prior to deformation. The elastically distorted state of the same lattice plane as a result of the elastic deformation is shown in part b) for a tensile force and in part c) for a compressive force.

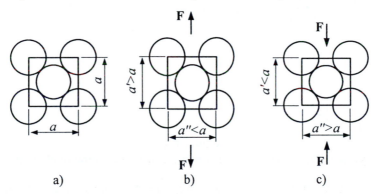

Figure 6.9
Schematic sketch to illustrate the elastic deformation
a) equilibrium state b) loaded with tensile force c) loaded with compressive force

The relationship between the deformation and the stresses during elastic deformation is given by *Hooke's law of elasticity*, which for a pure shearing load can be written in the following form:

$$\tau = G\gamma \qquad\qquad (6.22)$$

where G is the shear modulus of elasticity, γ is the shear strain and τ is the shear stress. In an equilibrium state, the atomic attractive and repulsive forces are equal and the atoms are located at the lattice points – see part a) of Figure 6.9. When the external load is removed, they return to their original locations.

Starting from the above considerations, we can determine the elastic limit stress for an ideal crystal (having no lattice defects). Above this limit, plastic deformation occurs. Consider for this analysis Figure 6.10. In the lower part of this figure, the shear stress (τ) required to slide neighboring crystal planes over each other is shown as a function of atomic displacement (x).

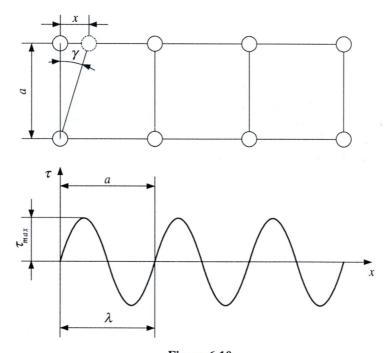

Figure 6.10
Sketch illustrating determination of the theoretical limit of elasticity

The periodic change of the shear stress exhibits a direct relationship with the periodic alteration of the atomic attractive and repulsive forces: it can be characterized by a sinusoidal function with respect to the distance of movement. Taking this into consideration, the function of the shear stress can be expressed as follows:

$$\tau = \tau_{max}\, \sin\frac{2\pi}{\lambda}x, \tag{6.23}$$

where the periodicity (λ) of the function can be taken equal to the lattice parameter (a). In the relationship (6.22), which is valid for elastic deformation, the angular distortion can be replaced by the following expression (see Figure 6.10):

$$\gamma = \frac{x}{a}. \tag{6.24}$$

As a result, the relationship (6.22) may be written in the form:

$$\tau = \frac{x}{a} G. \tag{6.25}$$

The function (6.23), describing the periodic change of the shear stress (τ) in the vicinity of point $x = 0$ (that is for small elastic deformations), can be replaced by the tangent of the function, thus:

$$\frac{\partial \tau}{\partial x} = \tau_{max} \frac{2\pi}{a} \cos \frac{2\pi}{a} x, \tag{6.26}$$

which at $x = 0$ leads to the value:

$$\frac{\partial \tau}{\partial x} = \tau_{max} \frac{2\pi}{a}. \tag{6.27}$$

Taking this into consideration, the change in the shear stress in the vicinity of $x = 0$ can be approximated by the following function:

$$\tau = \tau_{max} \frac{2\pi}{a} x. \tag{6.28}$$

(This approximation leads to $\pi/2$-times deviation in the value of τ_{max} compared to the value calculated with the relationship (6.23) describing the periodic change of τ. This deviation is negligible compared to the difference between the measured and calculated values which is several magnitudes greater.) The values of the shear stresses calculated from the relationships (6.25) and (6.28) should be equal for small deformations, that is:

$$\frac{x}{a} G = \tau_{max} \frac{2\pi}{a} x, \tag{6.29}$$

from which

$$\tau_{max} = \frac{G}{2\pi} \tag{6.30}$$

follows. The values calculated according to the relationship (6.30) are several magnitudes higher than the measured limit of elasticity in real crystals: this may be explained by the lattice defects and imperfections that are always present in real crystals. This will be analyzed in detail when the characteristics of plastic deformation are discussed.

6.2.2. Plastic deformation

6.2.2.1. The characteristics of plastic deformation

In contrast to elastic deformation, *plastic deformation* is characterized by the fact that the atoms are displaced for several hundreds (in certain cases for several thousands) of lattice constants under the effect of external forces. This means that the atoms do not remain in the neighborhood of the same atoms. A further essential difference – as a decisive feature of plastic deformation – is that on removal of the external forces, the atoms are unable to return to their original locations (due to the extensive displacement). Therefore, after removing the external forces, residual deformation remains, i.e. plastic deformation is an *irreversible process*.

6.2.2.2. The mechanisms of plastic deformation

The most characteristic and most frequent mechanism of plastic deformation is the so-called *slip mechanism*, which is realized by the movement of dislocations. However, under certain definite circumstances, plastic deformation may occur by *twinning, diffusion creeping, grain boundary sliding, grain rotation, deformation induced by phase transformation*, as well as by the combination of the mentioned deformation mechanisms.

Among the various deformation mechanisms, the *slip mechanism* is of decisive importance. *Diffusion creeping* and *grain boundary sliding* may occur in high temperature *hot working* and in *superplastic deformation* processes that have recently emerged in research activities. According to the above reasoning, the slip mechanism of plastic deformation will be discussed in detail, whilst the other mechanisms will be mentioned briefly.

The slip mechanism of plastic deformation

The slip mechanism is the most characteristic and most probable method of plastic deformation, which occurs by the slipping of crystal planes following well-defined rules of crystallography. These rules of crystallography mean that plastic deformation most likely occurs in the densest crystalline planes and, within these planes, in the densest directions.

Within a certain crystal system, the planes with the highest planar density are called *slip planes*, and the directions having the highest linear density are called *slip directions*. Slip planes and slip directions together form *slip systems*. In Table 6.2, the possible slip systems of various crystalline solids are summarized. In this table, the Miller indices of slip planes and the vector notation of slip directions are also given.

The number of *slip* systems can be directly related to *ductility*. In general, it is valid that metals having more slip systems are more ductile. From this we can conclude that metals with a face-centered cubic crystal structure are the most ductile, while those possessing the least number of *slip* systems belonging to the hexagonal system are the least deformable.

Table 6.2 Slip systems of various crystal structures

Crystal system	Slip planes		Slip directions		No. of slip systems
Notation	Notation	Number	Notation	Number	Number
bcc	{110}	6	<111>	2	12
fcc	{111}	4	<110>	3	12
hcp	(0001)	1	[11$\bar{2}$0]	3	3

Thus, in the slip mechanism external loading causes plastic deformation to occur in well-defined crystal planes (i.e. in the slip planes) and in well-defined crystal directions (i.e. in the slip directions). In Figure 6.11, the slipping of crystal planes is shown for an ideal crystal: in part *a*) the crystal is illustrated before slipping and in part *b*) after slipping. As can be seen from this illustration, for an ideal crystal all the atoms in the slip plane take part simultaneously in the slipping process. Considering this model, the resolved shear stress can be calculated from the Equation (6.30). This leads to values several magnitudes higher than the experimentally measured ones for real crystals. This contradiction can be resolved by considering lattice defects and imperfections that are always present in real crystals, i.e. more realistic figures can be obtained by calculations based on the *theory of dislocation.*

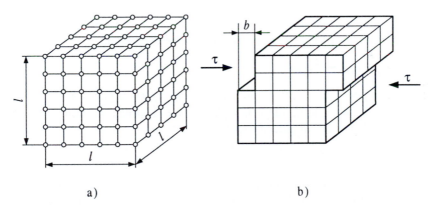

a)   b)

Figure 6.11
The slip mechanism of plastic deformation in ideal crystals

In real crystals, slip does not occur according to Figure 6.11, but rather in a way shown in Figure 6.12. From this figure, it can be seen clearly that in real crystals the slip mechanism is realized by the movement of dislocations. As we could see in Figure 6.11, in ideal crystals a displacement with a value of $\vec{b}$ can occur by the simultaneous movement of atoms on both sides of the slip plane. In contrast, in a real crystal, a displacement with a value of $\vec{b}$ is realized as a sequence of sliding steps by a continuous elementary displacement of the dislocations. It also follows from this, that the slip mechanism realized by the dislocation mechanism decreases the amount of stress required to cause the plastic deformation by several magnitudes.

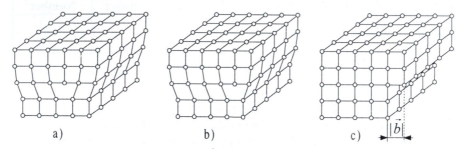

a)   b)   c)

Figure 6.12
The slip mechanism in real crystals

In order to analyze this, follow the next simple sequence of thoughts. Select a part of an ideal crystal having the size of $l \times l \times l$, as shown in Figure 6.11. Denote the shear stress required for sliding two adjacent planes over one another by τ_{id}. The external force required to realize a displacement with an elementary value of $\vec{b}$ can be calculated from the relationship:

$$F_{id} = A_s \tau_{id} = l^2 \tau_{id} \, , \qquad (6.31)$$

where $A_s = l^2$ means the area of the slip plane. The work done during the elementary slip with the value of $\vec{b}$ can be calculated from the expression:

$$W_{id} = F_{id} b = l^2 \tau_{id} b . \qquad (6.32)$$

From Figure 6.12, it can be seen that if the crystal contains a dislocation, a displacement with a value of $\vec{b}$ is realized as a sequence of sliding steps by a continuous elementary displacement of the dislocations. (For the sake of simplicity the size $l \times l \times l$ of the crystal in Figure 6.12 was selected so that an edge dislocation can be found within the crystal (this is called a *mosaic block*).

Denote the shear stress required for slipping of the slip planes of the real crystal over one another by τ_{real}. In this case, the value of the external force required to produce a slip with an elementary value of $\vec{b}$ can be calculated from the relationship:

$$F_{real} = A_s \tau_{real} = l^2 \tau_{real} . \qquad (6.33)$$

Thus, the amount of work done during the elementary slip in a real crystal can be determined from the relationship:

$$W_{real} = F_{real}l = l^2\tau_{real}l. \tag{6.34}$$

In order to realize an equivalent amount of displacements, the two values of work done in ideal and real crystals have to be equal, i.e. $W_{id} = W_{real}$. From this,

$$l^2\tau_{id}b = l^2\tau_{real}l \tag{6.35}$$

follows. From this expression, the value of the shear stress τ_{real} related to the real crystal can be expressed in terms of the ideal shear stress τ_{id} as follows:

$$\tau_{real} = \tau_{id}\frac{b}{l}. \tag{6.36}$$

Consider that the value of $\vec{b}$ (i.e. the magnitude of Burger's vector) falls into the order of magnitude of the lattice parameter ($|\vec{b}| \approx 10^{-7}$ mm) and the size l of the mosaic block can be found in the magnitude of 10^{-4} mm. Then, the ratio of the shear stress (τ_{real}) measured in a real crystal and calculated for an ideal crystal (τ_{id}) according to relationship (6.30) can be given as

$$\tau_{real} \approx 10^{-3}\tau_{id} \tag{6.37}$$

which already shows good agreement with experimental results.

Table 6.3 Theoretically calculated and experimentally measured critical shear stresses of some metallic single crystals

Metal	G [N/mm^2]	τ_{id} [N/mm^2]	τ_{real} [N/mm^2]	τ_{id}/τ_{real} [-]
Iron	81,420	12,965	2.700	4,802
Aluminum	26,400	4,204	0.981	4,285
Copper	47,090	7,498	0.982	7,635
Nickel	73,090	11,639	5.684	2,048
Magnesium	16,670	2,654	0.441	6,018

In Table 6.3, the values of theoretically calculated (τ_{id}) and experimentally measured (τ_{real}) critical shear stresses are summarized for some metallic single crystals of industrial significance. From this table, it can be seen that the difference is in the order of magnitude of 10^3 for all the listed materials.

It is obvious from the previously performed analysis that in the realization of the slip mechanism of the plastic deformation, the dislocation mechanism plays an outstanding role. During their movements, dislocations may interact with one another. Depending on the types of interactions, various dislocation reactions can occur. For example, in the case of edge dislocations:

— Dislocations can cease (extinguish each other); a condition where dislocations of opposing sign meet during slip, having a common slip plane.

— Dislocations may create empty lattice sites (vacancies) when the distance between the dislocations having opposite signs is equal to one lattice parameter.

— Dislocations can generate one or more rows of interstitial atoms provided there is an "overlapping" extending one or more lattice parameters between the dislocations of opposite signs.

Apart from the previously analyzed types of *dislocation reactions,* during plastic deformation typically such reactions occur that result in an increased number of dislocations. A great number of new dislocations are generated or, in other words, the *dislocation density is* increased. (This has been verified by the results of numerous measurements). One of the most probable mechanisms for the increase of dislocations is the multiplication of dislocations by the so-called Frank-Read *dislocation multiplication mechanism.* The essence of this *multiplication mechanism* is described hereunder.

According to the Frank-Read dislocation multiplication mechanism, closed dislocation loops are formed by the effects of shear stresses acting in slip planes. In order to analyze this, consider the movement and deformation of the dislocation line fixed at the A and B ends due to the effect of shear stress τ (see Figure 6.13). This type of "immobilization" of dislocation lines may occur for several reasons, for example when it crosses other dislocations not capable of moving in the case of a certain load, etc.

The straight dislocation line between points A and B becomes curved under the effect of shear stress τ (i.e. it "bends out" between the fixing points). When this curved line becomes semi-circular, the shear stress reaches its maximum value and the dislocation expands further on without any increase in the external loading. During further deformation, the dislocation line gradually takes up the shapes shown in Figure 6.13. The points marked by C_1 and C_2 in step 5 form screw dislocations of opposite sign, "extinguishing" each other when they meet during further expansion of the dislocation loop (step 6). Thus, a closed dislocation loop and another, new dislocation line are formed. The newly created dislocation line *AB* is capable of further plastic deformation. The closed dislocation loop may extend up to the grain boundaries, realizing an elementary slipping process with the magnitude of one Burger's vector. At the same time, the newly formed dislocation line becomes the source for the formation of further dislocations by repeating the above-described process.

Obviously, a single source of dislocation cannot generate an unlimited number of new dislocations. On the one hand, the dislocations may become immobile or become "blocked" due to several obstacles. On the other hand, the elastic stress field arising around the newly formed dislocations also forms an obstacle against creating an unlimited number of dislocations. Concerning this latter action, the operation of the dislocation source ceases when the level of stresses required for the further movement of the dislocation is exceeded by the stress level of the elastic stress field around the piled-up dislocations. The Frank-Read source of new dislocations is only one possible mechanism of the formation of new dislocations. The grain boundaries and phase boundaries can also act as sources of dislocations.

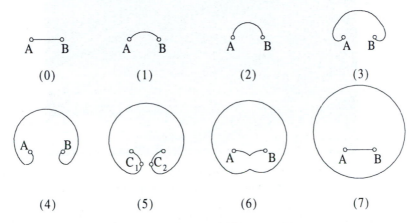

Figure 6.13
The Frank-Read dislocation multiplication mechanism

Concerning the slip mechanism, the formation of the so-called *slipping clusters* is worth mentioning. It is regarded as a strongly localized method of plastic deformation. This primarily can be observed in metals and alloys deformed at high strain rates. The slipping clusters can be considered as intensive slipping zones. This type of localization of plastic deformation plays an important role in the plastic fracture mechanism, since plastic fracture frequently occurs along such slipping clusters. This is the reason why the formation of slipping clusters is often considered as part of the fracture mechanism. However, it has been experimentally proved that the formation of slipping clusters under compressive stresses delays fracture rather than induces it. Therefore, a much more valid statement is that the formation of slipping clusters is the mechanism of high strain-rate deformation achieved without fracture.

The slipping clusters have two main types: the so-called *deformed slipping cluster* and the *phase transformation induced slipping cluster*. The former is rather characteristic of non-ferrous alloys (e.g. alloys of *Cu*, *Al* and *Ni*). The latter can be frequently observed in alloyed steels as well as in the alloys of titanium and uranium. In these alloys, the phase transformation induced plasticity plays an important role at high temperatures. The slipping clusters due to phase transformation are a few hundreds of micrometers wide and possess a very characteristic microstructure. The deformed slipping clusters are thin clusters that are etched to dark, visible under the optical microscope.

An example of deformed slipping clusters is shown in Figure 6.14, whilst in Figure 6.15 the formation of deformed slipping clusters are shown in the main rolling directions for a strongly hot-rolled copper slab.

Figure 6.14
Microscopic picture of deformed slipping clusters

Although there is no phase transformation in the deformed slipping clusters, the fine microstructure refers to a dynamically renewed microstructure within the slipping clusters. Plastic deformation in such slipping clusters may reach the value of φ=3 ~ 4, meaning a very significant degree of deformation. It is worth mentioning that slipping clusters should be considered as macroscopic features in the sense that the surface of slipping clusters does not necessarily consist of crystallographic slipping planes.

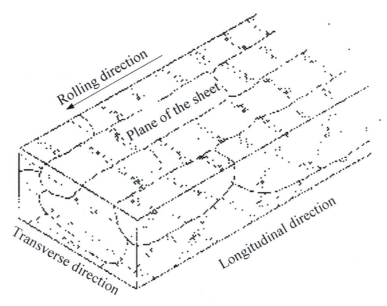

Figure 6.15
Slipping clusters formed during rolling

The twinning mechanism of plastic deformation

The twinning mechanism of plastic deformation differs essentially from the slip mechanism. In the slip mechanism the orientation of the crystals above and below the slip plane remains more or less unchanged, but the essence of twin formation is that the crystal orientation significantly changes in the deformation region undergoing plastic deformation. This difference between the two deformation mechanisms is illustrated in Figure 6.16.

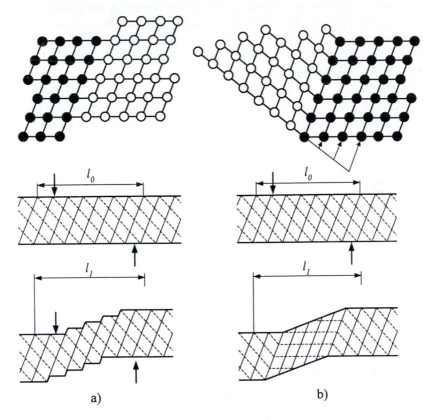

a)   b)

Figure 6.16

Comparison of the mechanisms of plastic deformation
a) slip mechanism b) twinning mechanism

The essence of twinning can be summarized as follows. In twinning, part of the atomic lattice is deformed so that it forms a mirror image of the undeformed part next to it. The crystal plane of symmetry between the deformed and undeformed parts is called the *twin plane*. Twinning – like slip – occurs in specific directions called t*win directions*. This mechanism is similar to the slip mechanism, and results in elongation of the crystal in the direction of the tensile force. Similarly to the slip mechanism, twinning can occur when the shear stress attains a certain critical value. However, in the slip mechanism, the atoms on one side of the slip plane all move an equal distances, whilst in twinning, the atoms move distances proportional to their distance from the twin plane. This also means that in twinning all the planes between the two twin planes parallel to them slip in the same direction and to the same extent with respect to the adjacent plane. This results in a change in lattice orientation towards a more favorable position from the point of view of plastic deformation with slip mechanism. This, in turn, provides a condition that is more favorable for further slipping. This is the most important consequence of twinning, particularly in metals having a small number of slip systems (e.g. hcp crystals). Twinning occurs very quickly, i.e. within a few microseconds. In certain cases, twinning is accompanied by a characteristic sound effect called "*tin noise*", named after the sound effect observed during the twinning of a tetragonal tin crystal.

In certain crystals – especially in metals crystallizing in the hexagonal system and having a small number of slip systems – the orientation of the slip plane can change to a more favorable position from the point of view of plastic deformation by twinning. Consequently, it can be deduced that twinning characteristically "operates" when the positions of the slip planes are unfavorable with respect to the external loading.

Additionally, deformation twinning can be observed in metals crystallizing according to the body-centered cubic system at high deformation rates and at low temperatures (for example in α-iron at a temperature of about $T = -200\ ^oC$). In face-centered cubic crystals, possessing high-energy stacking fault (as in aluminum), twinning is not characteristic at all. However, in certain alloys possessing low energy stacking faults (as for example in alloys of *Cu-Zn* or *Ni-Cu*) twinning can be observed despite the face-centered cubic crystal structure. Concerning *Cu-Zn* alloys, it should be noted that twinning can be observed in the α-phase under optical microscopes in both crystallization and recrystallization processes.

Grain boundary sliding

Grain boundary sliding is one of the most important deformation mechanisms in hot plastic forming. Grain boundary sliding represents a deformation mechanism mainly restricted to disordered grain boundaries. Grain boundary sliding primarily means slipping along grain boundaries that separate adjacent grains possessing orientation differences. However, this often leads to the operation of the so-called *grain rotation mechanism* as well. This latter mechanism may become dominant in *superplastic deformation processes*.

Analysis of the deformation of the so-called *slipping polycrystalline grain groups* led to remarkable results concerning the mechanism of grain boundary sliding. The traditional polycrystalline deformation model elaborated by Taylor, stipulates homogeneous deformation in the grain as a whole, as well as along the grain boundaries (see part a/ of Figure 6.17). In contrast to this, in grain boundary sliding (part b/ of Figure 6.17) significant kinematic discontinuities are experienced on the grain boundaries that disturb the uniform deformation within the grain as well.

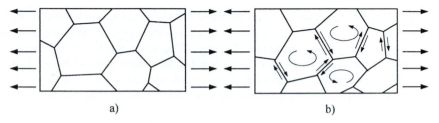

a)   b)

Figure 6.17
The so called grain boundary slip mechanism of plastic deformation

The polycrystalline deformation model based on grain boundary sliding has helped understand several phenomena, which were found previously difficult to explain on the basis of traditional deformation models. Thus, it is successfully applied in many fields starting with the high temperature creep processes, through the analysis of the mechanics of fiber-reinforced composites and to the examination of superplastic and superconductive materials.

The mechanism of phase transformation induced plastic deformation

The term *phase transformation induced plastic deformation* includes those deformation processes that are connected with internal microstructural changes due to the effects of mechanical or thermodynamic driving forces. The zone transformed during the phase transformation is separated from the mother phases (i.e. from the untransformed regions) by boundaries that have undergone rapid microstructural changes.

The *diffusional phase transformation* that is necessarily connected to the migration of material by the diffusion of atoms represents a relatively slow process – especially in low-temperature transformations. *Phase transformations* that are free from diffusion occur without the migration of material, therefore, they represent rapid transformations even at low temperatures, thus enabling fast propagation of the transformation boundary surfaces. The best known example of this kind of transformation is the *martensitic phase transformation of steels*. Moreover, from recent results of developments in materials, the ZrO_2-based ceramics, as well as certain alloys possessing *pseudo elastic properties* and the so-called *memory alloys* should be mentioned here.

Phase transformation induced plastic deformation may occur through the effects of mechanical loads (an example of this is *stress induced martensitic transformation of ZrO_2-based ceramics*), under thermal effects (the most typical example of this is the phase transformation of steels from austenite to martensite occurring during quenching); as well as under thermal and mechanical effects (good examples can be found among the so-called *shape memory alloys*).

6.2.2.3. Elementary processes of plastic deformation – The plastic deformation of single crystals

The elementary mechanisms and the basic rules of slipping can be studied most easily during the deformation of *single crystals*. In single crystals, the effects of the grain boundaries do not influence slipping. (The term *single crystal* means a crystalline material that consists of one crystal in the whole mass of material. It is grown by a *controlled crystallization procedure* described in Chapter 4. This requires a properly selected ratio between the rate of nucleation and the rate of crystallization.)

For the sake of simplicity, it is practical to choose a single crystal that possesses few slip systems. In this case, the simultaneous operation of complex slip systems can be excluded. Therefore, for this analysis the most appropriate materials can be found in the hexagonal system, which possesses only one slip plane according to Table 6.2. The crystallographic orientation of the slip plane can be determined by the *Laue X-ray diffraction method* described in Chapter 3.

For this analysis, consider Figure 6.18. In this figure, the initial angle between the slip plane and the axis of the specimen is denoted by χ_o. The test specimen is loaded with a tensile force denoted by F. (For the sake of simplicity, the tensile force F, the $\vec{n}$-normal vector of the slip plane and the slip direction within the slip plane have been selected in such a way that they are located in one plane.)

The stress arising in the cross section of the specimen (denoted by A_o) through the effect of the uniaxial tensile force (F) can be calculated from the following expression:

$$\sigma_o = \frac{F}{A_o}. \tag{6.38}$$

The tensile force F can be decomposed into two components: one of them is perpendicular to the slip plane (F_n), whilst the other is parallel to it (F_t). The value of these two components can be determined according to the relationships:

$$\begin{aligned} F_n &= F\sin\chi_o, \\ F_t &= F\cos\chi_o. \end{aligned} \quad \text{and} \tag{6.39}$$

Taking into consideration that the area of the slip plane can be calculated from the expression

$$A = \frac{A_o}{\sin\chi_o}, \tag{6.40}$$

the normal and tangential stress components can be calculated from the following expressions:

$$\sigma_n = \frac{F_n}{A} = \frac{F \sin \chi_0}{\dfrac{A_0}{\sin \chi_0}} = \frac{F}{A_0} \sin^2 \chi_0 = \sigma_0 \sin^2 \chi_0 \, ,$$

$$\tau_n = \frac{F_t}{A} = \frac{F \cos \chi_0}{\dfrac{A_0}{\sin \chi_0}} = \frac{F}{A_0} \sin \chi_0 \cos \chi_0 = \sigma_0 \sin \chi_0 \cos \chi_0$$

(6.41)

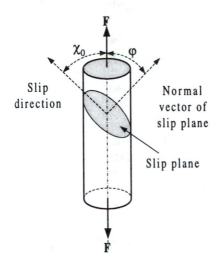

Figure 6.18
Theoretical scheme for studying the elementary processes of plastic deformation of single crystals

As the external loading force F is increased, the values of both components (i.e. F_n and F_t) are also increased.

For all materials, two characteristic stress values τ_{crit} and σ_{crit} can be determined, where τ_{crit} represents the critical shear stress required to initiate plastic deformation, and σ_{crit} represents the critical normal stress required to separate crystal planes.

If the value of the critical shear stress (τ_{crit}) is known, the value of the critical uniaxial stress $\sigma_{o,crit}$ can be determined. This critical value of the uniaxial stress $\sigma_{o,crit}$ is denoted by R_r and is called the *elastic limit stress*.

Using the second relationship of expression (6.41), the elastic limit stress can be calculated as:

$$R_r = \frac{\tau_{krit}}{\sin \chi_0 \cos \chi_0} \, .$$

(6.42)

Since τ_{crit} is a characteristic material property, for a given material it has a constant value. Using the relationship (6.42), the change in the elastic limit stress R_r can be determined as a function of the orientation of the slip plane (χ_0) of the specimen. This relationship is shown in Figure 6.19. From this figure, it can be seen that the most favorable position of the slip plane with respect to the initialization of the plastic deformation corresponds to the initial orientation $\chi_0 = 45°$. The initiation of plastic deformation requires the smallest value of external loading in this position.

If the initial orientation of the specimen falls into the range of $\chi_0 < 45°$, both the tangential force component F_t and the total area of the slip plane (A) will increase. As a result, the elastic limit stress (R_r) necessary to initiate plastic deformation will also increase compared to the case of a more favorable (i.e. $\chi_o = 45°$) initial orientation of the slip plane.

When $\chi_o > 45°$, both the tangential force component (F_t) and the total area of the slip plane (A) will decrease. Consequently, the elastic limit stress (R_r) necessary to initiate plastic deformation will also increase compared to the case of the more favorable initial orientation (i.e. $\chi_o = 45°$) of the slip plane. The changes in the elastic limit stress (R_r) as a function of the orientation of the slip plane are shown in Figure 6.19.

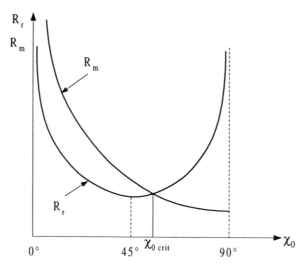

Figure 6.19
The change in the elastic limit stress and the ultimate tensile strength as a function of the crystallographic orientation

If the value of the critical normal stress (σ_{crit}) is known, another critical uniaxial stress $\sigma_{o,crit}$ can be determined. This critical value of the uniaxial stress is denoted by R_m and is called the *ultimate tensile strength*. Using the first relationship of expression (6.41), the ultimate tensile strength can be calculated as:

$$R_m = \frac{\sigma_{crit}}{\sin^2 \chi_o} . \tag{6.43}$$

σ_{crit} is a characteristic property of the material. Therefore, by using the relationship (6.43), we can determine how the ultimate tensile strength (R_m) required to separate crystal planes changes as a function of the initial orientation. This relationship is also shown in Figure 6.19. It can also be seen from Figure 6.19 that the value of the ultimate tensile strength monotonously decreases with an increase in the orientation angle. Simultaneously, it can be seen that by decreasing the value of χ_o the value of R_m rapidly increases, i.e. as $\chi_o \to 0$ the value of $R_m \to \infty$ is valid.

When the two functions – i.e. $R_r = f(\chi_o)$ and $R_m = f(\chi_o)$ – intersect, the condition $R_r = R_m$ is fulfilled, which leads to the relationship

$$\frac{\tau_{crit}}{\sin \chi_o \cos \chi_o} = \frac{\sigma_{crit}}{\sin^2 \chi_o} . \tag{6.44}$$

From the expression (6.44), a critical orientation angle ($\chi_{o,crit}$) can be determined using the expression:

$$tg \ \chi_{o,crit} = \frac{\sigma_{crit}}{\tau_{crit}} . \tag{6.45}$$

Analyzing Figure 6.19, it can be stated that when the angle between the slip plane and the plane of the external force is smaller than the previously determined critical value (that is $\chi_o < \chi_{o,crit}$), the stress required to initiate plastic deformation is always less than that required to separate crystal planes (i.e. the condition $R_r < R_m$ is fulfilled). Consequently, in this case, by increasing the external load, plastic deformation will always be initiated first.

It also follows from Figure 6.19, that when the angle between the slip plane and the plane of the external force is greater than the critical value determined by relationship (6.45) (i.e. $\chi_o > \chi_{o,crit}$), the crystal planes can be separated at a lower stress level than that required to initiate plastic deformation (i.e. $R_r > R_m$). Consequently, the crystal in this case breaks without plastic deformation. (In certain metals that have a tendency for twinning, the orientation may change to a favorable one, providing the possibility for further plastic deformation).

The translational movement of a hexagonal single crystal during plastic deformation under tensile load perfectly illustrates the mechanism of plastic deformation described earlier (Figure 6.20). When the magnitude of stress which initiates slipping is reached, the slip planes start to slip over each other. As the load increases, the slip planes gradually turn inward into the plane of the loading force (which means that $\chi_o \rightarrow 0$).

Due to this rotation, the magnitude of the force component parallel to the slip plane increases. As a result of these translational movements, the single crystal, originally being

Figure 6.20
Translational movement of a hexagonal single crystal during plastic deformation occurring in uniaxial tension

circular, gradually takes a flattened elliptical shape and the cylindrical test specimen becomes an ever thinning "strip". This stage of deformation is called the *principal elongation* of the crystal. This lasts as long as the angle between the slip plane and the direction of the loading force reaches a critical value depending on the metal examined and the temperature of deformation. Consequently, the so-called principal elongation is exhausted, and the crystal is not capable of further plastic deformation.

The deformation of metals belonging to the cubic system, having more slip systems, is a more complicated process. Due to the greater number of slip systems, there are several slip planes where slipping can occur by the effect of shear stresses of the same magnitude. In this way, for example, in the face-centered cubic crystals, there are four planes having the {111} orientation that are equivalent from the crystallographic point of view. Deformation starts in the plane most favorably oriented with respect to the loading force. In the course of deformation, the slip planes rotate, and subsequently, another (111) plane is positioned so that translation commences in it as well. Since this process may be repeated several times during deformation, the subsequent operation of several slip systems can be observed. Due to slipping movements that occur in the slip planes possessing various positions, it may result in complicated cross-slipping processes. However, these operating slip systems guarantee that the originally circular cross section remains more or less circular during the deformation.

6.2.2.4. Plastic deformation of polycrystalline metals

Although the plastic deformation of polycrystalline metals is significantly more complicated than that of the single crystals, it can be stated that basically it occurs by the same mechanisms. However, the essential difference is that the individual crystal planes are positioned in a well-defined manner in single crystals, whilst in polycrystalline metals, having more practical importance, crystal planes are positioned in a statistically disordered way. Consequently, there are crystallites oriented favorably to the loading force and there are unfavorably oriented ones from the point of view of plastic deformation. This is schematically illustrated in Figure 6.21.

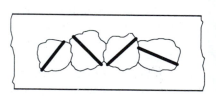

Figure 6.21
Statistically disordered distribution of slip planes in polycrystalline metals

A further significant difference is presented by the fact that in polycrystalline metals every crystal is surrounded by neighboring crystals, contrary to single crystals where the whole mass of the deforming metal consists of a single crystal. Obviously, plastic deformation commences in the slip planes and slip directions of the most favorably oriented crystallites (that is, in those slip planes forming a 45° angle with the external loading force). Plastic deformation of these types of metals starts when the shear stress achieves the critical value required for slipping.

However, considering that the favorably oriented crystallites are surrounded by significantly less favorably oriented crystallites, the deformation of favorably oriented crystallites is hindered by the unfavorably oriented ones. Consequently, high local strains may be generated that initiate plastic deformation even in the unfavorably oriented crystallites. Therefore, nearly all the crystallites take part in the process of plastic deformation, though the degree of deformation shows great local deviations.

Figure 6.22

Slipping lines on the surface of a plastically deformed polycrystalline metal

Any of the slip planes may extend from one boundary of the crystal to another. This is verified by the slip lines, which can be observed on the micrographs of such crystals (see Figure 6.22). The slip lines are virtually the "traces" of the slip planes generated by the plastic deformation of the specimen and visible on the polished surface of the specimen. The directions of slip lines are different in neighboring crystals. Initially, the distinguishable slip lines on the microscopic picture of polycrystalline metals are straight lines in the case of a small deformation. As the degree of deformation increases the number of slip lines increases in the crystal and the less favorably oriented slip planes start to operate. In some cases, the slip lines intersect. At the intersections of slip lines, these slip lines become step-like.

Figure 6.23

The crystallographically ordered structure of slip planes of polycrystalline metals formed during deformation

During deformation, the slip planes of the crystallites in poly-crystalline metals also tend to turn into the plane of the loading force. Consequently, the crystallites origi-nally positioned in a statistically disordered manner become crystallo-graphically ordered as can be seen in Figure 6.23. Thus, a *crystallographic order* is attained that may be characterized by a significantly greater number of slip planes with definite crystallographic orientations. This type of crystallographic order is called *deformation texture*. The consequence of this is that the originally quasi-isotropic mechanical properties change, and the deformed metals become anisotropic. The analysis of these consequences of plastic (residual) deformation will be discussed in the next subsection.

In polycrystalline metals possessing a small number of slip systems, residual deformation frequently starts with twinning. This process of the slip mechanism starts only following reorientation of the slip planes by twinning, obtaining an orientation more favorable to plastic deformation.

6.2.3. The consequences of plastic deformation

It is well known from the analysis of dislocations that they are essential for the realization of plastic deformation. It is also known that the speed of dislocation movement in metallic materials varies in the magnitude of 100 m/s. Though, this is quite a high speed, the movement of dislocations and thus the plastic deformation itself takes a definite amount of time. This feature of plastic deformation can be considered the reason for certain phenomena summarized under the term *the consequences of plastic deformation*. Some of these consequences can be observed even in the range of the so-called *apparently elastic deformations,* whilst most of the consequences appear only in the range of *real plastic deformation*. The former, i.e. the consequences observed in the range of *apparently elastic deformations,* cause certain irregular phenomena that cannot be explained and verified by the classical rules of elastic deformation. Therefore, before analyzing the main consequences of plastic deformation, we shall briefly discuss those phenomena that are collected under the term *consequences of apparently elastic deformation.*

6.2.3.1. The consequences of apparently elastic deformations

Conventionally, deformation is considered elastic if it is caused by a stress that is smaller than the *elastic limit strength* of the given material. During the analysis of the plastic deformation of single crystals (see subsection 6.2.2.3), we could see that the elastic limit strength of metallic materials varied to a great extent depending on the position of the slip plane with respect to the external loading. It obviously follows from this that every single crystallite of polycrystalline metals consisting of statistically disordered crystallites possesses different limits of elasticity. We also saw that the smallest stresses to start plastic deformation are required in those crystallites that are most favorably positioned (i.e. in which the slip plane forms an angle of nearly 45° with the direction of the loading force). Obviously, greater stresses are required to start plastic deformation in those crystallites that are less favorably positioned.

It also follows from the previous statements, that the elastic limit strength in polycrystalline metals is a macroscopic feature of the material: it is connected to the commencement of plastic deformation, which is measurable macroscopically using materials testing methods. Obviously, in certain crystallites plastic deformation occurs even under smaller stresses, thus in microscopic dimensions it is regarded as a residual (plastic) deformation though macroscopically it is not measurable. Among others, one of the consequences of this phenomenon is that the rules of classical elasticity are only approximately valid for polycrystalline metals. Similarly, those phenomena that are summarized under the term consequences of apparently elastic deformation can be explained by these reasons. These phenomena include elastic repercussion, relaxation, elastic hysteresis and the Bauschinger effect.

Elastic repercussion

The most characteristic manifestation of elastic repercussion is that the degree of elastic deformation of polycrystalline metals depends not only on the magnitude of the applied stress but also on the time of its application. Under the effect of a slowly increasing load, a greater degree of deformation may occur than under the effect of the same stress during fast loading.

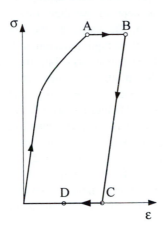

One illustrative manifestation of this phenomenon is the apparent increase in the modulus of elasticity (E) during rapidly increasing loading. Another manifestation of the same phenomenon is that the deformation of metallic materials continues under certain conditions though the applied stress is lower than the elastic limit strength but it is maintained at a constant level for a while. This is illustrated by the segment AB in Figure 6.24. It can also be observed in this figure that a portion of the deformation of the test specimen loaded by a stress lower than the elastic limit ceases only some time after the load has been removed (segment CD).

Figure 6.24
The phenomenon of elastic repercussion

Relaxation

It follows from the nature of elastic repercussion, that above a certain level (but still below the elastic limit strength), the stress should be gradually decreased if we wish to maintain a constant value of deformation. Although this phenomenon is the direct consequence of the elastic repercussion, due to its great practical importance it is denoted by the distinguishing term *relaxation*. It is very important to consider this phenomenon when we design structures operating under prestressed conditions. As follows from the Figure 6.24 as well, above a defined stress level and after a certain amount of time, part of the elastic deformation results in residual deformation. The occurrence of plastic deformation will lead to an undesirable decrease in the stress level adjusted by pre-stressing.

The Bauschinger effect

The *Bauschinger effect* is also a kind of elastic repercussion, although this phenomenon can be observed only in metals and alloys already subjected to prior plastic deformation. The substance of the *Bauschinger effect* can be briefly summarized as follows. When a metallic test specimen is subjected to a load causing a few percent of residual deformation, the value of the elastic limit strength determined by a tensile test will be different depending on the type of previously applied loading (i.e. whether the prior load was tensile or compressive load). This is illustrated by the σ–ε curves of Figure 6.25. The section a) refers to a previously annealed specimen, section b) to a formerly elongated one causing 2 % of residual deformation, and section c) indicates a test specimen previously subjected to compressive load causing 2 % of residual deformation.

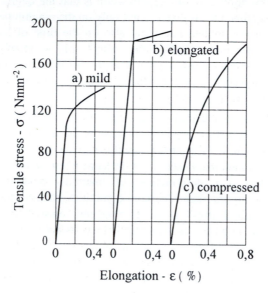

Figure 6.25
Illustration of the Bauschinger effect

This can be explained as follows. The low degree (1-2 %) of residual deformation does not cause plastic deformation in all the crystallites of the tested metal. At the boundary surfaces of the plastically deformed crystallites and only elastically deformed ones, elastic stresses arise that exert a restoring effect after deformation. Accordingly, if the former deformation were initiated by a tensile load, the subsequent load in a uniaxial tensile test – applied to determine the limit of elasticity following the prior deformation – would work in the same sense as the former deformation. Consequently, the repeated slipping will occur only under a stress greater than the former one. Therefore, in this case, a higher elastic limit strength is measured than the stress required, providing the former small (1–2 %) deformation. However, when the prior deformation was achieved by a compressive load, the effect of the uniaxial tensile test applied to determine the limit of elasticity is added to the restoring effect trying to restore the previous condition. Therefore, the same amount of residual deformation can be achieved by a stress lowered by the magnitude of the internal returning forces arising from the prior compressive loading. This is the reason why after prior compressive loading a lower tensile load is required to produce residual deformation. Due to this effect, in some cases the almost complete absence of the elastic (straight) section of the stress – elongation curve may be observed after prior compressive loading (see Figure 6.25).

6.2.3.2. The consequences of real plastic deformation

Almost all properties of metals and alloys change due to plastic deformation. The chemical and physical properties (with the exception of some electrical characteristics) change to only a small degree; however, the changes in the mechanical properties (including strength and deformation characteristics) are significant.

The consequences of plastic deformation, however, can be cancelled – i.e. the original properties can be restored – by heat treatment following the deformation process. Consequently, the effects of the plastic deformation can only be observed after deformation if the deformation was carried out below the so-called *recrystallization temperature* characteristic of the material. This is one of the bases for distinguishing cold and hot forming. Namely, we can speak about *cold deformation* when the consequences of the plastic deformation can be observed after the deformation process. In *hot deformation,* the consequences of plastic deformation are cancelled at the temperature of deformation, since the changed properties are restored immediately due to the high temperature of the forming process. It is known that the changed properties are restored above the recrystallization temperature. For most of the metals and alloys, the *recrystallization temperature* is roughly equivalent to $0.4\ T_{melt}$ (where T_{melt} is the temperature of the melting point expressed in degrees Kelvin). If we introduce the concept of the *homologous temperature* as the ratio of actual temperature and the melting point of the alloy expressed in degrees Kelvin (i.e. $T_{hom}=T/T_{melt}$), the recrystallization temperature can be given as $T_{hom}= 0.4$. It follows from this statement that the temperature distinguishing between cold and hot forming is considered as the so-called 0.4 homologous temperature.

It obviously follows from the above statements that cold deformation does not necessarily mean a deformation process performed at room temperature. The deformation of metals having low melting points (such as tin, lead, cadmium, etc) can already be considered *hot forming* at room temperature from the point of view of physical metallurgy. At the same time, the deformation performed in the temperature range of 800-1000 °C on high melting point metals (such as tungsten or molybdenum) may be regarded as cold forming. Since the essential mechanism of cold plastic deformation is the slip mechanism, in the forthcoming subsections the consequences of cold plastic deformation realized by the slip mechanism will be studied.

Lattice energy, lattice distortions

Plastic deformation is realized by external forces. These forces do work during the deformation. A major portion of the work done (according to some measurements 95-98 % of the work done) during plastic deformation is transformed into heat, the remaining few percent is stored in the material, increasing its energy level. The amount of stored energy depends primarily on the magnitude of deformation, and increases proportionally with it. (The magnitude of stored energy depends to a small extent on the microstructure of the deformed material as well. The same degree of deformation results in a little more stored energy in materials having a fine microstructure than in those possessing coarser grains.) The fact that a portion of the work done during deformation is stored in the material can be proved by the so-called *asterism* of the Laue X-ray diffraction examination.

In the Laue shots of non-deformed metals, a diffraction image of sharp dots can be seen. The deformed metals show elongated, "tail-like" reflections (see Figure 6.26). The asterism of the Laue-image also verifies that, during plastic deformation, the crystal planes existent prior to the deformation (that are considered to be virtually perfect planes apart from certain lattice defects) become distorted and bent. (This can be understood by a simple experiment, as follows. When a light beam is reflected by a plane surface toward some kind of a receptor surface, a sharp point-like image is obtained. When, however, this light beam is reflected by a bent surface, the image on the receptor surface will show a tail-like patch of light. This phenomenon is called *asterism*.) Thus, plastic deformation causes distortion of the crystal lattice by bending the theoretically perfect crystal planes. This lattice distortion, as the consequence of the internal forces, is related to the formation of lattice defects and disorders arising during deformation. These lattice defects represent a well-defined amount of surplus energy as was shown in Chapter 5. The portion of the work done accumulated in the deformed material correlates with this energy surplus.

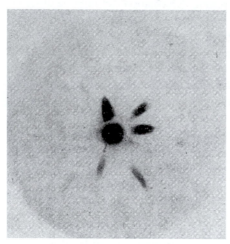

Figure 6.26

Asterism observed on the Laue-image of plastically deformed metal

The energy accumulated in the deformed material also means that the energy level of the crystals is higher compared to the undeformed state. This energy level, being higher than the equilibrium one, tends to reduce. However, this usually cannot occur spontaneously, or if so, it is an extremely slow process of low intensity.

In practice, this process is accelerated by a higher temperature. On increasing the temperature, the changes in properties that have occurred during plastic deformation gradually diminish and at the same time the energy level changes towards the equilibrium state. Under defined circumstances, the original state (the one preceding deformation) can thus be restored. In the following chapters, these processes and the metallographic reasons for these changes will be discussed.

Strain hardening

As we noted in the analysis of the slip mechanism, plastic deformation is realized essentially through the dislocation mechanism by the movement of dislocations. We also noted that during these movements, dislocations interact with each other. During these reactions, some of the dislocations disappear, but it is more characteristic that a great number of new dislocations are generated. Thus, the dislocation density increases in the deformed metal. As the number of dislocations increases, the energy level of the crystals also increases. This is due to the high elastic stress field around dislocations, as already shown in Figure 5.6. The elastic stress field around the dislocations hinders the movement of dislocations besides increasing the energy level of the crystals. Dislocations cannot approach each other within a certain limit due to the elastic stress field around them. Therefore, dislocations pile-up behind each other at a certain distance.

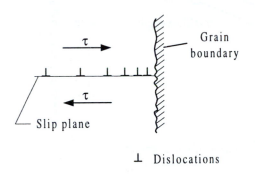

Figure 6.27
Schematic sketch illustrating
dislocation pile-up at grain boundaries

However, the movement of dislocations is hindered not only by other dislocations, but also by other obstacles, such as grain boundaries and foreign atoms. Figure 6.27 shows a schematic sketch of dislocation pile-up at the grain boundaries.

As a result of these effects, the further movement of dislocations, that is the continuation of plastic deformation, can be realized only by continuously increasing the stress. This phenomenon is called *strain hardening* of cold-deformed metals and alloys. It is characterized simply by the fact that an increasing degree of deformation can be achieved only by an ever-increasing magnitude of stress.

Strain hardening or the effect of cold plastic deformation can be indicated by changes in the mechanical properties of the deformed metal. Of these mechanical properties, the strength characteristics, such as yield point ($R_{p0.2}$), ultimate tensile strength (R_m) and hardness (HB) are increased, whereas the ductility parameters, such as specific elongation (A), contraction (Z), are decreased.

The effects of strain hardening are illustrated in Figure 6.28, where the changes in the ultimate tensile strength, the yield strength and the specific elongation of pure copper are given as a function of cold deformation.

In this figure, the degree of deformation is characterized by the specific reduction in the cross section determined from the following expression:

$$q = \frac{S_o - S}{S_o} \times 100. \qquad\qquad (6.46)$$

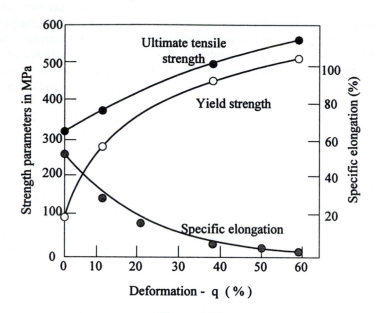

Figure 6.28
Illustration of the consequences of strain hardening, showing the changes in
strength and deformation characteristics of pure copper

The degree of strain hardening besides the actual deformation is also a function
of the crystal structure. This is illustrated in Figure 6.29 by the change in the
critical shear stress required to cause increasing plastic deformation.

As can be seen from this figure, the degree of strain hardening of metals
crystallizing in the hexagonal system and possessing only one parallel slip plane
{0001} (e.g. Cd, Zn, Mg), is significantly lower than that of metals crystallizing in
the face-centered cubic system possessing four slip planes {111} (e.g. Au, Al, Ag,
Cu, Ni). In metals crystallizing in the face- and body-centered systems and
possessing several slip planes, due to the intersections of the slip planes operating
simultaneously, dislocations hinder the mutual movements and thus these metals
harden to a greater degree.

Strain hardening as the consequence of plastic deformation can be evaluated as
beneficial in one sense and as harmful in another. Considering the increase in
strength properties, this is regarded as a favorable consequence: the increasing
degree of plastic deformation increases the loadability of metals and alloys. It is
important to note that *cold plastic deformation is the only way to increase the
strength of single-phase metallic materials*.

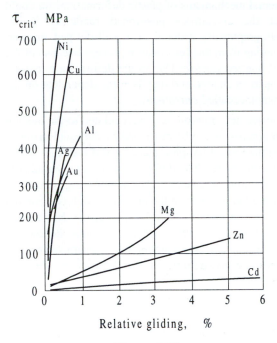

τ_{crit}, MPa

Relative gliding, %

Figure 6.29

The change of the critical tangential stress in metals having various crystal structures

Simultaneously, the unfavorable consequence of strain hardening is that the formability of the deformed material is exhausted after a certain degree of deformation. Therefore, when the forming process requires a high degree of cold deformation to produce a final component, intermediate heat treatments are included between the forming stages. Intermediate heat treatment is necessary to restore the properties changed during cold plastic deformation and in order to avoid cracks and failures.

A disadvantageous consequence of cold plastic deformation is also that the increasing degree of deformation can be achieved only by an ever-increasing value of stress. In other words, this means that deformation requires an ever-increasing forming force and energy. This has disadvantageous consequences with regard to the loading, the wear of forming tools, as well as the requirements set up concerning the performance of forming machines.

Strain hardening is frequently characterized by the true stress (σ') - true strain (φ) curves describing the change of the true stress as a function of true strain. For most metals applied in engineering practice, this can be expressed by the so-called Nádai's exponent law i.e.:

$$\sigma'= K\varphi^{n}, \tag{6.47}$$

where K is the coefficient of hardening and n is the exponent of hardening.

Deformation texture, anisotropy

In the normal (i.e. the mild) state, polycrystalline metals may be considered more or less isotropic, i.e. they possess essentially the same properties in any direction. The basis of this phenomenon is that the crystallites of polycrystalline metals are oriented randomly, there is no order whatsoever in their orientation.

From the features of the fundamental mechanisms of plastic deformation, we could see (sub–section 6.2.2.4) that the crystallites positioned randomly before deformation may become metallographically ordered in a well-defined manner. The crystallographic order originating from the deformation process because of the plastic deformation is called *deformation texture*. Due to the deformation texture, the originally quasi-isotropic properties change and they become dependent on the crystal orientation. This phenomenon is called *anisotropy*.

Anisotropy, as the result of plastic deformation, is closely related to the method and character of the deformation process. It can be described by the usual methods of crystallography applied to denote crystallographic planes and directions, i.e. by Miller indices and crystal vector components. The deformation texture is gradually formed as the degree of deformation increases. The neighboring grains of polycrystalline metals disturb each other not only in the process of deformation but also in the formation of the deformation texture. Therefore, a perfect deformation texture cannot be formed in polycrystalline metals even at extreme degrees of deformation. The deformation texture can be unambiguously characterized by using the indices of crystal planes and orientations only in those cases when the deformation texture is fully developed.

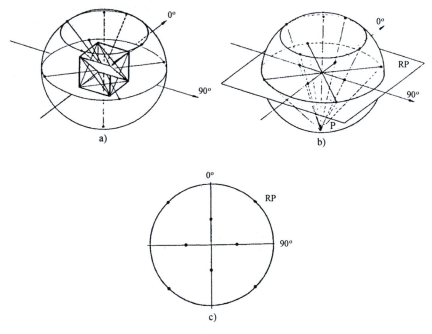

Figure 6.30
The procedure of creating a pole diagram

To describe the developing texture (non-fully developed texture), the so-called *pole diagram* determined by the methods of *stereographic projection* is applied (Figure 6.30). The pole diagram is obtained in such a way that all the crystals of the deformed metal (or at least a great number of them) are placed in the center of a sphere, keeping their orientations formed in the deformed metal. Then, the intersection points of the normal vector of the selected crystal planes with the surface of the sphere are determined. These points are projected to the so-called equatorial plane by using the opposite pole point P as the stereographic center. These projected points are the *pole points* of the selected plane, and the pole points together with the equatorial plane form the *pole diagram*. The above-described procedure of drawing the pole diagram is illustrated in Figure 6.30.

It follows from the above analysis that the pole points of metals consisting of crystallites with completely random orientation cover the total area of the circle with a uniform distribution. The pole diagram of crystalline materials possessing a fully developed texture is made of points of defined symmetry. It is very similar to the pole diagrams of single crystals, since the fully developed texture means that all the crystallites have exactly the same orientation. The pole diagrams of polycrystalline metallic materials – following from the previously described non-fully developed texture – always represent transitions between the two extreme cases. The more perfect the texture is, the more concentrated are the pole points around the point representing the completed texture. In Figure 6.31, the pole diagram of the {100} plane of a mild steel sheet rolled with a reduction of q = 90 % is illustrated. In the pole diagram, the plane of the rolled steel (RP) was selected parallel to the equatorial circle and the direction of rolling (RD) was selected parallel to the diameter of the equatorial circle.

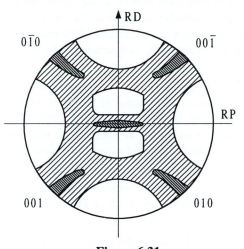

Figure 6.31
Polar diagram of steel rolled at a high degree of decrease in cross section

The residual stress of deformation

In macroscopically distinguishable sections of products manufactured by cold forming, the stresses originating from the non-uniform deformation are called *residual stresses of deformation*. The residual stress state due to plastic deformation can be considered beneficial in some cases and disadvantageous in some others.

Certain forming processes are applied with the direct aim of producing favorable residual stresses in the deformed workpiece. It is generally true that a residual stress of compressive nature is beneficial from several points of view. For example, the residual compressive stresses are definitely beneficial as they increase the fatigue limit in parts subjected to cyclic loads. For this purpose, complete families of forming processes have been developed that are collectively described as *surface strengthening processes.* (Further aims of surface strengthening processes are to increase the hardness of the surface layer, to ensure dimensional stability and a high-quality surface finish.)

Residual stresses arising from plastic deformation may also have unfavorable consequences. The undesirable changes of shape and sizes of the formed components observed immediately after deformation can be listed in this group (for example, spring back after bending). Similarly, the residual deformation of formed parts, occurring slowly during a long period and resembling a relaxation type process, may also be ranked as an unfavorable consequence of residual stresses.

The increased tendency of certain cold-formed products to undergo corrosion should also be mentioned as a disadvantageous consequence of residual stresses. This often leads to stress-corrosion occurring in the simultaneous presence of mechanical stresses and corrosive media. For example, in components made of brass, this effect may be experienced even under atmospheric corrosion conditions.

The disadvantageous consequences of residual stresses are process-specific in certain forming processes. Thus, for example, in drawn wires and rods, central bursting as a typical error of over-drawing is not only a technological failure, but is often attributed to the unfavorable distribution of residual stresses.

The unfavorable residual stresses can quite frequently be reduced by properly adjusting the technological parameters, appropriately dimensioning and designing the forming tools. In certain cases, an additional deformation of the product may also be advantageous to reduce the residual stresses. A frequently applied method of modifying the distribution of residual stresses is the straightening process.

One of the most effective methods for elimination of disadvantageous residual stresses is the *stress-releasing heat treatment* performed after deformation. The purpose of this annealing is to maintain the hardening caused by deformation with the simultaneous and significant reduction of residual stresses. The *recrystallizing annealing* performed at a temperature higher than that of the stress-releasing annealing restores all the properties modified by cold plastic deformation and simultaneously leads to the complete removal of the residual stresses. This will be discussed in detail in the next section.

6.3. Restoration of the modified properties of cold-deformed metals

Analyzing the consequences of cold plastic deformation, we noted in the previous section that the most important common feature of all changes occurring during plastic deformation is the increasing level of energy in the deformed crystals. The question arises why the higher energy level of the deformed crystal is not reduced by a spontaneous process, and why the atoms do not spontaneously return to their equilibrium state representing the lowest energy level.

The reduction or elimination of the changes representing the higher energy level of the crystal may occur only through the diffusion movement of atoms in the solid state. This type of diffusion is called *self-diffusion*. The *driving force for self-diffusion* is provided by the difference in the energy levels of deformed crystallites; however, the *activation energy* needed to initiate diffusion is usually insufficient to start the process at a measurable rate at room temperature. On the other hand, the rate of diffusion processes essentially depends on the value of the *diffusion coefficient* that increases exponentially with increasing temperature.

Essentially, there are two reasons explaining why the higher energy level of cold-deformed metals cannot spontaneously be reduced at the temperature of cold deformation. One of the reasons is the absence of the necessary activation energy to initiate the process, and the other one is the low value of the diffusion coefficient. It is well known that self-diffusion processes can occur at significant rates only above the recrystallization temperature of the deformed metal. At elevated temperatures, the activation energy for initiation of the process is available in the form of thermal energy and the value of the diffusion coefficient is sufficiently high.

At the same time, the above mentioned reasons unambiguously justify the fact that the property changes caused by cold deformation cancel in certain metals even at room temperature. For some metals having a low melting point, room temperature already represents a temperature above that of recrystallization. This also explains why the typical consequences of cold plastic deformation cannot be observed after hot forming. (When the condition $T_{forming} > T_{recr}$ is fulfilled, all the necessary conditions are available to restore the properties modified during deformation immediately at the temperature of the forming process. A more detailed analysis of this will be discussed when examining the metallographic aspects of hot forming.)

6.3.1. Thermally activated processes

It follows from the previous statements that as soon as the temperature is high enough to activate diffusion-type processes, the change in the properties modified by plastic deformation is initiated toward the original state (the state preceding deformation). It also follows from the former discussion that these processes occur at different rates depending on the temperature. In addition, certain processes can only occur above a certain temperature.

In Figure 6.32, the property changes of a cold-deformed metal are illustrated as a function of temperature. The diagram can be divided into three main sections, i.e. the stages of recovery *(I)*, recrystallization *(II)* and grain growth *(III)*. From this diagram, it can be seen that the properties modified by plastic deformation change toward their original values with an increase in temperature, and that significant changes can occur only in a well-defined range of temperature. In the forthcoming part, we shall analyze the processes occurring in cold-deformed metals and alloys at continuously increasing temperatures.

6.3.1.1. Recovery

During *recovery,* the energy accumulated in cold-deformed metals is reduced in a sub-microscopic extent. It follows from this that in the stage of recovery those properties of cold-deformed metals that are affected by lattice defects will change. However, these effects can differ significantly from each other with respect to the properties upon which they exert their effects. As can be seen from Figure 6.32, certain electrical and magnetic properties change significantly in the *recovery* stage (for example, electrical conductivity, the so-called thermoelectric force and magnetic susceptibility, as shown in region I in Figure 6.32). The mechanical properties (strength and deformation properties) also change to a slight degree but there are no changes in the microstructure that can be observed by an optical microscope.

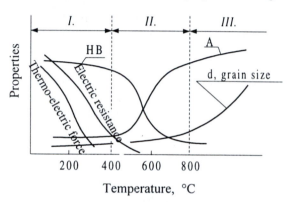

Figure 6.32
Change in the properties of cold-deformed metal with respect to temperature

Summarizing these changes, the *recovery* process can be characterized as follows. The properties modified by cold plastic deformation change towards their initial (pre-deformation) values without a significant change in the crystalline structure and without a significant change in the microstructure. In the *recovery* stage, the change in certain physical (electrical and magnetic) properties is much more significant than the change in the mechanical properties. Accordingly, the occurrence of the *recovery* process can primarily be observed by measuring the electrical and magnetic properties.

Recovery is inversely proportional to the degree of the preceding cold deformation meaning that a greater degree of cold deformation lowers the speed of recovery. It is also an experimentally verified fact that recovery is hindered by grain boundaries. Practically, this is the reason why, under identical conditions, recovery is more significant in single crystals than in polycrystalline metals.

6.3.1.2. Polygonization

Polygonization is a special kind of recovery. This means the "straightening" of the crystal planes bent and distorted by cold deformation, and the rearrangement of originally randomly positioned dislocations, which results in a characteristic polygonal structure, as shown in Figure 6.33.

Essentially, polygonization is the migration of dislocations to line up over one another in small-angle tilt boundaries. These small-angle tilt boundaries form internal sub-grain boundaries that surround sub-crystals, which are virtually free of dislocations.

Figure 6.33
Schematic illustration of polygonization

Polygonization is often called *structural recovery*. A greater degree of preceding deformation – under otherwise identical conditions – results in smaller sub-crystals. During polygonization, the mechanical properties also change, though to a lesser degree, but the hardness of the deformed metal is reduced, especially due to the increasing size of the sub-crystals as the temperature rises.

The formation and growth of sub-crystals as well as the accompanying changes in the mechanical properties are rarely observed in practice in polycrystalline metals under ordinary circumstances because of the recrystallization process and the concurring softening.

6.3.1.3. Recrystallization

Recrystallization is a process during which the crystallites distorted and loaded with stresses as a consequence of deformation behave as crystallization nuclei for the formation of new, stress-free, proportional crystallites. *Recrystallization* and the resulting *mechanical softening* completely cancel the consequences of plastic deformation.

Primary recrystallization

The temperature range, in which the formation of new, proportional grains from the distorted crystallites of the deformed metal is completed in the whole mass of the metal, is regarded as the *primary stage of recrystallization* (section II of Figure 6.32). Obviously, it follows from this that in the primary stage of recrystallization, the new, already mild crystallites (having a low dislocation density) coexist with the distorted crystal segments until the end of this stage of the process. The process of recrystallization is illustrated in Figure 6.34.

The stage of primary recrystallization is considered to be completed when the mechanical properties of the deformed metal regain their original values (the values they had prior to the deformation process). Recrystallization, similarly to all other crystallization processes, requires *crystallization centers* or *nuclei*. The segments of the deformed metal, possessing different energy levels, can serve as crystallization centers and the driving force required for recrystallization is provided by the energy level differences existing between the crystallized sections and the segments deformed to different degrees.

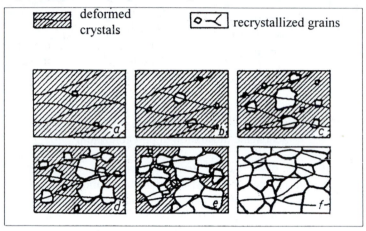

Figure 6.34
The scheme of recrystallization

Deformation of the metal is more unevenly distributed among the crystallites subjected to a greater degree of deformation than among the crystallites subjected to a lower degree of deformation. Consequently, greater differences in the energy level exist which provide for an increased amount of driving force for recrystallization. Therefore, a greater number of crystallization nuclei can form that lead to finer crystalline structures and the recrystallization process occurs faster. Further, it is important to note that:

— For the commencement of the recrystallization process, a certain minimum degree of deformation is necessary. (This is called the *recrystallization threshold*). The formation of crystallization nuclei necessary for recrystallization requires the existence of energy level differences between the differently deformed parts of the metal.

— The lower the degree of deformation, the higher the temperature required to start recrystallization, and vice versa, i.e. the higher the degree of deformation, the lower the temperature at which recrystallization starts. This is shown in Figure 6.35, illustrating the change in hardness vs. annealing temperature for copper specimens deformed to different extents.

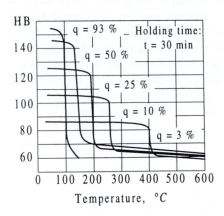

Figure 6.35

Effect of the preceding deformation on the temperature of recrystallization

— A similar relationship can be formulated between the temperature of recrystallization and the holding time, namely: under otherwise identical conditions, recrystallization occurs in a shorter period at a higher temperature.

The grain size obtained after recrystallization is determined by temperature, holding time and the degree of the preceding deformation. This relationship is shown in Figure 6.36, illustrating the effect of the degree of deformation on the size of ferrite grains of mild steel at constant temperature $(T = 870\ ^\circ C)$ and at constant holding time $(t = 30\ min)$. It can be seen from the diagram that, following a small degree of deformation $(q < 7 \sim 8\ \%)$, recrystallization does not commence because a sufficient amount of energy is not available (insufficient magnitude of driving force) to start the process. Increasing degrees of deformation already result in a sufficient energy differences in certain sections of the deformed metal to start the formation of crystallization nuclei. Because of the low number of crystallization nuclei and the high speed of recrystallization (at the temperature $T = 870\ ^\circ C$ the activation energy is high), the microstructure will consist of coarse grains.

Under defined conditions, the growth of the grains can reach an extreme magnitude. One of the manifestations of this is the so-called *recrystallization grain coarsening that* requires the coincidence of certain values of recrystallization temperature, holding time and degree of preceding deformation. The degree of deformation leading to the sudden and drastic growth of the grains is called *critical degree of deformation*. With the parameters of Figure 6.36, the critical degree of deformation is 8 %. By increasing the degree of deformation, ever-increasing energy differences are generated that lead to a rapid increase in the number of crystallization nuclei and to a finer recrystallized microstructure.

In Figure 6.36, the change in grain size is shown as a function of preceding deformation at a fixed recrystallization temperature. In practice, we need to follow the changes occurring in metals deformed to various degrees at various

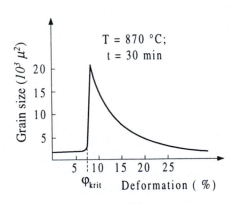

Figure 6.36

Recrystallization grain growth

temperatures. This is illustrated by the so-called complete *recrystallization diagram*. The complete recrystallization diagram can be determined by the examination of a great number of specimens. Deformation is performed by the same type of process (e.g. rolling, drawing) to various degrees of deformation, and then the specimens are annealed at different temperatures for the same holding time. In Figure 6.37, a complete recrystallization diagram is shown for mild steel. The axes of the base plane of the diagram represent the degree of deformation and the temperature of annealing, while the vertical axis represents the grain size.

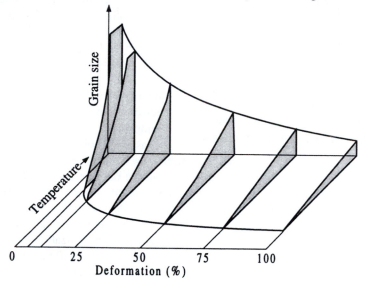

Figure 6.37
Complete recrystallization diagram of mild steel

Several important conclusions can be drawn from the complete recrystallization diagram. In the base plane of the diagram one can see the *curve of the initial temperature of recrystallization,* indicating at what temperature recrystallization starts with respect to the degree of deformation. It is clear from the diagram that under otherwise identical conditions the higher the degree of deformation the lower the temperature at which recrystallization starts. It is important to note that the recrystallization of the non-deformed metal does not usually occur at a temperature below its melting point. Therefore, in cast metals, recrystallization usually does not take place.

It is also seen from the diagram that the greater the degree of deformation, the finer the recrystallized microstructure. Additionally, above a certain temperature the lower the temperature of annealing, the finer the grain sizes. With regard to the grain size of test specimens deformed to various degrees and then heated to the same temperature, the following can be observed: every temperature corresponds to a degree of deformation below which recrystallization does not commence. This is called the *recrystallization threshold* value. When this value is reached, grain size suddenly increases. Following this, a finer microstructure is gradually obtained. The recrystallized grain size may even be finer than the original one after recrystallization, following a sufficiently great degree of deformation. It is also seen in the diagram that the recrystallization threshold value represents a lower degree of deformation with increasing the heating temperature.

Coarsening of grains, secondary recrystallization

However, the grains formed by recrystallization are not constant. They can grow further if the metal is held at the temperature of recrystallization for a longer period than required for primary recrystallization or when it is heated to a temperature higher than required for recrystallization. This process, causing the gradual growth and coarsening of grains, cannot be sharply distinguished from the primary section of recrystallization. This section of recrystallization is called *secondary recrystallization*, in other words, the *section of gradual grain coarsening*.

The growth of grains is related to the fact that the grain boundaries represent transient zones between the neighboring grains of different orientation. The atoms positioned at the grain boundaries do not belong definitely to the lattice of any of the grains and, therefore, their energy levels are higher. The increase in temperature – behaving as activation energy – moves the atoms (disorderly positioned at the grain boundaries) toward the lattice points representing lower energy levels. Figure 6.38 assigns disordered atoms to grain lattices.

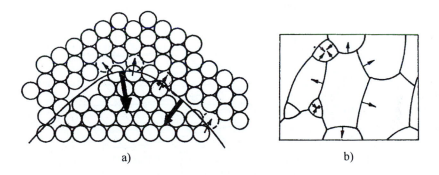

a)   b)

Figure 6.38
a)   The mechanism of grain growth by grain boundary migration
b)   Illustration of duplex microstructure

It can be seen in the figure that the attraction of the atoms of the concave grain boundary to the atoms located on the grain boundaries is greater than the attraction of the atoms of the convex grain boundary. Therefore, the positions of the atoms on the concave boundaries are more stable than the positions of the atoms on the convex grain boundaries. A consequence of this is illustrated in Figure 6.38, where the grain boundary migration of single-phase metallic materials occurs under the effect of heat. It can be observed that the grain boundaries move toward their centers of curvature.

According to the mechanism of grain growth described so far, certain grains may grow to become extremely large, while the other grains in their vicinity remain small as long as they "dissolve" into one of the grains growing large. In this type of microstructure, there is a period when coarse and fine grains exist simultaneously. This type of microstructure is called the *duplex microstructure.* This microstructure, consisting of mixed (coarse and finer) grains, can be extremely disadvantageous as regards certain mechanical properties (see section b/ of Figure 6.38).

Figure 6.36, shows the so-called *recrystallization grain coarsening* occurring due to coincidence of the recrystallization temperature, the holding time and the critical degree of deformation during the recrystallization process. A similar degree of grain coarsening may happen during the secondary recrystallization, too, as shown in Figure 6.39, illustrating the complete (three-dimensional) recrystallization diagram of wrought aluminum.

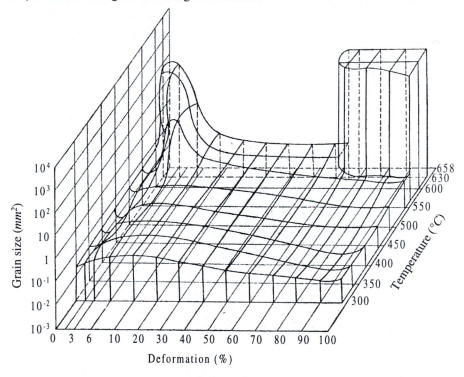

Figure 6.39
Complete recrystallization diagram of aluminum

The recrystallization texture

When we discussed the consequences of plastic deformation, the formation of the deformation texture and its consequences were analyzed. However, a change in microstructure causing a defined crystalline order may occur not only during deformation, but also during the recrystallization annealing. This is called *recrystallization texture* in order to distinguish it from the deformation texture and to refer to its origin.

The recrystallization texture can be explained mainly as follows. The recrystallization nuclei are formed from the deformed crystallites that possess a defined crystalline order. Therefore, the positions and orientations of the new crystallites depend on the crystalline order from which they have been formed. The growth of the crystallization nuclei is affected mainly by the holding time and the temperature of annealing, in addition to the preceding deformation. Consequently, the recrystallization microstructure may be equivalent to the deformation microstructure, but it may also differ from it, largely depending on the parameters of the recrystallizing annealing.

The texture of the rolled sheets of metals and alloys belonging to the face-centered cubic system can be characterized either by the ordered *cubic texture* having the index of $\{100\}$ <001> or by the so-called *silver texture* having the index of $\{110\}$ <$\bar{1}$12>. In rolled sheets having a cubic texture one of the hexahedron planes of the crystallites is parallel to the surface of the sheet and one of the crystalline axes is parallel to the direction of rolling. In sheets having a silver texture one of the rhomb-dodecahedron planes of the crystallites is parallel to the surface of the sheet and their <$\bar{1}$12> direction is parallel to the direction of rolling.

6.4. Hot forming

In the previous subsections, the consequences of cold forming and the restoration of properties changed during cold plastic deformation were discussed. In this subsection, the metallographic processes occurring during hot deformation are briefly discussed with a comparison of hot and cold forming.

In physical metallurgy, the term *hot forming* is used when deformation is performed at a temperature higher than the recrystallization temperature of the deformed material. It follows from this that the consequences of plastic deformation (hardening, lattice distortions, etc.) cease at the temperature of deformation, virtually simultaneously with the deformation (or between two operations of the process) due to the recrystallization occurring in the meantime.

The time requirement of the two essential but opposing changes – namely deformation causing the accompanying hardening and recrystallization resulting in mechanical softening – taking place during hot forming, are significantly different from each other. When we analyzed the mechanism of deformation hardening, we could see that hardening (as well as the dislocation reactions causing hardening) represented an extremely fast process, practically with no time requirement. In contrast to this, recrystallization is a diffusion process with a significant time requirement. Essentially, the significant difference between the required times results in the difference between cold and hot forming.

The strain hardening during cold deformation can be characterized by Nádai's exponent law, establishing the correlation between true stress (σ') and true strain (φ) i.e.

$$\sigma' = K\varphi^n ,$$ (6.48)

where K is the coefficient of hardening, n is the exponent of hardening. The value of the hardening exponent is equal to the true strain corresponding to the uniform elongation limit, i.e.: $n = \varphi_m$.

In contrast to this, in hot deformation the true stress is rather the function of the rate of deformation ($\dot\varphi$) in accordance with the time requirements of the recrystallization and annealing processes. Therefore, it is described by the relationship:

$$\sigma' = M\dot\varphi^m$$ (6.49)

According to the earlier discussion, cold and hot forming may be distinguished by the so-called true stress – true strain curves shown in Figure 6.40.

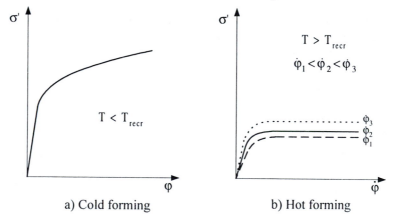

a) Cold forming   b) Hot forming

Figure 6.40
Distinguishing cold and hot forming based on the true stress - true strain curves

During cold forming, true stress increases with the increase in true strain according to Nádai's exponent law. The strain hardening – as a consequence of residual deformation – can be well seen on the true stress-true strain curve (in section a/ of Figure 6.40). In contrast to this, the true stress does not depend only on the true strain, but depends rather on the rate of deformation (see section b/ of Figure 6.40).

6.4.1. Static and dynamic changes in microstructures during hot deformation

It follows from the previous discussions that *recovery* and *recrystallization* occur not only during the annealing of cold-deformed metals, but also during hot forming at the temperature of deformation. The recovery and recrystallization processes occurring during recrystallization annealing after cold forming or during any other heating, holding or cooling processes are called *static recovery*, *static recrystallization*.

In contrast to this, we speak about *dynamic recovery* and *dynamic recrystallization* during hot forming when the recovery and recrystallization occur in a changing microstructure during the process itself, which is affected by deformation. Obviously, in hot forming the deformed workpiece may – and in many cases does - undergo simultaneously static recovery and static recrystallization during cooling following the hot forming accompanied by dynamic recovery and dynamic recrystallization. It also follows from this statement that the final microstructure and thus the mechanical properties of the deformed workpiece can be modified in a wide range by adjusting the conditions of hot deformation and the consequent cooling - heating. This is utilized in many recently developed processes called *thermo-mechanical treatments*.

6.4.1.1. Dynamic recovery

The consequences of *dynamic recovery* can be observed on the so-called *hot yield curves* determined by hot forming experiments (part a/ of Figure 6.41). The region denoted by (I) in the hardening section is the range of increasing dislocation density. As the stationary section of the yield curve is reached (section II in part a/ of Figure 6.41) the formation of subgrains typical of *recovery* commences (see section 6.3.1.2). The size, uniformity and orientation of the subgrains are a function of strain rate and deformation temperature.

The values of the main parameters of subgrains – the density of dislocations, the average length of the free movement of dislocations, and the mutual orientation of subgrains – do not change along section II of the hot deformation yield curve. This means that a dynamic equilibrium exists between the formation and the termination of new dislocations. Essentially this stabilizes the values of dislocation density and yield stress, as shown in section II of part a/ in Figure 6.41.

The size of the subgrains formed during dynamic *recovery* significantly influences the microstructure during the recrystallization following hot deformation. It is also well known that dynamic *recovery* occurs much more easily in metals characterized by high energy stacking faults (as, for example, in the face-centered cubic aluminum, γ-iron, etc.). In contrast to this, metals crystallizing in the body-centered cubic system are characterized by low-energy stacking faults having a lower tendency to dynamic *recovery*.

6.4.1.2. Dynamic recrystallization

The occurrence of *dynamic recrystallization* appears characteristically in the hot yield curves (part b/ of Figure 6.41). In most hot forming processes – carried out at a relatively high deformation rate – after reaching the maximum point of the yield curve, one can observe a decrease in the yield strength. This is due to the dynamic recrystallization which takes place continuously and rapidly.

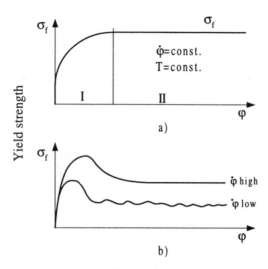

a)

b)

Figure 6.41
Yield curve for hot forming to illustrate
a) dynamic *recovery* b) dynamic
recrystallization

Apart from this, at lower deformation rates, "waves" can appear on the lower curve of part b/ in Figure 6.41. These waves are due to rapidly repeated dynamic recrystallization.

Dynamic recrystallization can be characterized by the formation of the steady section following the decrease in yield stress. In this section, another dynamic equilibrium evolves between the formation of new dislocations and the cessation of dislocations due to dynamic recrystallization. This region is further characterized by the dynamic equilibrium of a more or less steady grain size.

Based on experience gained during practical research experiments, deformation exceeding a certain critical value is a condition for dynamic recrystallization. This critical value of deformation is equivalent to about 20 % of the deformation corresponding to the maximum yield stress. If the degree of deformation is below the critical value, usually only dynamic recovery occurs during hot deformation. If the deformation exceeds the critical value, both dynamic recovery and dynamic recrystallization occur. At an even higher degree of deformation, dynamic recrystallization becomes dominant.

In hot forming, usually the same type of texture is produced as in the heat treatment following cold forming. However, the texture of hot deformed metals is not as defined, and is less perfect. This may be explained by the occurrence of dynamic recovery and dynamic recrystallization and as a consequence, the conditions – characteristic of cold forming – to form a defined texture are absent.

According to the metallographic processes occurring during hot forming, besides the temperature of forming and the true stress - true strain curves, the hot and cold forming can also be distinguished based on changes in microstructure.

The proportional, statistically disordered grains, having roughly the same sizes in all three dimensions of space (section a/ of Figure 6.42) – due to the slip mechanism of deformation – form a crystallographically ordered microstructure, elongated in the direction of the forming force (section b/ of Figure 6.42). It should be noted that the volumes of the grains before and after deformation – in accordance with the law of volume conservation – are equal, i.e.: $V_o = V_1$.

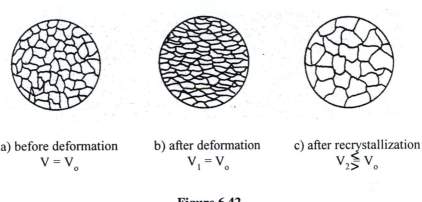

a) before deformation
$V = V_o$

b) after deformation
$V_1 = V_o$

c) after recrystallization
$V_2 \lessgtr V_o$

Figure 6.42
The change in microstructure during cold deformation and
during the subsequent recrystallization

The distorted, crystallographically ordered microstructure can be restored to proportional grains similar to the state preceding cold deformation by recrystallizing annealing (part c/ of Figure 6.42). However, the size of the grains may be smaller, equal to or greater than the size of the grains before deformation depending on the parameters of the recrystallizing annealing.

After hot deformation, proportional grains can be observed under the optical microscope as a result of the dynamic recrystallization occurring at the temperature of deformation (Figure 6.43). Compared to cold forming, it can be seen that after hot forming only the a) and c) stages are observed.

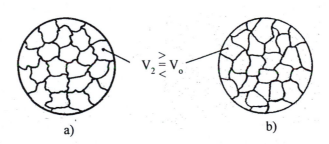

$V_2 \gtreqless V_o$

a)

b)

Figure 6.43
The change in microstructure during hot forming

The recrystallization occurring during hot forming results in a finer microstructure at lower temperatures and at higher degrees of deformation. Grain size may be considerably reduced by a well-designed and carefully performed hot deformation. When hot deformation is carried out at elevated temperatures and at low degrees of deformation, a coarse microstructure is formed, as was experienced during the recrystallization performed with similar parameters following cold deformation.

7. METALLIC ALLOYS AND PHASE DIAGRAMS

7.1. Definition of alloys and the mode of alloying

In most cases, pure metals consisting of a single component are unable to fulfill certain requirements. This statement is particularly valid considering the dynamic development of recent years and the consequent changes. To match the increasing requirements special materials, in many cases special alloys are needed. From the mechanical engineering point of view, metallic alloys are of decisive significance, and therefore, in this section we shall exclusively deal with metallic alloys.

Metallic alloy means at least an apparently homogeneous metallic material that is produced by combining two or more components in various ways. The important characteristic feature of metallic alloys is that the fundamental component (i.e. the basic component) is in all cases a metallic element.

The *purpose of alloying* is to establish such definite physical, chemical, mechanical or other special properties that cannot be achieved with pure metals. The components of metallic alloys are mainly metallic elements, but at least the basic element should be metallic (for instance *Fe, Al, Cu*, etc.). The further components may be metals (for instance *Cr, V, Ni*, etc.), so-called metalloid elements having metallic nature (such as *C, Si, Sb*), non-metallic elements (for instance *S, P*) and even gases (as for instance N_2) in certain cases.

The *production of alloys* is most frequently achieved by melting the components in the liquid state. In many cases, this is the most obvious and simplest way of alloying since most metals dissolve each other without limitations in the liquid state. Thus, the homogeneous structure – important from the point of view of alloys – can be ensured in the simplest way by melting. In this respect, aluminum (*Al*) and lead (*Pb*) are rare examples, since these two metals do not dissolve each other even in the liquid state. When these two metals are mixed together in the liquid state, after a sufficient amount of time, they separate due to their different densities.

There may be other difficulties in alloying by melting together the liquid components. When the difference between the melting points of the components is great, alloying by melting of the components is not suitable. The reason is that a major part of the component with a lower melting point may burn out (oxidize) before the component having the higher melting point melts. For instance, this is the case in the production of high-speed steel having iron as the fundamental metal. Iron has a melting point of $T_{FE} = 1536\ ^{\circ}C$ and the other components (besides carbon present in all steels) are tungsten ($T_W = 3410\ ^{\circ}C$), molybdenum ($T_{Mo} = 2625\ ^{\circ}C$), chromium ($T_{Cr} = 1890\ ^{\circ}C$) and vanadium ($T_V = 1735\ ^{\circ}C$). The differences between the melting points of these components are so great that this alloy cannot be produced in the classical manner, by melting the components. In such cases, so called *master alloys* are produced consisting of the higher melting point components, which in turn are added in predetermined quantities to the liquid of the base metal. (For instance, to produce the high speed steel, the master alloys of the higher melting point components made with iron, i.e. the so-called *ferro-alloys,* are utilized in the form of *FeV, FeCr, FeMo* and *FeW.*

A further method of alloying is the combination of the components in the solid state using the *process of powder metallurgy.* This alloying process is applied when extremely high melting point components are used. One of the most typical examples is the production of hard metals. *Hard metals* are mixtures of powders of high melting point carbides (for instance, *WC, TiC, TaC, NbC*) and the so-called embedding material or matrix (*Co*). The production of hard metals includes the pressing of the powder mixture into the desired shape followed by subsequent annealing (sintering). The sintering temperature is slightly lower than the melting points of the main constituents. During the sintering process, a metallic bond is established between the adjoining metal particles due to diffusion. The hard metals produced in this manner are not "real" alloys according to the original definition of alloys. Therefore, they are often referred to as *pseudo-alloys.*

One of the special methods of alloying in the solid state is performed by annealing the component to be alloyed in a reactive gas media or salt bath at the appropriate temperature. During this procedure, the required alloying element diffuses into the surface layer of the material to be alloyed. Utilizing this process, the carbon content of the surface layer of low carbon steels can be increased to improve hardenability: this process is called *carburization.* Similarly, the process known as *nitriding* is aimed at changing the composition of the surface layer. During nitriding, metal nitrides of high hardness and wear resistance are created in the surface layer. In certain cases, the aim is to increase both the carbon and the nitrogen content of the surface layer: this process is called *carbo-nitriding.* The diffusion of other alloying elements (such as boron, aluminum, chromium, etc.) can be achieved similarly. These processes are summarized under the name of *surface alloying processes,* or *thermo-chemical treatments.*

It is an obvious consequence of the previous statements that the structure of alloys is much more complex than that of the pure metals. The properties of the alloys are determined to a significant extent by the types of relationships between the components. These relationships will be analyzed in the following section.

7.2. The relationships between the components of metallic alloys

In metallic alloys, the following main relationships are possible among the components:

a) the components can dissolve each other in the solid state and form a *solid solution*;

b) the components react chemically with each other and produce various *metallic compounds*;

c) the components form neither solid solutions nor metallic compounds with each other: during crystallization these alloys usually solidify as a mixture of the components which is called a *eutectic*.

Both solid solutions and metallic compounds represent homogeneous, single-phase structures: their components cannot be observed separately. However, the eutectic represents a two-phase, heterogeneous microstructure: its constituents can be well distinguished under an optical microscope. The characteristics of the above mentioned types of alloys will be discussed in detail in the next sections.

7.2.1. Solid solutions

It is well known that, generally, various materials in the liquid state dissolve each other without any limit. If the particles of the solvent and the solute materials cannot be distinguished in the solution, it is called a homogeneous, single-phase liquid solution.

Similarly to liquid phases, solid solutions can occur in the solid state, also when the two components (i.e. the solvent and the solute material) cannot be distinguished in the homogeneous solid solution. One of the common characteristics of solid solutions is that the crystal lattice of the solid solution is the same as that of the solvent material. If the atoms of the solute material substitute the atoms of the solvent at the lattice points, a *substitutional solid solution* is formed. If the atoms of the solute material are located in the interstices of the crystal lattice, an *interstitial solid solution* is formed. Solid solutions are denoted by the letters of the Greek alphabet.

7.2.1.1. Substitutional solid solutions

Substitutional solid solutions are formed when the atoms of the solvent material are substituted by the atoms of the solute material at the lattice points. In Figure 7.1, a sketch of a substitutional solid solution can be seen. In this figure, the atoms of the solute material substitute the atoms of the solvent material randomly, but always at the lattice points of the solvent crystal.

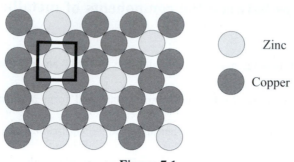

Figure 7.1
Sketch of a substitutional solid solution

There are several conditions for the formation of substitutional solid solutions: the sizes of the atoms of the solvent and the solute materials should be close to each other and their electron structure should be similar. For example, for the solid solution formed by zinc and copper shown in Figure 7.1, the difference between the atomic radii is minimal (approximately 8 % $r_{Cu} = 0.1278$ *nm*, $r_{Zn} = 0.139$ *nm*) and they have a similar electron structure (the electron configuration of *Cu* is $3d^{10}4s^1$, while the electron configuration of *Zn* is $3d^{10}4s^2$).

Under well-defined conditions, the two materials can form an *unlimited solid solution*. The conditions of forming an unlimited solid solution obviously impose stronger requirements regarding both the solvent and the solute material, namely:

- The two materials should have the same crystalline structure.

- The difference between the sizes of the atoms of the solvent and the solute material should not exceed 15 %.

- The number of the valence electrons of the solvent and solute materials should be the same, and the chemical affinity between the elements should not exceed a certain value. (Otherwise, they could form a metallic compound instead of a solid solution).

If all the above listed conditions are fulfilled, the two components form an unlimited solid solution. This is characterized by the fact that the atoms of one of the components (denoted by *A*) are continuously replaced by the atoms of the other component (denoted by *B*). As can be seen from Figure 7.2, starting from the pure *A* metal, a continuous transition to the crystal lattice of pure metal *B* can be obtained.

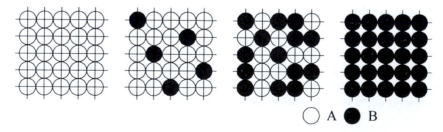

Figure 7.2
Illustration of the formation of an unlimited solid solution

Such an unlimited solid solution is formed by *Cu* and *Ni*. However, all the conditions of unlimited solid solubility are rarely fulfilled and in these cases, the components can be characterized by limited solid solubility. For example, in the case of *Cu* and *Zn* shown in Figure 7.1, the atoms of *Zn* can substitute a maximum of 37 % of the atoms of *Cu*, thus forming a limited solid solution.

The majority of substitutional solid solutions can be characterized as disordered solid solution, as shown in Figure 7.1. This means that the atoms of the solute material are located randomly in the crystal lattice of the solvent material. However, in certain cases so called *ordered solid solutions* are formed. In ordered solid solutions, the substitution occurs according to strictly defined geometrical rules (see Figure 7.3).

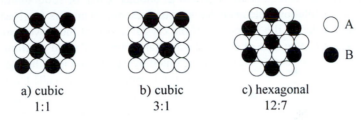

a) cubic	b) cubic	c) hexagonal
1:1	3:1	12:7

Figure 7.3
Schematic illustration of ordered solid solutions

In an ordered solid solution, the ratio of the atoms of the solvent and the solute materials can be expressed by simple integer numbers. Thus for example, in the cubic crystalline system $N_{solvent} : N_{solute} = 1{:}1$, or $3{:}1$, while for metals in the hexagonal system, the ratio $N_{solvent} : N_{solute} = 12{:}7$ is typical of ordered solid solutions.

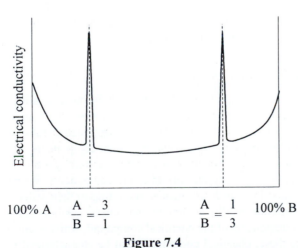

100% A $\quad \dfrac{A}{B} = \dfrac{3}{1} \qquad\qquad \dfrac{A}{B} = \dfrac{1}{3} \quad$ 100% B

Figure 7.4
The change in electric conductivity vs. composition in an alloy system containing an ordered solid solution

One of the features of ordered solid solutions is that certain properties change suddenly at the composition ratios corresponding to the ordered solid solution. This can be seen in Figure 7.4, where the change in electrical conductivity is shown for metals having a cubic crystal structure.

The order in solid solutions may be perfect when all the atoms are located in well-defined places of the crystal lattice according to the previously described ratios. Perfect order in solid solutions can be achieved by extremely slow cooling.

Non-perfect order can also occur in solid solutions. This may be characterized by the fact that certain atoms are located at lattice points as defined by the ratios of ordered solid solutions, while the rest of the atoms are located randomly. Concerning certain characteristics, the ordered solid solutions may be considered as intermediate phases between solid solutions and metallic compounds to be discussed in Section 7.2.2.

7.2.1.2. Interstitial solid solutions

Interstitial solid solutions are regarded as another group of solid solutions. In interstitial solid solutions, elements having small atomic diameters (such as *H, O, N, C, B*) are located in the interstices of the crystal lattice of the solvent material. In engineering practice, one of the most important interstitial solid solutions is the

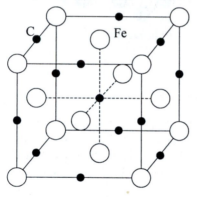

Figure 7.5
Interstitial solid solution of
C and *Fe*

one formed by carbon and iron. In Figure 7.5, the crystal lattice of the so-called γ-iron can be seen. The γ-iron has a face-centered cubic lattice. In γ-iron, *Fe* atoms are located at the lattice points (i.e. at the corner points and at the face centers), whereas *C* atoms are located in the largest interstices of the lattice (i.e. at the body center and at the edge centers). In Figure 7.5, all the possible locations of the *C* atoms in the γ-lattice are shown. In reality, the carbon atoms can be found only at some of these locations, since the *C*-solubility of γ-iron is limited. This will be discussed in detail in Chapter 8 dealing with iron-based alloys.

7.2.2. Chemical compounds in metallic alloys

There are two basic rules concerning chemical compounds in metallic alloys. One of them is that the chemical valence rule should be valid between the components of the chemical compounds. This means that the stochiometric ratio of components should correspond to the ratio of the valences of the components. The other important rule is that chemical compounds can be characterized by a unique crystal lattice that is independent of the crystal lattices of the components. Within this lattice, the atoms are located in an ordered manner according to their valence ratios. The chemical compounds are usually denoted by the general symbol $A_m B_n$, where *A* and *B* stand for the components of the chemical compound, while *m* and *n* are integer numbers. Besides these, chemical compounds can be characterized by the fact that their crystallization – similarly to pure metals – occurs at a definite, constant temperature. Furthermore, their composition changes may result in extreme changes in some of their properties. Three subgroups of chemical compounds are distinguished in metallic alloys: the so-called *ionic compounds*, the *electron compounds* and the *interstitial chemical compounds*.

7.2.2.1. Ionic compounds

Strong metallic elements (for instance *Na, Ca*, etc.) form ionic compounds with non-metallic elements (for instance *Cl, F*, etc.). The rules for forming ionic compounds are the same as the rules for forming ionic bonds discussed earlier. The metallic material losses its valence electrons from the outermost electron shell and becomes a positive ion. At the same time the non-metallic material picks up the lost electrons and thus becomes a negative ion. Accordingly, ions are located at the lattice points of the crystal and the ionic compound is kept together by ionic bonds.

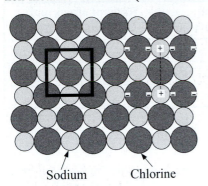

Sodium   Chlorine

Figure 7.6
Schematic illustration of *NaCl* ionic compound

One of the best examples of ionic compounds is *NaCl* made up of Na^+ and Cl^- ions. In Figure 7.6, the ordered structure of the *sodium chloride* ionic compound is shown.

7.2.2.2. Electron compounds

Typically electron compounds are formed by high-melting-point metals (for instance *Cu, Ag, Au, Fe, Co, Ni*) with lower-melting-point metals (for instance *Cd, Al, Sn, Zn, Be*). These chemical compounds are often named *Hume-Rothery phases* – particularly in the English-speaking world – in honor of the British scientist who discovered them.

The most important feature of electron compounds is that the ratio of atoms and valence electrons ($N_A : N_E$) taking part in the chemical bond can be expressed as the quotient of simple integers.

Similarly to solid solutions, electron compounds are denoted by the letters of the Greek alphabet. Metallic compounds crystallizing according to the body-centered cubic crystal system are denoted by β and are characterized by the ratio $N_A : N_E = 2 : 3$. Metallic compounds crystallizing according to the complex cubic crystal system are denoted by γ and are characterized by the ratio $N_A : N_E = 4 : 7$, while metallic compounds crystallizing according to the hexagonal close-packed crystal system are denoted by ε and are characterized by the ratio $N_A : N_E = 13 : 21$.

Similarly to other types of chemical compounds, electron compounds have a homogeneous, single phase structure possessing a single crystal lattice. Electron compounds can be found in many metallic alloys of industrial importance.

A further special feature of electron compounds is that if the two metals form an electron compound, then all three types of electron compounds can be found in the alloy system. An excellent example of this is the *Cu - Zn* alloy system. The characteristics of the electron compounds in this alloy system are summarized in Table 7.1.

Table 7.1 Electron compounds in the Cu - Zn alloy system

Symbol of compound	Symbol of phase	Crystal structure	Ratio of atoms and electrons, N_A:N_E
CuZn	β	Bcc	2 : 3
$CuZn_3$	γ	Complex cubic	4 : 7
Cu_5Zn^8	ε	Hexagonal	13 : 21

7.2.2.3. Interstitial chemical compounds

Usually, *interstitial chemical compounds* are formed by high-melting-point metals crystallizing in the cubic system (for example *Fe, Cr*) with metalloids of a small atomic radius (for example N_2, *C*). The condition for forming interstitial chemical compounds is that the relationship $r_{metalloid}/r_{metal} = 0.55 ... 0.66$ be fulfilled for the radius of the metalloid ($r_{metalloid}$) and the radius of the metal atom (r_{metal}). In interstitial chemical compounds, the metal is located at the lattice points and the metalloid in the interstices of the lattice.

It is also valid for the interstitial compounds that the ratio of the components can be expressed by the simple formula M_xN_y, where *M* stands for the metallic element and *N* for the metalloid. The crystal lattice of interstitial chemical compounds differs from the original crystal lattice of the metallic component. This provides the basic difference compared to an interstitial solid solution, which is also formed typically by high-melting-point metals and elements of a small atomic radius, but preserves the original crystal lattice of the metal.

Interstitial chemical compounds are characterized by high hardness and resistance to wear, therefore they are important constituents in tool steels. It is also worth mentioning that *Fe* and *C* form an interstitial solid solution as well as an interstitial chemical compound. The interstitial chemical compound of *Fe* and *C* is iron-carbide (*cementite)* with the chemical symbol Fe_3C.

7.2.3. Eutectic and eutectoid

If the components form neither a solid solution nor a chemical compound with each other, then during crystallization the alloy will solidify into a mixture of the two components. This heterogeneous, two-phase structure directly solidifying from the liquid phase by primary crystallization is called *eutectic*. The eutectic structure – in contrast to homogeneous solid solutions and chemical compounds discussed in the previous two sections – is a heterogeneous, two-phase microstructure.

In a eutectic, the two phases can be well distinguished under an optical microscope. This type of heterogeneous microstructure may also be formed during recrystallization in the solid state. It is called a *eutectoid* to distinguish it from the eutectic, which is the product of primary crystallization. These two types of microstructures (the eutectic and the eutectoid) have many common characteristics. The microscopic structure of a typical eutectic microstructure (called *ledeburite* and existing in iron-carbon alloys) is shown in Figure 7.7.

Depending on the crystallization process, the eutectic may have a lamellar or a granular microstructure. Its constituents may be pure metals, solid solutions or metallic compounds, but regardless of the phases of the components, the eutectic is characterized by a defined composition. Similarly to pure metals, a binary eutectic crystallizes at a constant temperature. The alloy with eutectic composition solidifies at the lowest temperature in a given alloy system. (Its name originates from this feature: *eutectic* is a Greek word meaning *well melting*).

Figure 7.7
Microscopic picture of a pig iron with eutectic composition, C=4.3 %,
(N = 500×, etched with HNO_3)

The properties of the eutectic are determined by the properties of the constituent phases. If both components are rigid, the eutectic will also behave rigidly. If both phases are ductile, the eutectic will also be ductile. If one of the phases is rigid and the other one is ductile, the properties of the embedding phase (the *matrix*) will be of primary importance. When the embedding phase is rigid, the eutectic will be rigid, while in the case of a ductile embedding phase, the eutectic will also be ductile, at least to some extent. The ductility of metallic materials containing rigid and plastic phases essentially depends on the quantity, shape, size and distribution of the rigid phase. Beside these parameters, the strength of the rigid and ductile phases, the crystallographic relationship between the two phases and the energy of the phase boundaries also play an important role. Separation of the effects of all these factors is hardly possible. If the rigid phase forms a network or is present in the form of large lamellas within the ductile embedding phase, then a microstructure with relatively poor ductility is obtained. The most favorable case occurs when the rigid phase is present in the form of dispersed particles. This type of microstructure possesses good toughness, thus good resistance to dynamic loads.

7.2.4. Phases and microstructural elements in metallic alloys

When we summarized the thermodynamic terminology of crystallization, it was stated that the *phase* in alloys represents an independent part of the system, separated by a boundary surface from other phases. Within a phase, the composition and the various properties are homogeneous. It follows from the previous definition and from the preceding sections that in a metallographic system, the following phases can be found:

- *liquid phase* (homogeneous, single phase melt of a metal or an alloy).
- *solid phase*, which can further be classified as:
  * pure metal,
  * solid solution of two or more components, forming a substitutional or an interstitial solid solution,
  * metallic compound which can be an ionic, electronic or interstitial chemical compound.

The term *microstructural element* (or, simply, *microstructure*) has to be distinguished from the term phase. The *microstructural element* means a distinguishable part under the optical microscope that was formed during crystallization or recrystallization and has an independent boundary surface. Consequently, the microstructural elements may be the same as the phases listed above (except the liquid phase because it cannot be observed under an optical microscope), but microstructural elements may be multi-phase heterogeneous structures as well, like the eutectic and the eutectoid.

7.3. Two-component (binary) alloy systems

Before analyzing the most important types of binary alloy systems, it is necessary to give some basic definitions. This will be detailed in the next subsection.

7.3.1. Fundamental terminology

The term *two-component (binary) alloys* means alloys containing only two components. In metallic alloys, the fundamental component must be a metallic element, called the *base metal* of the alloy. Besides the base metal, the other components may be metals, metalloids (elements of metallic nature), non-metallic elements and even gases.

The crystallization and transformation of alloys at various temperatures can be studied in so-called *binary phase diagrams* (often termed *equilibrium* or *constitutional diagrams*). The phase diagram of a two-component alloy can be presented as a planar diagram that has the temperature (T) on its vertical axis and composition (concentration - c) along the horizontal axis. Capital letters A and B symbolize the two components; the left boundary line of the diagram represents the pure A metal (100 % A) and the right boundary line represent the pure B metal (100 % B). This is illustrated in Figure 7.8.

Binary phase diagram may be defined accordingly as a planar diagram that *gives the quantity and quality of phases being in equilibrium* of any alloy within the alloy system at any arbitrary composition and any selected temperature.

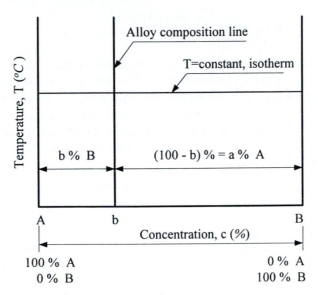

Figure 7.8
The phase diagram and interpretation of markings

Phase diagram is also often termed *equilibrium diagram*. The word *equilibrium* has a double meaning: it refers to the equilibrium of phases, but also relates to the *equilibrium cooling conditions*. Equilibrium cooling conditions mean a cooling procedure in which, during cooling, sufficient time is provided at every temperature to complete all diffusion processes. This implies ever increasing time with lowering of the temperature; therefore, the realization of a completely equilibrium cooling process is not feasible in practice. Thus, "*equilibrium*" diagrams applied in practice are to be considered as *quasi equilibrium diagrams* in reality.

Any horizontal line in the diagram represents a constant temperature (*T = constant*). Constant temperature lines in phase diagrams are called *isotherms*. Any vertical line in the diagram represents an alloy having a defined composition (for example denoting an alloy having *a % of A metal* and *b % of B metal*). Therefore, these lines are called *alloy lines,* or simply *composition lines*. It follows from the diagram that for any alloy the relationship

$$a + b = 100\,\% \tag{7.1}$$

is valid.

Taking into consideration the number of elements having great practical importance, the number of possible two-component, binary systems can be determined from the following relationship:

$$N = \binom{n}{2} = \frac{n!}{(n-2)!\,2!} = \frac{n(n-1)}{2} \;, \tag{7.2}$$

where n denotes the number of possible constituents and N is the number of possible binary systems. Substituting $n \approx 90$ (that is, the number of natural elements of great industrial importance) into Equation (7.2), $N \approx 4005$ is obtained. This means that the number of possible binary systems exceeds 4000. If we consider only the most widely used 50 elements, the number of possible binary systems is equal to $N \approx 1200$. This means that we need to analyze more than one thousand various combinations.

Thanks to the activity of Gustav Tammann (an Estonian scientist dealing with metals and alloys), the great number of binary alloy systems can be classified into 8 basic groups. These basic diagrams of binary alloy systems are called *ideal phase diagrams*. (The name Tammann phase diagrams is also used in appreciation of the systematization work done by Gustav Tammann.) Before starting the analysis of ideal binary phase diagrams, we briefly summarize the thermodynamic background of the phase diagrams, since the crystallization procedures of two- and multi-component systems are also controlled by the rules of thermodynamics.

As we noted in the analysis of the crystallization of pure metals, the existence of any phase can be connected to the thermodynamic potential (i.e. with the free energy of the phases). As we mentioned, binary phase diagrams relate to the equilibrium conditions of phases in a two-component system. Therefore, it is obvious that the phase diagrams can be determined based on the free energies of the phases, in a purely theoretical manner, too.

7.3.2. Theoretical determination of binary phase diagrams

For the sake of simplicity, an alloy system will be analyzed that consists of two components (A and B metals) having an identical crystal structure and matching all the conditions required to form unlimited solid solutions.

If the free energy functions of the solid and liquid phases of the two components are known at any arbitrary temperature for the whole composition range, the phase diagram can be determined without experiments, on a purely theoretical basis. (However, the amount of calculations required to determine the phase diagrams theoretically is so large that it can successfully be performed only by applying digital computers.)

For this analysis, consider first Figure 7.9. In this figure, the free energy curves for all possible compositions of the solid and liquid phases of the two components valid at the temperature $T_1 = constant$ are shown. The temperature T_1 is selected so that considering any possible composition of the two components, the free energy

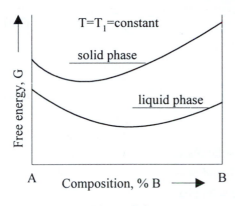

Figure 7.9
Free energy curves at an arbitrary
temperature $T_1 > T_{Liqvid}$

of the liquid phase will always be lower than that of the solid phase. Consequently, for any composition of the alloy system, the liquid phase is the stable one.

As the temperature decreases, the free energy curves change in such a way that the difference between the free energies of the liquid and the solid phase decreases continuously. At a temperature of $T = T_2 = T_{melt}^A$, the curves representing the free energies of the liquid and the solid phases intersect at 100 % A content (in other words, at the composition of pure A metal, see Figure 7.10). It follows from the knowledge of crystallization processes, that this temperature is the melting point of the pure metal A. This is the temperature at which, in equilibrium cooling, crystallization starts and also finishes theoretically (i.e. starts and finishes at the constant temperature $T = T_{melt}^A$).

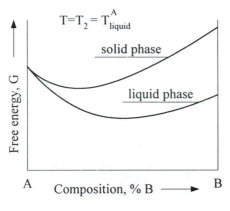

Figure 7.10
The change of free energy curves at the
temperature $T_2 = T_{melt}^A$

On further decrease of the temperature, the relative displacement of the energy curves occurs in such a way that, in the increasing composition range, the free energy of the solid phase become lower than the free energy of the liquid phase.

In Figure 7.11, the change in the free energy of the solid and the liquid phases are shown at a selected temperature $T=T_3$. At this temperature, the two curves intersect at point X. In the vicinity of point X, the lowest value of the free energy for any alloy can be obtained when the alloy is present in the form of a heterogeneous mixture of the solid phase with a composition p and the liquid phase with a composition q.

Compositions marked by points p and q can be determined on the basis of the condition that the minimum of the free energy of the two-phase heterogeneous range is indicated by the mutual tangent of the two curves (the curves describing the free energies of the liquid and the solid phases). Considering this, the positions of points p and q can be determined from the following relationships:

$$\left(\frac{\partial G_s}{\partial c}\right)_{c_s} = \left(\frac{\partial G_l}{\partial c}\right)_{c_l}, \tag{7.3}$$

as well as

$$\frac{\partial G_s}{\partial c} = \frac{G_s - G_l}{c_s - c_l} \tag{7.4}$$

where

G_s the free energy of the solid phase,

G_l the free energy of the liquid phase,

c_s the composition of the solid phase,

c_l the composition of the liquid phase.

(The above parameters refer to the mutual equilibrium of the two phases, appropriately.)

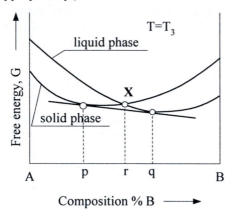

Figure 7.11

Free energy curves at temperature

$$T_{melt}^A > T_3 > T_{melt}^B$$

Furthermore, at the selected temperature $T = T_3$, we can see that for alloys having a composition in the range between pure A metal and the composition marked by point p, the free energy of the solid phase is lower at this temperature. Therefore, the stable state of these alloys is represented by the solid crystalline phase. However, for alloys having a composition in the range between the composition marked by point q and pure B metal, at this temperature the free energy of the liquid phase is still lower. Therefore, the stable state of these alloys is represented by the liquid phase.

A further decrease in temperature results in further displacement of the free energy curves. Naturally, together with this shift of free energy curves, the position of point X indicating the intersection of the two curves, as well as the position of the mutual tangent, will change, as can be seen in Figure 7.12. Similarly to the previous situation, the composition values u and w, meaning the mutual equilibrium of the two phases, can be determined. At the temperature $T=T_4$, these u and w values determine unambiguously the composition of the phases being in equilibrium. These remain unchanged as long as the temperature is unchanged.

Repeating the previously described procedure at a great number of temperatures, we can determine the compositions of the liquid and the solid phases being

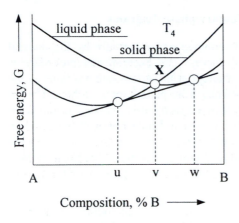

Figure 7.12

The change of free energy curves at a selected temperature $T_{melt}^A > T_4 > T_{melt}^B$

in equilibrium with each other at every temperature. Connecting the points that mark the equilibrium compositions of the liquid phase, we obtain the so-called *liquidus curve*, while connecting the points that mark the equilibrium compositions of the solid phase, we obtain the so-called *solidus curve*. This procedure is shown in Figure 7.13.

Above the liquidus curve, at any temperature the alloy consists of a homogeneous single-phase liquid and below the liquidus curve, a single-phase solid solution can be found. The zone bordered by the liquidus and solidus curves is a heterogeneous, two-phase zone containing solid and liquid phases. At any temperature, the compositions providing the equilibrium state of the liquid and the solid phases are indicated by the intersections of the isotherm representing that temperature and the solidus and liquidus curves. The section of the isotherm between the liquidus and the solidus curves is called the *tie-line*. Considering this, we can say that the compositions of the phases being in equilibrium are indicated by the end points of the tie-line. The end point of the tie-line on the liquidus curve indicates the composition of the liquid phase, while the end point of the tie-line on the solidus curve indicates the composition of the solid phase. The above statement comprises in itself one of the most important laws of phase diagrams, called the *quality rule* or the *first lever rule*.

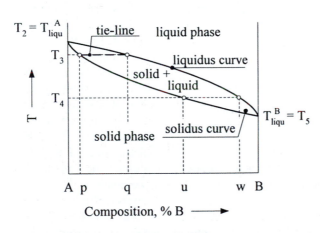

Figure 7.13

Determination of the solidus and liquidus curves of phase diagrams based on the free energy curves

7.3.3. Experimental determination of binary phase diagrams

The phase diagrams can also be determined experimentally by the thermal analysis of a great number of cooling curves. The experimental procedure is briefly summarized below. First, a set of alloys with various compositions of the given alloy system is to be produced, starting from pure A metal, followed by an increasing amount of B metal and finally finished with pure B metal. The alloys are melted, then continuously and slowly (near to equilibrium cooling) cooled down from the liquid state down to room temperature. During cooling, the so-called cooling curves of the alloys are recorded in a temperature-time coordinate system. As an illustration, a series of cooling curves is shown in Figure 7.14 for various compositions of *Cu-Ni* alloys.

The pure metals – as explained in Section 3 – crystallize at constant temperatures, which are indicated by a straight horizontal line between the beginning and the end points of crystallization on the cooling curves. In contrast to this, alloys typically do not solidify at constant temperatures but rather in a given temperature range, as can be seen in Figure 7.14. On the second, third and fourth cooling curves in Figure 7.14, this is illustrated by the sections with different slopes between points L_i and S_i. The points L_i indicate the starting temperature and the points S_i the finishing temperature of the solidification. This behavior, differing from that of the pure metals, can be proved by the application of Gibbs' phase rule. For binary alloys, the number of components, $C=2$, the number of phases present

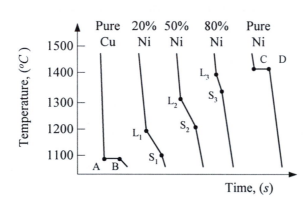

Figure 7.14
Cooling curves of Cu-Ni alloys with various compositions

in the system during crystallization is $P=2$ (the liquid and the solid phases). Substituting the above values into the relationship (4.2) of Gibbs' phase rule, $F=1$ is obtained for the degree of freedom. Since for a given alloy the composition is fixed, it means that the temperature can be changed during the crystallization of an alloy.

Using the cooling curves obtained according to the previously described experimental procedure, the phase diagram of the alloy system can be constructed as described below. First, we draw the alloy lines in accordance with the compositions of the examined alloys (see Figure 7.15). Then the temperatures corresponding to the breakpoints of the curves for each alloy are projected to the alloy line. (For example, on the cooling curve of pure copper the breakpoints marked by letters A and B are projected to the alloy line corresponding to 100 % Cu content, that is onto the left boundary line of the diagram. The breakpoints L_1 and S_1 on the cooling curve of the alloy containing 20 % Ni are projected to the alloy line representing 20 % Ni content, and the procedure is repeated for all the alloys.) Subsequently, the points having the same feature are connected with a continuous curve. The curve connecting the L_i breakpoints representing the beginning of crystallization is the *liquidus* curve, the curve connecting the S_i points representing the end of crystallization is the *solidus curve*, as can be seen in Figure 7.15. (Naturally, to obtain a phase diagram with high accuracy, a much greater number of cooling curves has to be recorded. In complex phase diagrams, the composition of selected alloys in certain composition ranges is changed by one tenth of a percent.)

The liquidus and solidus curves drawn according to the previously described procedure divide the area of the phase diagram into three zones. The area above the liquidus curve is the *zone of the homogeneous liquid phases* where the alloys are in a liquid state at any composition.

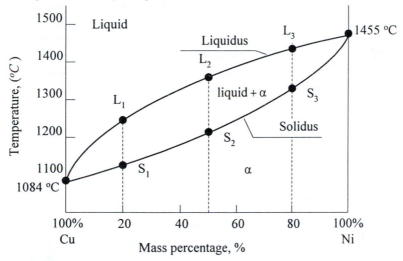

Figure 7.15
Experimental determination of a binary phase diagram using the cooling curves

Below the solidus curve, only the solid phase can exist for any alloy: this is the zone of *homogeneous solid phases*. In the zone bordered by the liquidus and the solidus curves, the crystallization of alloys has already started, but it has not yet been completed. In this zone, both solid and liquid phases can be found for any alloy composition, thus this is a *heterogeneous two-phase zone*.

7.3.4. The basic rules of analyzing phase diagrams

We already saw in the analysis of the theoretical fundamentals of binary phase diagrams (see Section 7.3.2) that the main principles of phase diagrams are determined by the rules of thermodynamics. Applying Gibbs' phase rule, it was already proved that for a given composition of an alloy in the heterogeneous zone, the temperature could be changed between the solidus and liquidus curves without any changes taking place in the equilibrium of the system. If for a given alloy, even the temperature is fixed, the quality (the composition) and the quantity of the phases being in equilibrium in the system can be unambiguously determined. To determine the quality of the phases, the *first lever rule (the* so-called *quality rule)* is used, while to determine the quantity of the phases the *second lever rule* (the *quantity rule*) is applied. It is worth mentioning that both the quality and the quantity rules can only be applied in heterogeneous zones. (In homogeneous zones, the alloy itself unambiguously determines the state of the system.) The composition of the phase present must be equal to the composition of the alloy examined and its quantity (there being no other phase present) should be equal to the total amount. The total amount of alloy is usually taken as a unit mass (for example 1 kg) which in percentage values equals 100 %.

7.3.4.1. The first lever rule – The quality rule

Consider Figure 7.16, in order to analyze the quality of the phases being in equilibrium in the heterogeneous zone. Select an alloy with an arbitrary composition (which remains fixed after selection) of a % of A metal and b % of B metal, $(a + b = 100 \%)$. Select a temperature T_1 such that the isotherm line marking this temperature intersects the alloy line in the heterogeneous zone (the zone bordered by the liquidus and the solidus curves).

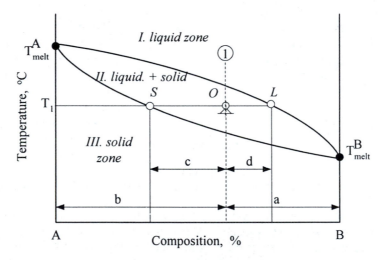

Figure 7.16
Sketch for the interpretation of first lever rule

According to the first lever rule, the intersection of the isotherm with the liquidus curve (L) gives the composition of the liquid phase, while the intersection of the isotherm with the solidus curve (S) indicates the composition of the solid phase being in equilibrium. The compositions may be determined by projecting the intersections to the composition line. Accordingly, and in conformity with the notations of Figure 7.16, the composition of the liquid phase is characterized by $(b + d)\%$ of B metal and that of the solid phase by $(b - c)\%$ of B metal.

Naturally, the first lever rule may be applied for much more complicated cases than the one shown in Figure 7.16. Summarizing, the first lever rule may be formulated in a general manner as follows. The quality of the phases (i.e. the types of the phases) being in equilibrium in a two-phase, heterogeneous zone are determined by the intersections of the isotherm corresponding to temperature T with the curves bordering the heterogeneous zone. The intersections (i.e. the end points of the previously mentioned tie-line) show the types of phases being in equilibrium. The composition of the phases can be determined by projecting the end points of the tie-line to the base line representing the compositions.

7.3.4.2. The second lever rule – The quantity rule

For the derivation of the second lever rule, consider Figure 7.16. Select a unit mass (for example 1 kg) for the purpose of examination. In conformity with the notations of Figure 7.16, the alloy contains $a\%$ of A metal and $b\%$ of B metal. Thus, the quantities of the two components in the original alloy are as follows:

$$\frac{a}{100} \times 1 \text{ kg is the amount of A metal,}$$

$$\frac{b}{100} \times 1 \text{ kg is the amount of B metal.}$$

$$(7.5)$$

Denote the unknown quantity of the solid phase at the temperature T_l by x kg. Consequently, the quantity of the liquid phase is equal to $(1 - x)$ kg. Therefore, the amount of B metal in the liquid phase can be given by the expression

$$\frac{b + d}{100}(1 - x) , \qquad (7.6)$$

and the amount of metal B in the solid phase is given by the expression

$$\frac{b - c}{100}x , \qquad (7.7)$$

respectively.

It follows from the rule of mass conservation that the quantities of the individual components (A and B metals) present in the alloy must be equal to the sum of their quantities in the liquid and the solid phases. Writing this relationship with respect to B metal, the following equation can be obtained:

$$\frac{b + d}{100}(1 - x) + \frac{b - c}{100}x = \frac{b}{100} \times 1 . \qquad (7.8)$$

Following some elementary transformations, we obtain the equation for the quantity of the solid phase:

$$x = \frac{d}{c+d}.$$ (7.9)

Expressing the value of $(1-x)$ from expression (7.8), after some transformation, for the quantity of the liquid phase we obtain the expression

$$1-x = \frac{c}{c+d}.$$ (7.10)

Using the relationships (7.9) and (7.10) for the heterogeneous zone, the quantities of the solid and the liquid phases in equilibrium can be determined at any temperature.

The *(c+d)* section of the isotherm in Figure 7.16 means exactly the length of the tie-line (since, according to the definition, *the tie-line is the section of the isotherm between the lines bordering the heterogeneous zone*). Define the intercept c of the tie-line between the solidus point S and the point O on the alloy line as the arm of the solid phase. Similarly, denote the intercept d of the tie-line between the liquidus point L and the point O on the alloy line as the arm of the liquid phase. Then the previous relationships can be formulated as follows: *at any temperature the quantity of the solid phase (x) can be calculated as the ratio of the arm of the liquid phase (d) and the total length of the tie-line (c+d) and the quantity of the liquid phase can be calculated as the ratio of the arm of the solid phase (c) and the total length of the tie-line.* This is actually the definition of the second lever rule, or the quality rule, valid for binary phase diagrams.

The above definition of the second lever rule unambiguously indicates that the quantities of the phases being in equilibrium are inversely proportional to the arms belonging to the respective phases. This is seen more clearly if the relationship (7.9) is divided by the relationship (7.10), which leads to the expression

$$\frac{x}{1-x} = \frac{d}{c}.$$ (7.11)

This relationship can be formulated so that the quantity of the solid phase (x) is proportional to the arm of the liquid phase *(d)*, and the quantity of the liquid phase *(1-x)* is proportional to the arm of the solid phase (c). This is the reason why the second lever rule is often referred to as *the rule of inverse arms*.

Consider now the tie-line as a two-armed lever. Denote the intersection point of the two-armed lever and the alloy line by point O. This point (O) is the center of the moment for the two-armed lever. The lever will be in equilibrium if a weight of x kg equivalent to the quantity of the solid phase is placed at point (S) and a weight of $(1-x)$ kg equivalent to the quantity of the liquid phase is placed at point (L), respectively. (This is the main reason why the quantity rule is also referred to as the *lever rule*.) The equilibrium condition of the lever may be written in the form

$$cx = d(1-x).$$ (7.12)

The first and second lever rules, introduced to determine the quality and quantity of phases, play an extremely important role in the analysis of binary phase diagrams. Using these rules, both the qualities (types and compositions) and the quantities of the phases being in equilibrium can be determined.

7.3.5. Analysis of ideal binary phase diagrams

The ideal binary phase diagrams are analyzed according to the systematization performed by Tammann. During these analyses, a systematic approach will be used. In each case, first we summarize the most important characteristics of the given phase diagram, then the analysis of the crystallization processes of some characteristic alloys will be performed.

7.3.5.1. The simple eutectic system (Tammann-1)

The first binary phase diagram – according to the systematization performed by Tammann – is the so-called *simple eutectic system*. The most important characteristics of this diagram are summarized below.

- The two components dissolve each other without any limitation in the liquid state (this means that the liquidus contains only curved sections).
- At the same time, the two components do not dissolve each other in the solid state at all. (The consequence is that the solidus may contain only straight, horizontal sections).
- Since the two components form neither a solid solution nor a metallic compound, a eutectic is formed consisting of a mixture of pure A and B metal during crystallization. (At a defined composition the total mass of the alloy solidifies as a eutectic.)

In Figure 7.17, the binary phase diagram characterized by the above listed features is shown.

The most important features of the phase diagram may be studied through the crystallization processes of typical alloys. Select first the alloy marked by the alloy line 1. The composition of this alloy is between pure A metal and the eutectic point defined by point E. Above the liquidus curve, denoted by L_1, the alloy can exist in a homogeneous, single-phase liquid state. The composition of the liquid corresponds to the composition of the alloy. (*Note*: this is a homogeneous, single-phase zone). Crystallization of the alloy starts when the alloy line intersects the liquidus curve (or more precisely, in accordance with the rules of crystallization discussed earlier, at a temperature that is lower than the liquidus temperature by a ΔT undercooling). Since crystallization starts as the liquidus curve is intersected, the alloy passes over from the so far homogeneous, single-phase liquid zone to the two-phase, heterogeneous zone containing both liquid and solid phases. In the heterogeneous zone, the first lever rule can be used to determine the composition of the first crystal to solidify. Drawing an isotherm at a temperature lower than the liquidus by ΔT, it intersects the line of pure A metal bordering the heterogeneous zone. Therefore, the first solidifying crystal is pure A metal.

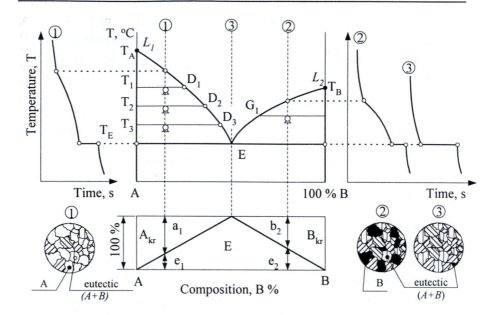

Figure 7.17
The simple eutectic system (Tammann-1 ideal phase diagram)

The crystals solidifying directly from the liquid are called *primary crystals*, as distinct from crystals formed later during further transformations in the solid state. These crystals are called secondary and tertiary crystals in the sequence of their formation. (Secondary as well as tertiary crystallization processes will be seen during the analysis of the crystallization of iron alloys.)

When crystallization starts, a breakpoint appears in the cooling curve of the alloy that can be attributed to the release of the latent heat that was "introduced" during melting. This results in a change of the slope on the cooling curve during crystallization.

On further decrease in the temperature, the types and compositions of the phases being in equilibrium at an arbitrary temperature of $T = T_2$ can also be determined using the first lever rule. Drawing the isotherm at temperature $T = T_2$, it can be seen that its end points still lie on the border line of pure A metal (solid phase), while the other end is on the liquidus curve (liquid phase). Because of this, the composition of the solid phase is the same (i.e. pure A metal), while the composition of the liquid phase is determined by point D_2 of the liquidus curve. From the above, it also follows that the A content of the remaining liquid continuously decreases since, during the crystallization process, continuously pure A metal solidifies. It also means that the liquid is becoming increasingly rich in the other component (i.e. the B content of the liquid is increasing). Accordingly, at various temperatures the composition of the liquid phase can be characterized by the points D_1, D_2, D_3 moving on the liquidus curve. It is also clearly seen in the figure that the lengths of the tie-lines belonging to various temperatures, as well as the lengths of the sections representing the quantity of the solid and liquid phases, change continuously. These changes may be characterized by the continuous increase in the quantity of the solid phase with the simultaneous decrease of the liquid phase (see: the changes in the lever sections). It follows from the above that pure A metal crystallizes continuously until the temperature $T = T_E$ is reached. The composition of the remaining liquid changes to the composition characterized by the eutectic point E. The reaction occurring at this temperature may be better understood by considering first the following. Whatever alloy composition is selected within the composition range determined by the pure A metal and the eutectic point E, the alloy line intersects first the liquidus section marked by L_1. This means that the crystallization starts with the solidification of pure A metal. From this, we can draw the conclusion that the liquidus section marked by L_1 represents a *solubility limit curve* with respect to component A. Thus, intersecting this liquidus section, the liquid phase becomes oversaturated in component A. Therefore, to maintain the equilibrium state, pure A metal has to solidify from the liquid phase.

Without performing a detailed analysis of crystallization within the composition range bordered by the point E and the composition line of pure B metal, the following can be concluded. It is clear from the previous analysis that the alloy lines in this composition range intersect the liquidus section marked by L_2, and therefore, the first solidifying crystal is pure B metal. Thus, the liquidus section marked by L_2 represents a *solubility limit curve* with respect to component B. Since the point E lies on both liquidus sections, consequently the liquid phase with the composition E can be considered as a saturated solution with respect to both components. Therefore, during the solidification of the liquid phase with the composition E, both A and B crystals have to solidify simultaneously to maintain equilibrium.

Figure 7.18 is the schematic illustration of the mechanism of the eutectic reaction. In Figure 18.a, the neighborhood of point E is enlarged for the sake of better demonstration. The solidification of the eutectic liquid starts after a ΔT amount of undercooling according to the well-known rules of crystallization. This moment is marked in the figure by point 1. However, at this temperature, the equilibrium condition would be represented either by point 2 or 5, lying on the extension of the liquidus sections marked by L_1, or L_2. In order to reach point 2 on the liquidus section L_1, the solidification of an A crystal is necessary. Similarly, to reach point 5 on the liquidus section L_2, the solidification of a B crystal is necessary. It is a matter of a random phenomenon, whether the solidification of A or B crystals will start. Therefore, consider one of the possible events. If the crystallization of the eutectic liquid starts with the solidification of an A crystal, the equilibrium condition is fulfilled at point 2. The latent heat released during the solidification of the A crystal increases the temperature of the liquid to point 3. However, at this temperature, the composition of point E corresponds to the equilibrium conditions. Therefore, a B crystal has to solidify, following the earlier crystallization of A crystal. With the solidification of a B crystal, the composition may not match the one defined by point E, but it may "overstep" to point 4, as can be seen in Figure 18.c. Due to the continuous heat removal, the solidification of the eutectic is more likely to occur in a concentration range of Δc and a temperature range of ΔT as an oscillating process, as shown in the figure. The progress of the solidification of the eutectic in time is shown in the part c) of Figure 7.18.

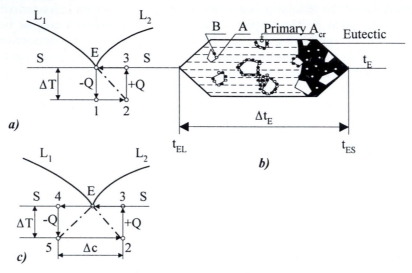

Figure 7.18
Schematic illustration of the solidification of a eutectic liquid

This process of crystallization, in which two new solid phases are formed from a homogeneous, single-phase liquid, is called a *eutectic reaction*. It can be described by the following reaction equation:

$$liqv_E \rightleftharpoons A + B. \tag{7.13}$$

In the reaction equation, the arrow sign showing in both directions means that the reaction is reversible, i.e. it can also occur in the opposite direction during heating. In this reaction, three phases (the liquid with the composition E as well as A and B pure metals) take part; therefore, the degree of freedom can be calculated from the Gibbs' phase rule as follows:

$$F = C + 1 - P = 2 + 1 - 3 = 0. \tag{7.14}$$

Consequently, none of the state factors can be changed until one of the phases disappears from the system. Therefore, the eutectic reaction is a *non-variant one*. This means that the temperature should remain constant until the total amount of the liquid solidifies. It appears as a horizontal straight section on the cooling curve of the alloy at the temperature $T = T_E$.

On continuing the cooling of the alloy down from the temperature of the eutectic reaction, no additional phase transformation occurs. Accordingly, the microstructure of alloy 1 at room temperature contains primary A metal crystals and a two-phase eutectic containing the mixture of *A* and *B* metals. The previously described procedure of crystallization can be followed in Figure 7.19, where the so-called *"family tree of crystallization "* is shown for alloy 1.

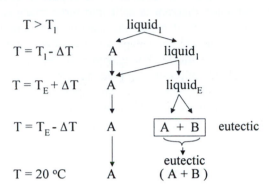

Figure 7.19
Crystallization family tree of alloy 1 in the Tammann-1 system

The *family tree of crystallization* is practically a flow chart of crystallization. In this flow chart, the quality and composition of the phases being in equilibrium are shown for all the characteristic temperatures. The indices of the phases identify the composition and the arrows indicate the direction of the process. This method of analysis was elaborated in the beginning of the 70's at the Department of Mechanical Engineering *of the University of Miskolc.*

The crystallization of alloy 2 is more or less similar to the crystallization of alloy 1. The basic difference is that the alloy line intersects the L_2 liquidus section. Therefore, the primarily crystallizing phase is pure *B* metal. The further procedure of crystallization is the same as described for alloy 1. Accordingly, in the microstructure of this alloy, pure *B* metal and the eutectic can be found at room temperature as shown in Figure 7.17.

Alloy 3 was selected so that its composition is exactly equivalent to the composition of point *E* (i.e., it is equivalent to the eutectic composition). The crystallization of this alloy proceeds according to the eutectic reaction described before. The two-phase eutectic consisting of the mixture of *A* and *B* metals solidifies from the liquid. Since the eutectic reaction is non-variant, the total amount of the liquid solidifies at the constant (eutectic) temperature of $T = T_E$. It can be seen from Figure 7.17 that the cooling curves of pure eutectic alloys are similar to those of pure metals.

In Figure 7.17, the so-called microstructure diagram of the alloy system is also constructed. The *microstructure diagram* for a binary system is a planar diagram that has the percentage values of the microstructure elements on the vertical axis and the percentage values of composition on the horizontal axis. The construction of the microstructure diagram is performed as follows. First, we have to find those composition values of the alloy system for which the whole alloy contains only one microstructural element. For the phase diagram shown in Figure 7.17, this means the concentrations of the two pure metals (i.e. *A* and *B*), as well as the concentration of the eutectic alloy. It follows from the first lever rule that in those composition ranges where two microstructural elements can be found, the quantities of the microstructural elements are changing linearly. In the examined phase diagram, this means that a straight diagonal line from the boundary line of pure A metal to the composition of the eutectic should be drawn. Similarly, another straight diagonal line from the eutectic point to the boundary line of pure B metal should be drawn. Thus, the microstructure diagram is split into three triangular zones indicating the respective fields of pure A metal, the eutectic and pure B metal. Based on the microstructure diagram, the percentage quantity of any of the microstructural elements can be easily determined. For example, in the case of alloy 1, projecting the alloy line down to the microstructure diagram intercept a_1 is proportional to the quantity of pure A metal, intercept e_1 is proportional to the quantity of the eutectic. Since the vertical axis of the microstructure diagram indicates 100 % of the quantity of the microstructural elements, the actual quantities can be calculated based on similar triangles.

7.3.5.2. Phase diagram with stable metallic compound (Tammann-2)

The most important features of this alloy system can be summarized as follows:

- The two components dissolve each other to an unlimited extent in the liquid state, and the number of phases primarily crystallizing is three (that is, the liquidus has three sections).
- The two components do not dissolve each other in the solid state at all.
- A stable metallic compound is formed at the composition $A_m B_n$. (A metallic compound is considered stable if it similarly to pure metals solidifies directly from the liquid. Therefore, a stable metallic compound primarily crystallizes at a defined, constant temperature and during heating it melts directly to become a liquid at the temperature corresponding to the temperature of crystallization.)
- A further feature of the diagram is that two eutectics are formed with different compositions.

The phase diagram having the above features is shown in Figure 7.20. In this alloy system, crystallization occurs basically in the same way as in the Tammann-1 system, but with consideration of the following features:

- There are three liquidus sections in this diagram:
  * If the alloy line intersects the liquidus section marked by L_1, primary crystallization starts with the solidification of pure A metal and finishes with the solidification of the eutectic with the composition E_1.
  * If the alloy line intersects the liquidus section marked by L_2, primary crystallization starts with the solidification of the metallic compound $A_m B_n$. If the composition of the alloy is in the range defined by the composition of the eutectic E_1 and the composition of the metallic compound $A_m B_n$, crystallization finishes with the solidification of the eutectic with the composition E_1. If the composition is in the range defined by the composition of metallic compound $A_m B_n$ and the composition of the eutectic E_2, crystallization finishes with the solidification of the eutectic with the composition marked by E_2.

* Finally, if the alloy line intersects the liquidus section marked by L_3, primary crystallization starts with the solidification of pure B metal and finishes with the solidification of the eutectic with the composition of E_2.

- The crystallization of the alloy corresponding to the composition of the metallic compound A_mB_n occurs exactly in the same way as the crystallization of pure metals: it proceeds with the solidification of a stable metallic compound at a constant temperature.

It is important to note that the points marked by empty circles in Figure 7.20, (on the composition lines of pure A and B metals, and that of the metallic compound A_mB_n) do not belong to the corresponding solidus sections. These points should be considered *singular points*.

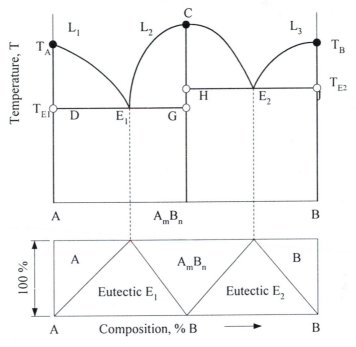

Figure 7.20
Binary phase diagram with a stable metallic compound

The microstructure diagram of the alloy system is shown in Figure 7.20 as well. It can be seen from this figure that there are five different microstructural elements in this alloy system. Three of them are single phases (pure A and B metals and the A_mB_n metallic compound) while the other two are heterogeneous, two-phase microstructural elements (the eutectics E_1 and E_2).

7.3.5.3. Phase diagram with unstable metallic compound (Tammann-3)

The most important characteristic features of this binary alloy system (shown in Figure 7.21) are summarized in the following:

- The two components dissolve each other without any limit in the liquid state: the number of phases crystallizing primarily is equal to three. (These are the pure A and pure B metals, as well as the A_mB_n metallic compound).

- The two components do not dissolve each other in the solid state, but they form a eutectic at a defined composition (point E).

- At a defined composition, an unstable metallic compound is formed in the alloy system. The metallic compound is considered unstable because it is not melted directly to liquid during heating: first, at a defined temperature ($T=T_{CDF}$ in the phase diagram) it decomposes into a solid and a liquid phase.

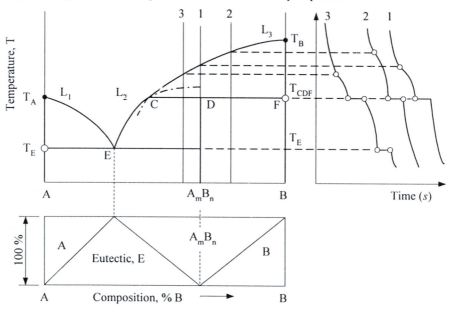

Figure 7.21
Binary phase diagram with an unstable metallic compound

The most characteristic reaction of this alloy system is the reaction resulting in the formation of the metallic compound at the temperature $T=T_{CDF}$. To analyze this reaction, select an alloy with a composition equivalent to that of the A_mB_n metallic compound (alloy line 1 in Figure 7.21). Crystallization of the alloy starts with the solidification of pure B metal when the liquidus curve is intersected. During cooling, pure B metal continues to crystallize. Thus the B content of the remaining liquid decreases. At a temperature that is higher than the temperature T_{CDF} by an elementary ΔT value, the liquid has the composition corresponding to point C. Thus, at this moment a liquid phase having the composition of C is in equilibrium with pure B metal. At the temperature $T=T_{CDF}$, a non-variant reaction occurs which is characterized by the reaction equation

$$liquid_C + B \rightleftharpoons A_m B_n .$$ (7.15)

In this reaction, a new solid phase, the $A_m B_n$ metallic compound is formed at a constant temperature $(T=T_{CDF})$. This is called a *peritectic reaction*. The name reflects the fact that A atoms diffuse towards the inside of the previously crystallized pure B metal through its perimeter. As a result of this process, first the surface layers of B crystals and a portion of the liquid transform into the metallic compound $A_m B_n$. Then by further diffusion, the whole crystal will be transformed into the $A_m B_n$ metallic compound, provided the ratios of the quantities of the liquid phase and the primarily crystallized pure B metal are appropriate for completing the reaction. This condition for the ratio of the liquid and solid phases – according to the second lever rule – can be given by the following relationship:

$$\frac{liquid_C}{B} = \frac{\overline{DF}}{\overline{CD}} .$$ (7.16)

If this condition is fulfilled, a pure peritectic reaction occurs. At the end of the reaction, there is only one homogeneous, single phase metallic compound in the system with the composition of $A_m B_n$.

If we have an alloy with the composition between points D and F (alloy 2), then also in this case liquid of composition C and pure B metal can be found in the system prior to the peritectic reaction. However, the ratio of the phases differs from the value given by expression (7.16). In these alloys, the quantity of pure B metal is greater and the quantity of the liquid phase is smaller than the respective quantities in expression (7.16). Therefore, besides the new phase (the $A_m B_n$ metallic compound) formed in the peritectic reaction, part of the pure B metal remains after the reaction, since the quantity of the liquid phase is insufficient to transform the total quantity of pure B metal to the metallic compound $A_m B_n$. Accordingly, the non-variant reaction occurring at the temperature $T=T_{CDF}$ can be described by the equation:

$$liquid_C + B \rightleftharpoons A_m B_n + B .$$ (7.17)

Studying the two sides of equation (7.17), we can observe that the left side of the equation contains the phases being in equilibrium at the temperature $T=T_{CDF} + \Delta T$, while the right side of the equation contains the phases being in equilibrium at the temperature $T=T_{CDF} - \Delta T$. This provides an essential basis to formulate the reactions occurring at a constant temperature (non-variant reaction). A non-variant reaction may be analyzed in the most reliable way by first determining the phases being in equilibrium before the reaction at a temperature above the isothermal temperature by a value of ΔT (the left side of the equation). Then, we determine the phases being in equilibrium after the reaction at a temperature below the isothermal temperature by a value of ΔT (this is the right side of the equation).

The ΔT temperature always represents an elementary small amount of temperature where $\Delta T \to 0$ (ΔT approaches zero).

If an alloy line is drawn between points C and D (alloy 3), a liquid with composition C and a pure B metal are present in the system prior to the peritectic reaction. However, the ratio of the quantities of these differs in this case as well from the values formulated by expression (7.16). In this case, the quantity of pure B metal is smaller and the quantity of the liquid phase is greater than the quantities stipulated by the conditional expression (7.16). Therefore, besides the new phase (the metallic compound A_mB_n) formed in the peritectic reaction, a portion of the liquid phase remains after the reaction since there is more liquid phase than necessary to transform the total amount of pure B metal into the metallic compound A_mB_n. Accordingly, the non-variant reaction occurring at the temperature $T=T_{CDF}$ for alloy 3 may be given by the following equation:

$$liquid_C + B \rightleftharpoons A_m B_n + liquid_C .$$ (7.18)

Since crystallization is not terminated with the peritectic reaction in this alloy, (there is still a liquid phase in the system) the analysis of crystallization must be continued. Upon further cooling, applying the quality rule, we can state that crystals of the A_mB_n metallic compound solidify from the liquid. Since the A_mB_n metallic compound has a high B content, the B content of the remaining liquid will also continuously decrease. Consequently, at the temperature $T=T_E$, the composition of the liquid corresponds to the eutectic composition of point E. At this temperature, a eutectic reaction occurs according to the following equation:

$$liquid_E \rightleftharpoons A + A_m B_n .$$ (7.19)

In accordance with the described crystallization process, at room temperature the microstructure of the alloy contains A_mB_n metallic compound formed by the peritectic reaction and a eutectic consisting of A pure metal and A_mB_n metallic compound formed by the eutectic reaction.

The cooling curves of the analyzed alloys have been drawn in Figure 7.21. It can be seen that both the peritectic and the eutectic reactions are indicated by horizontal straight sections on the cooling curve. For the other sections, the "*solidification during thermal change*" behavior is characteristic.

The microstructure diagram of the alloy system is also shown in Figure 7.21. From this diagram, it can be seen that pure A metal, eutectic, A_mB_n metallic compound and pure metal B can be found in the alloy system.

In Figure 7.22, the crystallization processes of the analyzed alloys are shown using the crystallization family tree method. This method clearly indicates the reactions occurring at the various temperatures. It also provides a good possibility to compare the three alloys of different compositions that have several similarities from the point of view of crystallization processes.

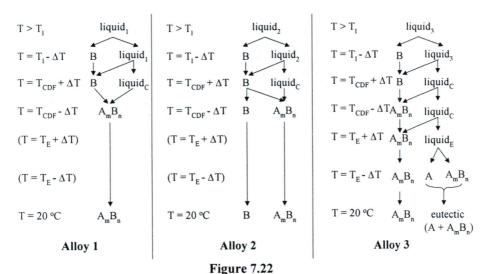

Figure 7.22
Crystallization of characteristic alloys in the Tammann-3 system

7.3.5.4. Phase diagram with monotectic reaction (Tammann-4)

From the point of view of mechanical engineering practice, the Tammann-4 phase diagram is of less practical importance. However, it includes a special type of the non-variant reactions; therefore, the main features of this diagram will also be summarized.

In the previously analyzed systems, the two components dissolved each other in the liquid state without any limitation. In this alloy system, in a defined composition range, the solubility is limited even in the liquid state. Consequently, as can be seen in Figure 7.23, a horizontal straight section can be found on the liquidus curve. The curve drawn with a dotted line above the liquidus section *FG* shows the composition changes in liquid phases separated by an independent phase boundary if the liquid is left undisturbed for an adequately long period. This section of the diagram also indicates that the limited solubility of the components is not absolute. Above the temperature defined by the dotted liquidus curve, the two separate liquid phases can dissolve each other, forming a homogeneous liquid phase.

In this alloy system, we find a new non-variant reaction at the temperature $T=T_{FGH}$. This non-variant reaction occurs purely in the alloy having a composition corresponding to point *G*. This is called a *monotectic reaction* and can be characterized by the following equation:

$$liquid_G \rightleftharpoons B + liquid_F. \tag{7.20}$$

In this reaction, a homogeneous liquid (*liquid_G*) is decomposed into a solid (pure *B* metal) and another liquid phase (*liquid_G*). This special reaction (not found in the previous phase diagrams) can be analyzed, for example, by the crystallization of alloy 1, shown in Figure 7.23.

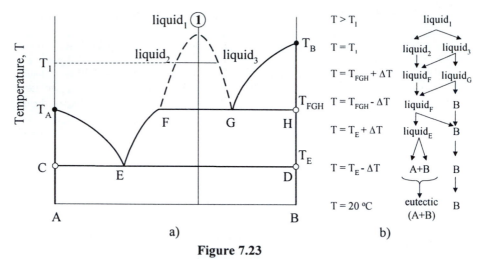

Figure 7.23
a) Phase diagram with a monotectic reaction
b) The crystallization tree of a characteristic alloy

Starting with the analysis of crystallization at a temperature above the dotted line, the liquid phase is in a homogeneous, single-phase state even in this system (this is *liquid$_l$*). During cooling, separation of the originally homogeneous, single-phase liquid occurs according to the dotted liquidus curve. Due to this separation, two liquid phases having the compositions of *liquid$_F$* and *liquid$_G$* are in equilibrium at the temperature $T=T_{FGH} + \Delta T$. At this temperature, the *monotectic reaction* takes place according to the equation:

$$liquid_F + liquid_G \rightleftharpoons liquid_F + B . \tag{7.21}$$

In alloy 1, the liquid phase marked by *liquid$_F$* on the left-hand side of Equation (7.21) does not take part in the monotectic reaction: even its quantity increases with the amount of *liquid$_F$* formed in the monotectic reaction. During further cooling, pure B metal crystallizes from the liquid phase. Due to this, the composition of the liquid changes to the composition of the eutectic point (*liquid$_E$*). This solidifies to a solid eutectic consisting of pure A and B metals by a eutectic reaction. Accordingly, at room temperature the alloy consists of a eutectic containing a mixture of pure A and B metals and the primarily crystallized pure B metal.

The pure monotectic reaction occurs in the alloy having the exact composition of point G. It can be described by the reaction equation:

$$liquid_G \rightleftharpoons liquid_F + B . \tag{7.22}$$

It clearly shows that a solid phase (pure B metal) and a new liquid phase of different composition (*liquid$_F$*) are formed by the monotectic reaction from the liquid phase of composition G (*liquid$_G$*).

7.3.5.5. Phase diagram with syntectic reaction (Tammann-5)

This ideal binary phase diagram is similar to the Tammann-4 diagram in the sense that the two components of this system do not dissolve each other to an unlimited extent in liquid state. In this system, in a defined composition range (marked by points F and H in Figure 7.24), the liquid separates into two liquid phases separated by an independent boundary surface.

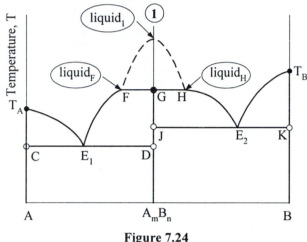

Figure 7.24

Phase diagram with syntectic reaction

The characteristic reaction of the alloy system (the so-called *syntectic reaction*) occurs in the alloy having the composition corresponding to point G at a temperature $T = T_{FGH}$. The non-variant syntectic reaction can be given by the following equation:

$$liquid_F + liquid_H \rightleftharpoons A_m B_n \ . \tag{7.23}$$

It can be seen from Equation (7.23) that in this reaction a new solid phase is formed from two liquid phases having different compositions. The further process of crystallization may be followed based on the crystallization procedures discussed previously.

7.3.5.6. Phase diagram with unlimited solid solution (Tammann-6)

During the theoretical calculation of binary phase diagrams based on thermo-dynamics (section 7.3.2), and also during the experimental determination of phase diagrams (section 7.3.3), we have already considered this basic type of phase diagrams. It may be characterized by unlimited solubility both in the solid and in the liquid state. Consequently, both the liquidus and the solidus lines are curved as illustrated in the diagram used for the derivation of the first and second lever (Figure 7.16).

Although the crystallization process in these types of alloy systems is extremely simple from certain aspects, the mechanism of the formation of the homogeneous solid solution reflects features that are important to recognize. Therefore, we shall briefly analyze the solidification procedure of an arbitrarily selected alloy shown in Figure 7.25.

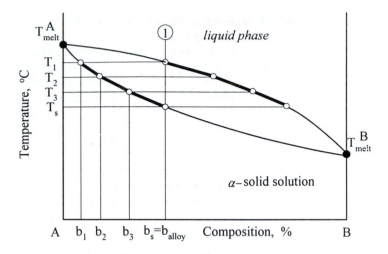

Figure 7.25
Analysis of the formation of a solid solution in the Tammann-6 alloy system

The alloy is in homogeneous, single-phase liquid state above the liquidus curve. Crystallization starts when the alloy line intersects the liquidus curve. The composition of the first solid crystal at temperature $T = T_1$ can be determined by projecting the intersection of the isotherm corresponding to this temperature and the solidus line onto the base line (i.e. the composition line). From this, we can see that α-solid solution, rich in pure A metal, solidifies. The composition of the α-solid solution is marked by b_1 % of B content. During further cooling, crystallization of the α-solid solution continues. At the temperature $T = T_2$, the composition of the total amount of α-phase can be characterized by b_2 % of B content. The composition of the liquid can be characterized with an ever decreasing A content since, from the beginning of the crystallization, an α-solid solution with high A content solidifies. (The change in the composition of the solid phase is illustrated by the bold section of the solidus curve, while the change of the composition of the liquid phase is illustrated by the bold section of the liquidus curve in Figure 7.25.)

During further cooling, the crystallization proceeds according to the previously described method, that is, an α-solid solution with a high A content solidifies (although the A content of the α-phase also decreases with the decreasing temperature). Consequently, the A content of the liquid phase also decreases. When intersecting the solidus curve, the total amount of the alloy has already been solidified to an α-solid solution. Its composition at this point already corresponds exactly to the composition of the alloy. Therefore, we can find only a homogeneous, single-phase α-solid solution under the solidus curve.

However, the previously described crystallization procedure is valid only for the so-called full equilibrium cooling. It requires holding the alloy at every temperature as long as the composition of the α-solid solution formed at the given temperature is equalized with the composition of the previously solidified α-solid solutions. (For example, at the temperature $T = T_2$, the α-solid solution contains b_2 % of B, while at the temperature $T = T_1$ the α-solid solution contains b_1 % of B. These different compositions should be equalized by a diffusion mechanism.) Only this infinitely slow, equilibrium cooling can lead to a perfectly homogeneous α-solid solution.

Core of the crystal solidifying first

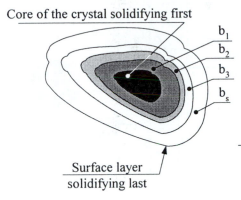

b_1
b_2
b_3
b_s

Surface layer
solidifying last

Figure 7.26
Structure of a solid solution

In reality, the requirement of equilibrium cooling described previously represents a cooling procedure which is so slow that it usually cannot be achieved in practice, since, due to the decreasing diffusion coefficient with lowering of the temperature, an ever-increasing time is needed. Therefore, significant differences can be observed in the composition of the crystals, as shown in Figure 7.26. It can be seen in the figure, that the core of the crystal is richer in the higher melting point component, while moving outwards, the composition is characterized by an increasing content of the lower melting point component. The inhomogeneous microstructure formed according to the above description is disadvantageous for the majority of mechanical properties. Therefore, the differences in composition following solidification should be ceased or at least decreased by the application of so-called diffusion annealing. This should be performed at a high temperature, close to the solidus, in order to achieve a fast diffusion process.

7.3.5.7. Phase diagram with limited solid solubility (Tammann-7)

In this alloy system, the two components dissolve each other to an unlimited extent in the liquid state and they form a limited solid solution in the solid state. This means that in certain composition ranges a homogeneous solid solution is formed. For example, in the composition range described by points $A \rightarrow G$ in Figure 7.27, a homogeneous α-solid solution can be found, but above this concentration, the two components are unable to dissolve each other in the solid state.

The solubility in the solid state can change in various ways with a decrease in temperature. The most typical case is a decrease in solubility with decreasing temperature, as shown in Figure 7.27 by the $\overline{DG}$-curve indicating the solubility of component B in an α-solid solution. In certain cases, the solubility does not change with the decrease in temperature. This is shown by curve $\overline{FH}$ in Figure 7.27. Consequently, this is represented by a vertical line in the diagram, which should not be confused with the vertical lines of the earlier diagrams marking the A_mB_n metallic compound! In some alloys, the solubility in the solid state may increase with the decrease in temperature. We can see an example of this in the *Cu-Zn* alloy system, where the Zn solubility of the so-called α-brass increases in the temperature range of $T = 903 - 454$ °C.

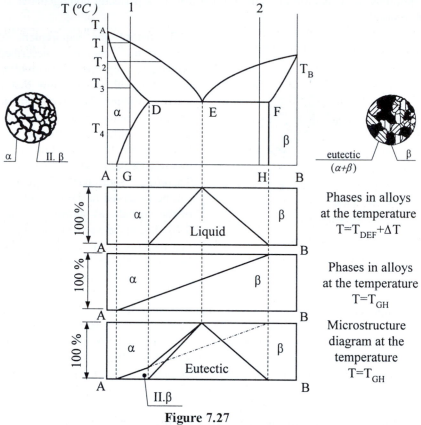

Figure 7.27
Phase diagram with limited solid solubility

Consider first the crystallization of alloy 1. The primary crystallization process is very similar to the solidification analyzed in the Tammann-6 system. This means that crystallization starts with the solidification of an α-solid solution when the liquidus curve is intersected and crystallization ends with the solidification of the α -solid solution at the temperature of the solidus curve. At this stage, the α-solid solution has the composition identical to that of the alloy. In the temperature range of $T_3 \rightarrow T_4$ (i.e. in the homogeneous zone of the α-solid solution), the whole alloy contains 100 % α-solid solution. Below the temperature determined by the solubi- lity limit curve indicated by the line $\overline{DG}$, the *B* solubility of the α-solid solution decreases along the solubility limit curve. Therefore, a phase with high *B*-content has to segregate from the α-solid solution to maintain the phase equilibrium. The composition of the segregating phase can be determined using the first lever rule. Drawing an isotherm from the intersection point of the alloy line and the curve indicating the limited solubility, the end point of the isotherm falls onto the curve $\overline{FH}$ indicating the boundary of β-solid solution. This means that the decrease of the B-content of the α-solid solution is ensured by the segregation of a β-solid solution having a high B-content. Since this segregation occurs in the solid state, this is denoted as a secondary crystallization process in order to distinguish it from the primary crystallization occurring from the liquid state. This segregation is also called *precipitation*. The segregated crystal is called a *segregate* or a *precipitate*.

Continuing the cooling of the alloy, the B solubility of the α-solid solution further decreases along the curve representing the limited solubility indicated by the $\overline{DG}$-curve. The equilibrium corresponding to this can be ensured by the continuous segregation of the secondary β-solid solution. Since this segregation occurs from the previously solidified crystals of the α-solid solution, the secondary β-solid solution precipitates along the grain boundaries of the primary α-crystals as can be seen on the left side of Figure 7.27. Thus, the alloy at room temperature consists of an α-solid solution and secondary β-crystals precipitated along the grain boundaries of the primary α-crystals.

Solidification of alloy 2 starts with the crystallization of a β-solid solution and this process lasts until the eutectic temperature marked by $T=T_E$ is reached. At this temperature, the β-solid solution and the liquid phase of the eutectic composition are in equilibrium. The eutectic liquid at a constant temperature ($T=T_E$) solidifies into a mixture of α_D and β_F solid solutions according to the reaction equation

$$liquid_E \rightleftarrows \alpha_D + \beta_F. \tag{7.24}$$

Since the A solubility of the β-solid solution does not change with temperature, the primarily crystallized β-solid solution and a eutectic consisting of a mixture of α_G and β_H solid solutions can be found in the alloy at room temperature. The compositions of the phases within the eutectic change according to the limited solubility. This means that the B solubility of the α-solid solution is reduced from the value marked by point D to the value marked by point G. This change means that the lamella of the β-phase within the eutectic becomes thicker but there are no other changes in the microstructure.

The construction of the microstructure diagram requires further considerations due to the secondary β-phase. Therefore, two further diagrams are shown in Figure 7.27 for two different temperatures (one slightly above the temperature of the eutectic reaction: $T=T_{DEF}+\Delta T$, the other at room temperature marked by $T=T_{GH}$). From these diagrams, both the types and the quantities of the phases being in equilibrium at the given temperatures can be determined. These diagrams are drawn at these two temperatures, since they enable assessment of what microstructural elements are formed and from which phases during crystallization.

It can be seen from the diagram drawn for the temperature $T=T_{DEF}+\Delta T$ that besides the primarily crystallized α- and β-solid solutions, a liquid of eutectic composition can be found. Since a eutectic can only arise from a liquid of eutectic composition, the zone of the eutectic in the microstructure diagram at room temperature should be the same as the zone of the eutectic liquid at the temperature $T=T_{DEF}+\Delta T$. Similarly, it can be easily understood that the quantity of the primarily crystallized β-phase (crystallizing in the composition range between points E and B) does not change during the eutectic reaction. Thus, the zone of the primary β-phase in the microstructure diagram drawn at room temperature should be the same as the zone of the β-phase in the diagram drawn at the temperature $T=T_{DEF}+\Delta T$.

From the analysis of crystallization processes, it is obvious that the secondary β-phase can precipitate only from the primary α-phase along the solubility limit curve marked DG. Since the B solubility of the α-phase at room temperature can be characterized by point G, in alloys with a B-content higher than this value, the secondary β-phase should appear. The maximum quantity of secondary β-phase can be found in the alloy corresponding to the composition marked by point D (since in this alloy the primary α-phase corresponds to 100 % at the temperature $T=T_{DEF}+\Delta T$). From the diagram drawn at the temperature $T=T_{DEF}+\Delta T$, it can be seen that the quantity of α-phase between the points D and E decreases linearly and reaches the value of 0 % at point E. Accordingly, the quantity of the secondary β-phase segregating from the primary α-phase has to decrease linearly between points D and E, too in such a way that its quantity equals zero in the alloy having the composition corresponding to point E.

7.3.5.8. Phase diagram with peritectic reaction in solid solution (Tammann-8)

In fact, this ideal, binary phase diagram does not convey significant new knowledge compared to the previously discussed basic types.

The reason for its inclusion as an independent basic type can be explained by the special peritectic reaction of this system. In the previously analyzed diagrams, in peritectic reactions metallic compounds were always formed, while in the alloys crystallizing according to Tammann-8, the result of the peritectic reaction is a solid solution (see Figure 7.28).

The most important characteristics of this diagram can be summarized as follows: the two components can dissolve each other to an unlimited extent in the liquid state. In the solid state, two solid solutions are

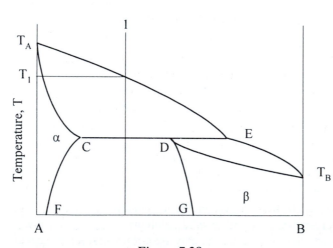

Figure 7.28
Phase diagram with a peritectic reaction in a solid solution

formed: both the α and β solid solutions are characterized by a decreasing limited solubility with a decrease in temperature. The most important feature of the diagram is that the α-solid solution is the product of the primary crystallization, while the β-solid solution is the result of the peritectic reaction.

Since the characteristic reaction of the alloy system occurs only during the crystallization of alloys within the composition range marked by points *CDE*, we have to select an alloy for the analysis in this range. Crystallization starts with the primary crystallization of α-solid solution when crossing the liquidus curve and it lasts until the temperature $T=T_{DEF}+\Delta T$ is reached. At this temperature, α_C and *liquid_E* are in equilibrium. At the temperature $T=T_{CDE}$, the peritectic reaction takes place according to the equation

$$\alpha_c + liquid_E \rightleftharpoons \alpha_C + \beta_D .$$ (7.25)

In this peritectic reaction, a solid solution (α_C) and a liquid phase (*liquid_E*) form a new solid phase (β_D). The peritectic reaction in its "pure" form occurs only in the alloy having the composition corresponding to point *D* as follows:

$$\alpha_c + liquid_E \rightleftharpoons \beta_D .$$ (7.26)

In the alloy 1, part of the α_C solid solution does not participate in the reaction, thus the peritectic reaction occurs according to Equation (7.25).

Further decreasing the temperature, the solubility of both solid solutions decreases. Therefore, secondary β-crystals segregate from the α-solid solution and secondary α-crystals segregate from the β-solid solution. Due to these precipitation processes, the alloy at room temperature contains α_F and β_G solid solutions, as well as secondary β-crystals precipitated on the grain boundaries of the α-phase and secondary α-crystals precipitated on the grain boundaries of β crystals.

7.3.5.9. Generalization of the basic rules of binary phase diagrams

The rules discussed during the analysis of ideal binary phase diagrams can be of great help in the analysis of more complex systems. Therefore, we shall briefly summarize the most important, general rules of the binary systems.

- The solubility in the liquid state may be characterized by the shape of the liquidus. If the liquidus consists of only curved sections, the two components dissolve each other to an unlimited extent in the liquid state. If the liquidus has horizontal, straight sections, the solubility is limited even in the liquid state in the composition range corresponding to the straight liquidus section.
- The number of liquidus sections equals the number of the primarily crystallizing phases.
- The solubility in the solid state may be characterized by the shape of the solidus. If the solidus contains only horizontal, straight sections, the two components cannot dissolve each other in the solid state at all. If the solidus contains only curved sections, the two components form unlimited solid

solutions. If the solidus consists of both curved and straight sections, limited solid solubility is characteristic of the two components.

– Drawing a continuous straight line into the phase diagram in any direction, the number of phases should change (decrease or increase) by one when intersecting any lines of the phase diagram. Actually, it follows from this rule that within a binary phase diagram a heterogeneous zone has to be followed by a homogeneous one and, vice versa, a homogeneous zone has to be followed by a heterogeneous one. There are two special exceptions: the horizontal and vertical boundary lines separating heterogeneous zones. (These exceptions are only virtual. In fact a horizontal, straight boundary line can be considered as a three-component heterogeneous zone having an infinitely small temperature interval. The vertical boundary line can be considered as a homogeneous zone having an infinitely small composition interval. In this sense, the intersection of these lines also fulfills the general rules.)

– In binary phase diagrams, there are six different non-variant reactions occurring at constant temperature. The reaction equations and the characteristic sections of phase diagrams are summarized in the following table.

Table 7.2 **Non-variant reactions in binary phase diagrams**

Type of reaction	The reaction equation	Typical section of phase diagram
Eutectic	$liquid_1 \rightleftharpoons solid_1 + solid_2$	$liquid_1$; $solid_1$; $solid_1+solid_2$; $solid_2$
Eutectoid	$solid_1 \rightleftharpoons solid_2 + solid_3$	$solid_1$; $solid_2$; $solid_2+solid_3$; $solid_3$
Peritectic	$liquid_1 + solid_1 \rightleftharpoons solid_2$	$solid_1$; $solid_1+liquid_1$; $liquid_1$; $solid_2$
Peritectoid	$solid_1 + solid_2 \rightleftharpoons solid_3$	$solid_1$; $solid_1+solid_2$; $solid_2$; $solid_3$
Monotectic	$liquid_1 \rightleftharpoons solid_1 + liquid_2$	sol_1; $liqu_1$; sol_1+liqu_1; $liqu_1+liqu_2$; $liqu_2$; sol_1+liqu_2
Syntectic	$liquid_1 + liquid_2 \rightleftharpoons solid_1$	$liqu_1$; $liqu_1+liqu_2$; $liqu_2$; sol_1+liqu_1; sol_1; sol_1+liqu_2

7.4. Three-component (ternary) alloy systems

Until now, we have dealt only with two-component, so-called binary alloys and phase diagrams. However, a great majority of alloys applied in practice are multi-component alloys. In many cases, alloys may be considered as a binary alloy of two dominant components (the fundamental metal and the most significant alloying metal) considering the effects of the rest of the components exerted on the phase diagram. However, there are many alloys of practical importance, which contain three equally important components. Therefore, in the following the most important rules of ternary phase diagrams will be discussed briefly.

The ternary phase diagrams can be illustrated in three dimensions: two dimensions are required to indicate the concentrations and the third dimension is used for the temperature. In order to indicate concentrations, an equilateral triangle is selected. This is called the basic (or composition) triangle of the ternary system. The apexes of the triangle correspond to 100 % of one of the components. Each side represents the composition of binary alloys of the respective two components.

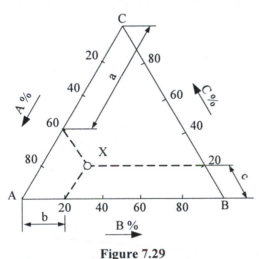

The composition of a three-component alloy corresponds to one of the points inside the triangle. It is a characteristic feature of equilateral triangle that the sum of the distances of an internal point from the sides is equal to the height of the triangle. The sum of the intercepts drawn parallel to the sides from the internal point equals the length of the side of the triangle.

Figure 7.29

Basic triangle of a three-component spatial phase diagram

Thus, the percentage value of the components of a three-component alloy can be obtained from the basic triangle as follows. First, a line parallel to the side opposite to the respective component (for example to A) is drawn from the point representing the composition (e.g. point X in Figure 7.29). This intersects the other two sides of the triangle at the percentage value corresponding to the respective component (in this case component A). For instance the composition of the alloy marked by point X in Figure 7.29 is 60 % A + 20 % B + 20 % C, altogether 100 %.

Points on the sides of the triangle represent binary alloys: the content of the component corresponding to the apex opposite the side is equal to zero. Beside the sides, two other characteristic lines inside the triangle are of utmost importance.

a) The lines parallel to the sides represent ternary alloys in which the quantity of the component of the opposite apex to the side is constant (Figure 7.30.a.).

b) The points of straight lines passing through an apex represent alloys in which the ratio of the other two components is constant (Figure 7.30.b.).

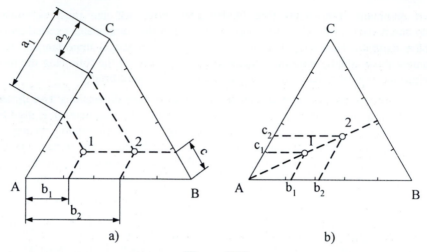

Figure 7.30
Characteristic composition lines in a ternary alloy system

Constructing the spatial phase diagram of a three-component system is quite a complicated procedure. Therefore, the spatial diagrams are used only in relatively simple cases. The planar sections perpendicular to the basic triangle or the horizontal sections corresponding to a constant temperature (the so-called *isothermal sections*) are used more frequently to study the crystallization of ternary alloys. In Figure 7.31, the isothermal section of the *Fe-Cr-Ni* ternary alloy system is shown. The *Fe-Cr-Ni* alloy system is of outstanding practical importance. One of the most generally applied corrosion resistant steels (the so-called austenitic *Cr-Ni* steel) can be considered basically as a three-component alloy containing 18 % *Cr* and 8 % *Ni* besides *Fe* as the fundamental metal).

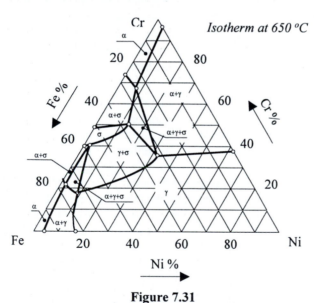

Figure 7.31
The isotherm section of a ternary alloy system of
Fe-Cr-Ni at a temperature of T=650 °C

In the following, a relatively simple ternary system will be shortly analyzed. It may be characterized as follows:

- The components dissolve each other in the liquid state without any limitation: this means that the phase diagram contains curved liquidus surfaces.

- The components form neither solid solutions nor metallic compounds in the solid state. This means a planar solidus surface in the diagram.

- The pairs of the components form binary eutectics (for example, components A and B form the binary eutectic marked by E_1 shown in the binary sections of Figure 7.32). In ternary systems – according to Gibbs' phase rule – the binary eutectics do not solidify at a constant temperature but in a temperature interval.

- The three components form a ternary eutectic. The ternary eutectic corresponding to point E crystallizes at a constant temperature. This also follows from Gibbs' phase rule, since during the formation of a three-component eutectic, the degree of freedom is zero. Consider that the number of components is $C=3$, the number of phases in the ternary eutectic is $P=4$ (the liquid and components A, B and C). Substituting these values into Gibbs' phase rule, $F=0$ is obtained for the degree of freedom.

In Figure 7.32, the binary phase diagrams are drawn in the plane of the basic triangle. The spatial diagram can be derived from this in such a way that the binary phase diagrams are turned to an upright position. A liquidus surface can be fit onto the respective two liquidus curves. Two liquidus surfaces intersect each other on curves representing the E_1E, E_2E, and E_3E binary eutectics, respectively. The three liquidus surfaces meet at point E representing the ternary eutectic. This point corresponds to the lowest point of the liquidus surfaces. If a plane is laid across this point, the planar solidus surface of the ternary system is obtained.

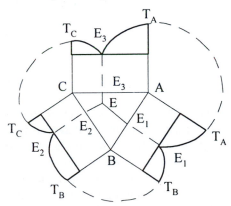

Figure 7.32

Binary diagrams of a three-component diagram spread to the plane

Let us analyze the crystallization of an alloy having a composition M' marked by the alloy line 1. The alloy line intersects the $t_A E_3 E E_1$ liquidus surface at point M. Crystallization starts at this temperature with the solidification of the component having the highest melting point. In this case it is the component A. Since only A crystals solidify from the melt, the B and C content as well as their ratios remain constant in the remaining liquid. These types of alloys fall onto such straight lines of the basic triangle that pass through points A and M'. Laying a projection plane onto the straight lines AM' and At_A, the intersection line of this with the liquidus surface (MK) gives the temperature of the liquid remaining during solidification and line $M'K'$ represents the composition of the liquid. The ratio of the liquid and the solid A crystals can be determined by using the quantity rule. This may be calculated from the following relationship:

$$\frac{M_1^{liquid}}{A} = \frac{m}{a}.$$

(7.27)

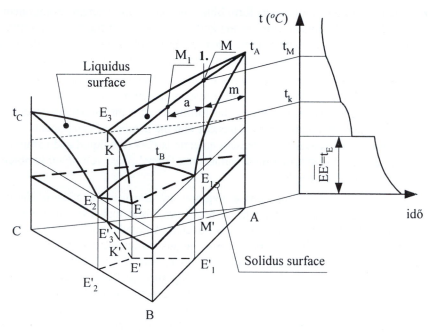

Figure 7.33
Three dimensional illustration of a ternary phase diagram

The crystallization of pure A metal lasts until the line t_AMK intersects the E_3E spatial curve (point K in the figure). The crystallization of the E_3 binary eutectic commences at this temperature, while the temperature of the liquid phase is changing along the KE curve towards point E and its composition is shifted along the $K'E'$ line towards point E'. When point E is reached, the liquid becomes saturated with respect to all the three components. Therefore, the total mass of it solidifies at this temperature as a ternary eutectic denoted by point E.

Thus, the following microstructural elements can be found in the alloy having a composition corresponding to M at room temperature: pure A crystals solidified by primary crystallization, the binary eutectic E_3 consisting of components A and C, as well as the tertiary eutectic "E" consisting of A, B and C components.

8. EQUILIBRIUM CRYSTALLIZATION OF IRON-CARBON ALLOYS

8.1. The industrial and practical significance of iron alloys

The terminology of *iron-carbon alloys* is used to designate the large group of metallic alloys in which the fundamental alloying element is iron (*Fe*). Despite the extremely rapid and dynamic development of materials science and technology in the past decades, iron alloys still represent the most important and most widely applied metallic materials in industrial practice. The quantity of iron alloys produced and utilized all over the world exceeds the quantity of all other metallic elements together by a factor of ten.

The utilization of iron alloys accompanies the history of mankind from the very beginning. Iron-based metallic materials were first utilized 1500 years B.C. Therefore, it is not a mere accident that a determinant period of the ancient history of mankind is called the *Iron Age*. The popularity of iron alloys prevailing even in present times can be explained by several factors:

- The ores that can be processed for the production of iron can be ranked as the most abundant minerals found in the Earth's crust. About 4.2 % of the Earth's crust consists of various ores of iron. From these ores, iron can be extracted by relatively simple and cheap processes compared to the extraction procedures used for other metals.
- The melting point of pure iron is $T = 1536\ ^\circ C$. Below this value iron is present in the forms of various allotropic modifications (*α-Fe*, *γ-Fe*, *δ-Fe*) in different temperature ranges. Consequently, there are many thermally activated pro‒cesses – in not too high temperature ranges – which can be applied to modify advantageously the properties of iron alloys in a very wide range.
- Below the temperature of $T = 769\ ^\circ C$ (the so-called *Curie point*) iron has *ferromagnetic properties* which makes numerous practical applications possible.

Iron alloys – besides iron (*Fe*) as the fundamental (or base) metal – always contain carbon (*C*), which is considered as the basic alloying element of iron alloys. The iron alloys also contain several other elements in smaller quantities. Due to the production processes of pig iron and steel, all iron alloys contain manganese (*Mn*), silicon (*Si*), sulfur (*S*) and phosphorus (*P*). These elements enter the metal as contaminants from the raw materials or from the products of combustion. In some cases, these elements are added to the raw materials for the sake of production.

It follows from this that iron alloys are multi-component metallic alloys. However, as long as the quantities of the elements listed above do not exceed the values unavoidably entering the alloy from the production process ($Si = 0,05 \div 0,4\,\%$, $Mn = 0,15 \div 0,7\,\%$, $S + P \leq 0,035\,\%$), these elements do not influence considerably the equilibrium diagram of the two-component *Fe-C* alloy system. Therefore, the metallographic concepts and processes connected to iron alloys can be studied in the two-component, i.e. the binary *Fe-C* equilibrium diagrams. However, prior to discussing the specific features of the equilibrium diagrams, it is useful to be acquainted with some essential characteristics of the two main components, i.e. some properties of iron and carbon.

The melting point of *pure iron* (*Fe*) is $T_{melt} = 1536\,^{\circ}C$. In the solid state, pure iron possesses three allotropic modifications (see Figure 4.17), namely: *δ-Fe* as a body-centered cubic crystal (existing in the temperature range of $1392\,^{\circ}C \leq T < 1536\,^{\circ}C$), *γ-Fe* as a face-centered cubic crystal (existing in the temperature range of $911\,^{\circ}C \leq T < 1392\,^{\circ}C$) and *α-Fe* as a body-centered cubic crystal (existing in the temperature range of $T < 911\,^{\circ}C$). Among the allotropic transformations, the $\alpha \rightarrow \gamma$, as well as the $\gamma \rightarrow \alpha$ phase transformations play important roles. These phase transformations form the theoretical basis for many heat treatment processes.

Iron forms solid solutions with many non-metallic elements. It forms substitutional solid solutions with chromium, nickel, cobalt and vanadium, while the most important interstitial solid solutions of iron are formed with carbon. However, while the carbon dissolving capability of *α-Fe* is very low (at room temperature *α-Fe* is capable of dissolving only 0.006 % carbon in the equilibrium state), the carbon dissolving capability of *γ-Fe* is higher by several magnitudes (the maximum carbon dissolving capacity of *γ-Fe* is 2.06 % at a temperature of $T=1147\,^{\circ}C$). The solid solution of carbon with *α*-iron is called *ferrite* and the one formed with γ-iron is called *austenite*.

However, iron forms not only solid solutions with carbon but also interstitial metallic compounds. The interstitial compound of iron and carbon is Fe_3C. This is called *iron carbide* or *cementite*. Further characteristics of iron carbide are its high hardness ($HV \approx 900$) and high rigidity. Practically, iron carbide is non-deformable. The melting point of iron carbide is $T_{melt} = 1250\,^{\circ}C$. Iron carbide cannot be considered an equilibrium phase. Under appropriate conditions, it can be decomposed into its components, *Fe* and *C*. This carbon (called graphite) is a stable, equilibrium phase.

8.2. The equilibrium phase diagram of iron – carbon alloys

The phase diagram of iron–carbon alloys for equilibrium crystallization is shown in Figure 8.1. This is often referred to as the *Heyn-Charpy dual phase diagram of iron-carbon alloys*. The original diagram is valid even today as regards its shape; however, during the time since its elaboration, the coordinates of some points (composition and temperature data) have been modified several times as a result of developments in measuring methods and instruments. In this book, the diagram containing the data published in Volume 8 of the *ASM Metals Handbook* is used.

The diagram is in fact a dual phase diagram. It relates both to the *equilibrium crystallization* (the so-called *stable crystallization*) and to the *quasi-equilibrium crystallization* (the so-called *metastable crystallization*). This duality is connected primarily to the form of carbon as exists in the alloys. Carbon can be found in the iron-carbon alloys in free form as graphite (C), as well as in a chemically bonded form as iron carbide (Fe_3C), i.e. a metallic compound.

The equilibrium state means the presence of carbon in free form (as graphite). This is indicated in the diagram of Figure 8.1. by the dotted lines (where we cannot find dotted lines besides the continuous ones, the continuous line is valid for the stable crystallization as well). However, in practice the metastable crystallization is much more general when carbon is present in the form of chemically bonded iron carbide as a metallic compound. The equilibrium diagram relating to this type of crystallization is shown in Figure 8.1. by the sections drawn with continuous lines.

The reason for the two types of existence of carbon can be found in the higher crystallization capability of carbide compared to that of graphite. The low carbon content alloys (steels) – under the cooling conditions common in practice – always crystallize according to the metastable system.

In practice, iron–carbon alloys are used with a limited carbon content. This is the reason why the iron–carbon phase diagram extends only to the value of $C = 6.687 \%$. This carbon content corresponds to the metallic compound Fe_3C in the metastable system. Therefore, the right hand side boundary line of the metastable phase diagram represents 100 % iron carbide (i.e. 100 % Fe_3C). This is the reason why the equilibrium diagram representing the metastable crystallization drawn with continuous lines is often referred to as the *Fe- Fe_3C* system.

Naturally, the right hand side border line of the diagram in the stable system represents 100 % C (a composition corresponding to pure graphite); therefore the stable system is frequently referred to as the iron-graphite (*Fe-C*) system.

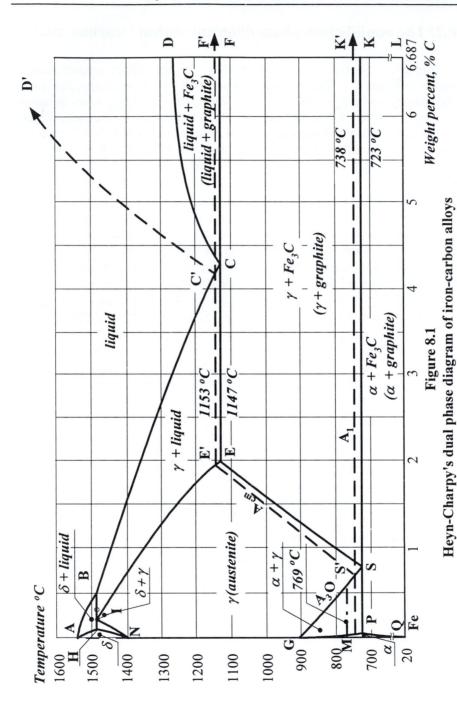

Figure 8.1

Heyn-Charpy's dual phase diagram of iron-carbon alloys

Table 8.1 Characteristic points of the Heyn-Charpy dual phase diagram
for iron–carbon alloys with the corresponding
temperature and composition values

Main points of the phase diagram	Temperature T ($^{\circ}$C)	Carbon content C (%)
A	1536	0.000
B	1493	0.510
C	1147	4.300
C'	1153	4.260
D	1250	6.687
D'	*	100.000
E	1147	2.060
E'	1153	2.030
F	1147	6.687
F'	1153	100.000
G	911	0.000
H	1493	0.100
I	1493	0.160
K	723	6.687
K'	738	100.000
L	20	6.687
L'	20	100.000
M	769	0.000
N	1392	0.000
O	769	0.512
P	723	0.025
P'	738	0.020
Q	20	0.006
S	723	0.800
S'	738	0.690

* The exact value is not known.

The most important features of the Heyn-Charpy dual phase diagram can be summarized as follows:

1. The components form an unlimited liquid solution: according to the general rules of equilibrium diagrams, this means that the liquidus line may consist of only curved sections. In Figure 8.1, it can be observed that the liquidus consists of three curved sections, meaning that the number of phases crystallizing from the liquid as primary crystals is equal to three. These phases are partly the same and partly different in the stable and the metastable systems.

   - The following phases crystallize primarily in the metastable system:
     * along the *AB* liquidus curve, δ-solid solution,
     * along the *BC* liquidus curve, γ-solid solution,
     * along the *CD* liquidus curve, Fe_3C metallic compound.
   - The following phases crystallize primarily in the stable system:
     * along the *AB* liquidus curve, δ-solid solution, again,
     * along the *BC'* liquidus curve, γ-solid solution,
     * along the *C'D'* liquidus curve, C-graphite.

2. The components form a limited solid solution; consequently, the solidus consists of both curved and straight sections. The previously mentioned three solid solutions can be found in the system. All the solid solutions are interstitial solid solutions of *Fe* and *C*. These are, in sequence, the following:
   - *α-solid solution*: a body-centered cubic crystal called *ferrite*,
   - *γ-solid solution*: a face-centered cubic crystal called *austenite*,
   - *δ-solid solution*, also a body-centered cubic crystal called *δ-ferrite*.

3. There are *three non-variant reactions* in this alloy system: *peritectic, eutectic* and *eutectoid* reactions according to the following:
   - In the *metastable system*:
     * At the temperature $T = 1493\ ^{\circ}C$, a *peritectic reaction* occurs in the concentration range of C = 0.10 - 0.51 % (between the points *H* and *B)*. The peritectic reaction may be described by the reaction equation:

$$\delta_H + liquid_B \rightleftharpoons \gamma_I . \tag{8.1}$$

     * At the temperature $T = 1147\ ^{\circ}C$, a *eutectic reaction* occurs in the concentration range of C = 2.06 - 6.687 % (between the points *E* and *F)*. The eutectic reaction may be given by the reaction equation:

$$liquid_C \rightleftharpoons \gamma_E + Fe_3C . \tag{8.2}$$

     This two-phase, heterogeneous product of the eutectic reaction is called *ledeburite*. It consists of the phases $\gamma_E + Fe_3C$ (austenite and cementite) at the temperature of its origin.

* At the temperature $T=723$ °C, a *eutectoid reaction* occurs in the concentration range of $C = 0.025 - 6.687$ % (between the points P and K). The eutectoid reaction may be given by the reaction equation:

$$\gamma_S \rightleftharpoons \alpha_P + Fe_3C. \tag{8.3}$$

This two-phase product of the eutectoid reaction in the metastable system is called *pearlite*. It consists of the phases $\alpha_P + Fe_3C$ (ferrite and cementite) at the temperature of its origin.

- In the *stable system*

  * At the temperature $T = 1493$ °C, a *peritectic reaction* occurs in the concentration range of $C = 0.10 - 0.51$ % (between the points H and B). This peritectic reaction may be given by the reaction equation:

  $$\delta_H + liquid_B \rightleftharpoons \gamma_I. \tag{8.4}$$

  * At the temperature $T=1153$ °C, a *eutectic reaction* occurs in the concentration range of $C = 2.03 - 100$ % (between the points E' and F'). This eutectic reaction may be described by the reaction equation:

  $$liquid_{C'} \rightleftharpoons \gamma_{E'} + C. \tag{8.5}$$

  This two-phase product of the eutectic reaction is called *graphite-eutectic*. It consists of the phases $\gamma_{E'}+C$ (i.e. austenite and graphite) in the stable system at the temperature of its origin.

  * At the temperature $T = 738$ °C, a *eutectoid reaction* occurs in the concentration range of $C = 0.02 - 100$ % (between the points P' and K'). This eutectoid reaction may be given by the reaction equation:

  $$\gamma_{S'} \rightleftharpoons \alpha_{P'} + C \tag{8.6}$$

  This two-phase product of the eutectoid reaction is called *graphite-eutectoid*. It consists of the phases $\alpha_{P'} + C$ (ferrite and graphite) at the temperature of its origin.

4. Three further phase transformations occurring in the alloy system are worth mentioning. These are connected to the crystallization and segregation of cementite and graphite. These processes have certain common properties, but also significant differences in the stable and metastable system; therefore they will be discussed separately.

   - In the metastable system:

     * As a result of a *primary crystallization* – i.e. crystallization from the liquid phase – the microstructure element *primary cementite* is formed along the liquidus curve *CD* in the concentration range of C = 4.3 - 6.687 %.

* Due to the decreasing carbon solubility of γ-austenite with decreasing temperature, *secondary cementite* is formed by a precipitation/segregation process in the solid state along the curve *ES*. This curve represents the limited solid solubility of carbon in austenite in the concentration range of C = 0.8 - 4.3 %.

* Due to the decreasing carbon solubility of α-ferrite, *tertiary cementite* is formed by a precipitation/segregation process in the solid state, too, along the curve *PQ*. This curve represents the limited solid solubility of carbon in ferrite in the concentration range of C = 0.006 - 0.8 %.

- In the stable system:

    * As a result of *primary crystallization, primary graphite* is formed along the liquidus curve *C'D'* in the concentration range of C = 4.26 - 100 %.

    * Due to the decreasing carbon solubility of γ-austenite, *secondary graphite* is formed by segregation in the solid state along the curve E'S'. This curve represents the limited solid solubility of carbon in austenite in the concentration range of C = 0.69 - 4.26 %.

    * Due to the decreasing carbon solubility of α-ferrite, *tertiary graphite* is formed by segregation in the solid state, too, along the curve *P'Q'*. This curve represents the limited solid solubility of carbon in ferrite in the concentration range of C = 0.006 - 0.69 %.

The consequent application of the previously summarized general rules makes the analysis of crystallization of various alloys significantly easier. Furthermore, as a basic rule, the following statement should always be taken into consideration during the analysis of crystallization processes: carbon never exists in the metastable system as graphite but as chemically bonded iron carbide (i.e. cementite). Moreover, carbon is always present in the stable system in the free form of graphite and it never exists as iron carbide.

In the following sections, the procedure of crystallization of typical alloys will be analyzed, first in the metastable system then in the stable system. It will provide an excellent opportunity to apply the knowledge acquired during the analysis of ideal, binary phase diagrams.

However, before starting the analysis of some characteristic alloys, it is useful to classify the iron-carbon alloys based on the equilibrium phase diagram since the terminology to be introduced here can be applied later in the analysis of crystallization.

8.3. Classification of iron-carbon alloys based on the equilibrium diagram

To perform the classification, consider Figure 8.2. In the upper part of the figure, the phase diagram with the characteristic points is shown, while the classification based on the diagram can be seen in its lower part.

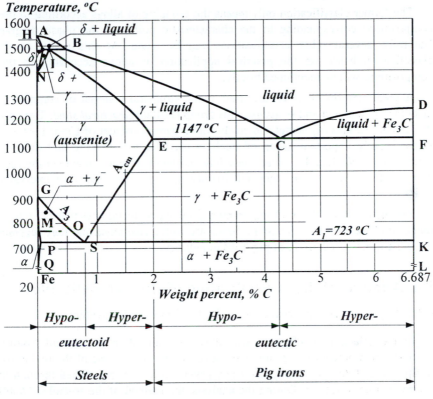

Figure 8.2
Theoretical classification of iron-carbon alloys

In industrial practice, only iron-carbon alloys having a carbon content of $C < 6.687\%$ are used. Point E ($C = 2.06\%$) in the equilibrium diagram divides the iron-carbon alloys into two main groups. The iron-carbon alloys containing less carbon than the composition of point E are called *steels,* while those containing more carbon than point E are called *pig irons.*

Further subgroups can be formed within the two main groups. In both cases, the typical point of further classification is connected to the non-variant reactions. In steels, the composition corresponding to point S represents the eutectoid reaction: therefore, the steel with this composition is called *eutectoid steel.* Steels containing less carbon than the eutectoid composition are called *hypoeutectoid steels* and steels with higher carbon content than the eutectoid composition are called *hypereutectoid steels*.

Similarly, for pig irons, the composition of point *C* represents the eutectic reaction: therefore, a pig iron with this composition is called *eutectic pig iron*. Pig irons containing less carbon than the eutectic composition are called *hypoeutectic pig irons*, and pig irons containing more carbon than the eutectic composition are called *hypereutectic pig irons*.

The former classification only means the grouping of alloys according to their compositions corresponding to the characteristic points of the phase diagram. Naturally, several other classifications can be made. At this point, another classification based on the practical application of alloys according to their compositions will be mentioned briefly.

A great majority of steels are used as the structural material of components, machines, equipment and structures: this group is called *structural steel*. Typically, this group includes *hypo*eutectoid steels with a carbon content of $C = 0.1 - 0.6$ %. A particularly important group of steels is used for tool making and these are called *tool steels*. Usually, these steels have a higher carbon content (in the range of $C = 0.6 - 2.06$ %), thus, they fall mainly into the *hypereutectoid* interval of steels.

Among pig irons, the most important are the hypoeutectic ones that fall into the range of $C = 2.5 - 4.0$ %. These pig irons are called *cast irons* and they represent the most widely used group of pig irons in industry.

8.4. Analysis of the crystallization of characteristic alloys

8.4.1. Crystallization processes in the Fe-Fe₃C metastable system

In the selection of characteristic alloys, we aimed to include all essential crystallization processes (e.g. non-variant reactions, etc.) in one of the real alloys. In this way, all the previously described general rules can be studied through the crystallization processes. During the analysis, the method of the so-called "*family tree of crystallization*" will be used and some qualitative analysis will be performed. The family tree of crystallization is a kind of *crystallization flow diagram*. In Figure 8.3, the alloy lines used to denote the alloys to be analyzed are also shown in the metastable phase diagram of the *Fe-Fe₃C* system.

Consider first *Alloy 1*, which can be characterized by a carbon content of $C = 0.12$ %. The crystallization process is shown in Figure 8.4. The solidification of the alloy starts where the alloy-line intersects the liquidus curve. The first crystal that solidifies is a δ-solid solution. The crystallization of δ-solid solution continues until the temperature $T = T_{HIB} = 1493$ °C is reached at which a peritectic reaction occurs according to the reaction equation:

$$\delta_H + liquid_B \rightleftharpoons \delta_H + \gamma_I .$$

(8.7)

Since, in this alloy, the phase δ_H is present in a quantity higher than required to a pure peritectic reaction, part of this phase will not participate in the reaction. In fact, the phase γ_I (with a composition corresponding to point I) is formed from the total quantity of the liquid (with the composition of point B) and part of phase δ_H (with the composition of point H). The quantities of the phases taking part in this reaction can be determined using the second lever rule in the following manner.

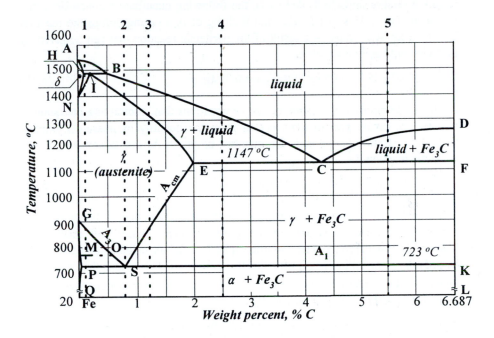

Figure 8.3

Sketch for analysis of the crystallization of various iron-carbon alloys

To perform the calculation of quantities, consider a unit mass (i.e. 1 kg) of the alloy. The quantities of the phases being in equilibrium before the reaction can be calculated from the ratio of the intercepts of the tie line drawn at the temperature $T=T_{HIB}+\Delta T$ using the following relationships:

$$\delta_H = \frac{0.51-0.12}{0.51-0,10} \times 1 = 0.95 kg \Rightarrow 95\%,$$

$$(8.8)$$

$$liquid_B = \frac{0.12-0.10}{0.51-0,10} \times 1 = 0.05 kg \Rightarrow 33\%.$$

The quantities of the phases being in equilibrium after the reaction can be calculated from the ratio of the intercepts of the tie line drawn at temperature $T=T_{HIB}-\Delta T$ using the following relationships:

$$\delta_H = \frac{0.16-0.12}{0.16-0.10} \times 1 = 0.67\,kg \Rightarrow 67\%,$$

$$(8.9)$$

$$\gamma_I = \frac{0.12-0.10}{0.16-0.10} \times 1 = 0.33\,kg \Rightarrow 33\%.$$

From the relationships (8.8) and (8.9), the following quantitative conclusions can be drawn. The 0.05 kg of *liquid$_B$* and 0.28 kg of δ_H existing before the reaction form 0.33 kg of γ_I-phase as a result of the peritectic reaction that can be given by the following reaction equation:

$$0.28\,kg\ \delta_H + 0.05\,kg\ liquid_B \rightleftharpoons 0.33\,kg\ \gamma_I\,.$$

$$(8.10)$$

It can also be seen clearly from the above equations that, subtracting the quantity of phase δ_H taking part in the reaction from the quantity of phase δ_H existing before the reaction, we obtain the quantity of phase δ_H not taking part in the peritectic reaction, i.e.:

$$0.95\,kg\,\delta_H - 0.28\,kg\,\delta_H = 0.67\,kg\,\delta_H\,.$$

$$(8.11)$$

Continuing the analysis of crystallization, it can be seen that as the temperature decreases the phase δ_H is transformed gradually into γ-phase. This results in a decrease in the carbon content of the γ-phase from its original composition γ_I towards the composition of the alloy itself (i.e. to $\gamma_{0.12}$). This means that the carbon content of the γ- phase below the curve $\overline{NI}$ (i.e. in the homogeneous γ-zone) will be the same as that of the starting alloy.

In the homogeneous γ-zone, there is no phase transformation until the temperature $T=T_{GOS}$, that is, until the line A_3 is reached. At this temperature, the nucleation of the so-called pro-eutectoid α-phase starts from the γ-phase. Since the α-phase has a much lower iron content than the γ-phase, the carbon content of the remaining γ-phase continuously increases until it reaches the eutectoid γ_S composition at the temperature $T=T_{PSK}=723$ $^\circ C$. The pro-eutectoid α-phase segregating from the austenite in the temperature range of A_3-A_1 is called *ferrite.* It can be observed as bright, polygonal crystallites under an optical microscope. The quantity of ferrite at the temperature $T_{PSK}+\Delta T$ is given by the equation:

$$\alpha_P = \frac{0.8-0.12}{0.8-0.025} = 0.877\,kg \Rightarrow 87.7\%\,.$$

$$(8.12)$$

At this temperature, the ferrite is in equilibrium with the γ_S-austenite phase. The quantity of γ_S-austenite can be calculated from the following equation:

$$\gamma_S = \frac{0.12-0.025}{0.8-0.025} = 0.123\,kg \Rightarrow 12.3\%\,.$$

$$(8.13)$$

At the temperature A_1 ($T = 723$ $^\circ C$), the following *eutectoid reaction* occurs:

$$\gamma_S \rightleftharpoons \alpha_P + Fe_3C \ . \tag{8.14}$$

The result of the eutectoid reaction is a two-phase, heterogeneous microstructure called *pearlite*. As can be seen from Equation (8.14), the pearlite is formed from the homogeneous, single-phase austenite (γ_S). The composition of pearlite corresponds to point S. As shown by the crystallization tree in Figure 8.4, the previously nucleated pro-eutectoid ferrite phase also remains in the alloy. Since pearlite was formed from the phase γ_S, its quantity is equal to the calculated quantity of γ_S (i.e. equal to 12.3 %).

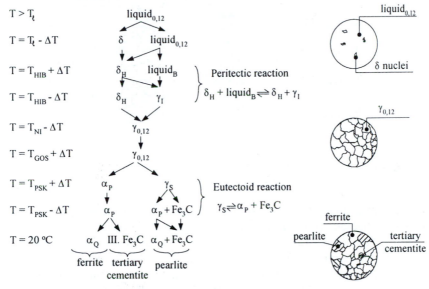

Figure 8.4
Crystallization tree of an alloy with a carbon content of $C = 0.12$ %

Following the eutectoid reaction, the carbon dissolving capability of the α-phase decreases along the curve $\overline{PQ}$ representing the solubility limit. This is realized by precipitation of the Fe_3C (cementite) phase having a high carbon content. This phase is called *tertiary cementite*. Due to the precipitation of tertiary cementite, the composition of ferrite containing 0.025 % of carbon at the temperature $T=723\,°C$ will decrease to 0.006 % carbon corresponding to the composition of α_Q. (As a matter of fact, the carbon content of ferrite within the pearlite will also decrease from 0.025 % to 0.006 carbon. This means that the ferrite components of pearlite will change from the α_P to the α_Q phase in such a way that the precipitating cementite will increase the thickness of the cementite plates within the pearlite. It also means that this cementite does not appear as an independent phase within the pearlite.)

Accordingly, at room temperature, the alloy consists of *ferrite + tertiary cementite + pearlite*. The maximum quantity of tertiary cementite is so insignificant (0.28 %) that in practice, it may be neglected.

The next alloy to be analyzed is shown in Figure 8.3 with the composition line 2. This alloy has a carbon content of $C = 0.8$ %. The process of crystallization can be seen in Figure 8.5.

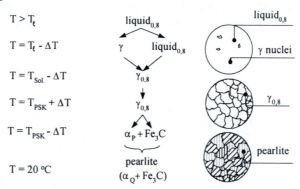

Figure 8.5
Crystallization process of an alloy with a carbon content of $C = 0.8$ % in the metastable alloy system

Solidification of this alloy starts with the crystallization of a γ-solid solution at crossing the liquidus curve. During further cooling, the γ-phase will continuously crystallize. The composition of the last crystal is equal to the composition of the alloy, that is, it contains exactly 0.8 % carbon. The alloy is further cooled in the homogeneous austenite zone from the temperature of the solidus curve until the temperature $T=T_{PSK}=723 \ ^oC$ is reached. Thus, in this temperature range, a homogeneous $\gamma_{0.8}$-solid solution can be found. At the temperature $T=T_{PSK}=723 \ ^oC$, the well known eutectoid reaction occurs according to the following reaction equation:

$$\gamma_S \rightleftharpoons \alpha_p + Fe_3C . \tag{8.15}$$

The result is a eutectoid microstructure called *pearlite* consisting of a mixture of $\alpha_p + Fe_3C$ phases. During further cooling down to room temperature, the carbon dissolving capability of the α-phase within the eutectoid decreases along the curve PQ. In accordance with the described process of crystallization, the alloy at room temperature consists of 100 % pearlite containing α_Q-ferrite and Fe_3C-cementite phases.

The next alloy to be analyzed is shown in Figure 8.3 with the composition line 3. This alloy has a carbon content of $C = 1.2$ %. The crystallization process is shown in Figure 8.6. Solidification of the alloy starts with the crystallization of a γ-solid solution when the liquidus curve is intersected. During further cooling, γ-solid solutions crystallize continuously from the liquid. At the end of the primary crystallization, the composition of the γ-solid solution is the same as that of the alloy, i.e. its carbon content is exactly 1.2 %. When the temperature is between the temperature of the solidus curve and the $T=T_{ES}$ curve, the alloy can be found in the homogeneous γ-zone. Therefore, in this zone, a homogeneous γ-solid solution can be found having a carbon content of $C = 1.2$ %. Since the ES curve indicates the carbon dissolving capability of the γ-solid solution, a high carbon content phase must precipitate when it is crossed.

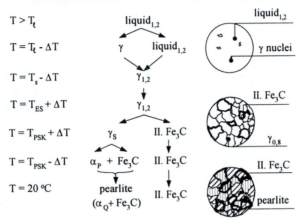

Figure 8.6
Crystallization process of an alloy having a carbon content of $C = 1.2$ % in the metastable system

This phase is the Fe_3C-cementite in the metastable system. Thus, secondary cementite (*II. Fe₃C*) precipitates continuously along the ES curve. Due to this segregation process, the carbon content of the γ-solid solution gradually decreases until it reaches the composition of the eutectoid ($C=0.8$ %) at the temperature $T=T_{PSK}=723$ °C. The well-known eutectoid reaction occurs at this temperature resulting in the formation of the eutectoid (i.e. pearlite) consisting of $a_P + Fe_3C$ phases. During further cooling to room temperature, the carbon dissolving capability of a_P-ferrite also decreases along the PQ-curve within the pearlite. No changes occur in the previously segregated secondary cementite. Accordingly, at room temperature the alloy consists of pearlite ($a_Q + Fe_3C$) and secondary cementite (*II. Fe₃C*).

For determination of the quantities of the constituents of the microstructure, the quantities of the constituents at the temperature $T=T_{PSK}-\Delta T$ should first be calculated. Applying the second lever rule, the percentage value of pearlite can be determined from the expression:

$$perlite \ \% = \frac{6.687 - 1.2}{6.687 - 0.8} \times 100 \ \% = 93.2 \ \% . \tag{8.16}$$

The quantity of the secondary cementite can be calculated from the equation:

$$II. \; Fe_3C \; \% = \frac{1.2 - 0.8}{6.687 - 0.8} \times 100 \; \% = 6.8 \; \% \; . \tag{8.17}$$

All the previously analyzed alloys belong to the group of steels. The alloy 4 $(C = 2.5 \%)$ belongs to the group of pig irons and possesses several properties differing significantly from those previously analyzed. The crystallization procedure of the alloy can be seen in the flow chart in Figure 8.7.

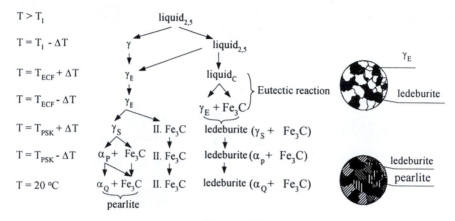

Figure 8.7
Crystallization process of an alloy having a carbon content of $C = 2.5 \%$ in the metastable system

Solidification of the alloy starts with the crystallization of a γ-solid solution when crossing the liquidus curve. During further cooling the γ-solid solution crystallizes continuously until the temperature $T=T_{ECF}=1147\,^{o}C$ is reached. At this temperature, the composition of the γ-solid solution corresponds to the composition indicated by point E $(C = 2.06 \%)$. The composition of the liquid corresponds to the composition of the eutectic point $(C = 4.3 \%)$. The liquid having the eutectic composition solidifies to a eutectic microstructure at constant temperature according to the following non-variant reaction equation:

$$liquid_E \rightleftharpoons \gamma_E + Fe_3C \; . \tag{8.18}$$

The eutectic has the composition corresponding to point E and consists of a γ-solid solution and Fe_3C-cementite phases. The microstructure name of the eutectic in the metastable system is *ledeburite*.

During further cooling, the carbon dissolving capability of the γ_E-solid solution having a composition corresponding to point E decreases along the *ES-curve.* Therefore, secondary cementite $(II. \; Fe_3C)$ having a high carbon content precipitates continuously. Due to this segregation process, the carbon content of the γ_E-solid solution decreases to the composition of γ_S corresponding to the eutectoid composition. (The same precipitation occurs within the ledeburite, i.e.

cementite segregates from the γ_E-phase within ledeburite. However, this cementite does not appear in the microstructure as an independent phase. It contributes only to the thickening of cementite plates within the ledeburite.) At the temperature $T=T_S$, the γ_S-solid solution corresponding to composition S decomposes according to the eutectoid reaction into pearlite consisting of the phases $a_P + Fe_3C$. The γ-phase of ledeburite also decomposes into $a_P + Fe_3C$ phases at this temperature; however, it does not form a new microstructure element: it is only a constituent of the ledeburite. This is shown in the flow diagram of crystallization, also by indicating the composition of the phases put in parentheses beside the name of ledeburite. In accordance with the previous description, at room temperature this alloy consists of pearlite, secondary cementite and ledeburite.

Since this alloy at room temperature consists of three microstructure elements, the quantities of these cannot be determined by the simple application of the second lever rule. In such cases, we must also consider those earlier phases from which the microstructures existing at room temperature originated. As can be seen in Figure 8.7, pearlite and secondary cementite originate from the primary γ_E-solid solution. It can also be seen in the Figure that $T=T_{ECF}=1147\,^\circ C$ is the temperature where γ_E and the liquid having the eutectic composition are in equilibrium. This is the liquid from which ledeburite is formed. Thus their quantities can be directly determined by the application of the second lever rule, as follows:

$$\gamma_E \% = \frac{4.3 - 2.5}{4.3 - 2.06} \times 100\,\% = 80.4\,\%,$$

$$ledeburite \% = \frac{2.5 - 2.06}{4.3 - 2.06} \times 100\,\% = 19.6\,\%. \tag{8.19}$$

According to the flow diagram of crystallization, it is obvious that the quantity of ledeburite does not change during further cooling, thus the value calculated with the aid of relationship (8.19) also gives the amount of ledeburite at room temperature. However, further considerations are required to determine the quantities of pearlite and secondary cementite. First, the maximum possible quantity of secondary cementite has to be determined. This can be done by analyzing the crystallization of the alloy having the composition of point E. (The carbon content of this alloy is equal to $C = 2.06\,\%$). The maximum quantity of secondary cementite can be found at the temperature $T=T_{PSK}=723\,^\circ C$ according to the following relationship:

$$(II.Fe_3C)_{max} = \frac{2.06 - 0.8}{6.687 - 0.8} \times 100\,\% = 21.3\,\%. \tag{8.20}$$

However, the alloy under examination contains only $80.4\,\%$ of γ_E-solid solution [see relationship (8.19)]; therefore, the actual value of secondary cementite can be calculated from the equation:

$$(II. \ Fe_3C)_{actual} = \gamma_E \times (II. \ Fe_3C)_{max} = 0.804 \times 21.3 \ \% = 17.2 \ \% . \qquad (8.21)$$

Since, originally, both pearlite and the secondary cementite are formed from the γ_E-solid solution, the quantity of pearlite can be calculated from the equation:

$$pearlite \ \% = \gamma_E \ \% - (II. \ Fe_3C)_{actual} \ \% = 80.4 - 17.2 = 63.2 \ \% \qquad (8.22)$$

The last alloy that we briefly analyze in the metastable system is denoted by the composition line 5 and has a carbon content of $C = 5.5 \ \%$. Solidification of this alloy starts with the crystallization of primary cementite ($I. \ Fe_3C$). As a result of this, the carbon content of the liquid gradually decreases. At the temperature $T=T_{ECF}=1147 \ ^oC$, the composition of the liquid corresponds exactly to the eutectic point ($C = 4.3 \ \%$).

The eutectic liquid solidifies into ledeburite with a non-variant reaction and accordingly, the alloy consists of ledeburite and primary cementite at room temperature. (Obviously, with the decrease in temperature the same changes occur within the ledeburite, as analyzed for alloy 4, but this does not change the fact that the alloy consists of ledeburite and primary cementite at room temperature.)

Following the analysis of the characteristic alloys, the microstructure diagram of the metastable iron–iron carbide (Fe-Fe_3C) alloy system, valid at room temperature has been drawn in Figure 8.8. This is performed according to the basic rules described for the analysis of ideal binary phase diagrams (Chapter 7).

In the microstructure diagram, – in accordance with the analysis of crystallization – the following microstructure elements can be found: ferrite, pearlite, ledeburite, as well as primary, secondary and tertiary cementite. (The maximum quantity of tertiary cementite is so insignificant that it is usually not indicated in the microstructure diagram.)

Another diagram is also drawn at room temperature ($T = 20 \ ^oC$) to illustrate the types and quantities of phases existing in the metastable system. It can be seen that there are only two different phases at room temperature in the alloy system, namely: ferrite and cementite.

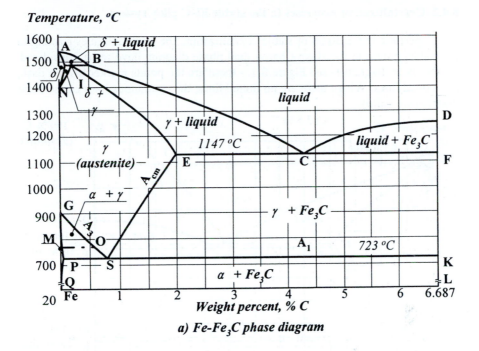

a) Fe-Fe₃C phase diagram

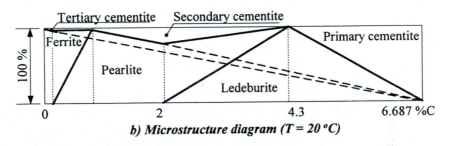

b) Microstructure diagram (T = 20 °C)

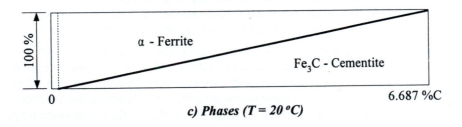

c) Phases (T = 20 °C)

Figure 8.8
Microstructure and phase diagram of the metastable *Fe-Fe₃C* alloy system

8.4.2. Crystallization processes in the stable Fe-C alloy system

To perform the analysis of stable crystallization, the point and lines relating to the stable system in the Heyn-Charpy dual phase diagram have to be considered. (In the basic diagram - see Figure 8.1 – these are the points marked with a dash, and dotted lines. Where there are no dotted lines, the continuous lines of the basic diagram are valid).

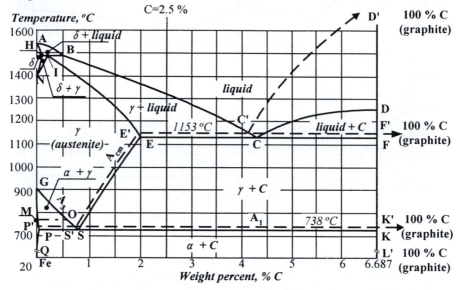

Figure 8.9

Sketch for the analysis of crystallization processes in the *Fe-C* stable system

In this case, the boundary line of the diagram on the right hand side can be found at 100 % carbon content, which corresponds to the composition of pure graphite. Therefore, to get a proportional diagram, the right hand side of the diagram is only indicated (see points *D'*, *F'*, *K'* and *L'* corresponding to 100 % of *C* content). This is justified by two reasons: firstly, drawing the complete diagram would make the important section of the diagram useless. Secondly, compositions having a carbon content higher than 6.687 % *C* are of no practical importance.

The analysis of stable crystallization is quite reasonable, since analyzing it, essential theoretical knowledge can be obtained that is indispensable when considering pig iron alloys having high carbon content. Among pig iron alloys, cast irons are of utmost importance in industrial practice. Therefore, the analysis of the stable crystallization is carried out only for one alloy in the composition range of *industrial cast irons*.

To perform the analysis, select an alloy having a carbon content of *C* = 2.5 %, which is equivalent to the one analyzed previously in the metastable system. This provides a good ground to compare the crystallization processes of the same alloy in the stable and the metastable systems. (The crystallization process is shown in

Figure 8.10. Note that pure graphite is marked by the letter C_{gr} in order to distinguish it from point C of the diagram.)

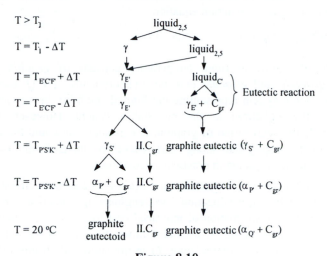

Figure 8.10
Crystallization process of an alloy having a carbon content of C = 2.5 %
in the stable system

Solidification of the alloy starts when the liquidus curve is intersected. The first solid crystal is a γ-solid solution in the stable system, too. During further cooling the γ-solid solution crystallizes continuously until the temperature $T=T_{E'C'F'}$ is reached. At this moment, the composition of the γ-solid solution corresponds to point E' ($C = 2.03$ %); the composition of the liquid corresponds to the composition of the eutectic ($C = 4.26$ %) denoted by point C'. At this temperature, we experience the first significant difference between the crystallization process in the stable and the metastable systems. The liquid having a eutectic composition solidifies into a eutectic consisting of γ-solid solution with a composition corresponding to point E' and C_{gr}-graphite at a constant temperature by a non-variant reaction according to the reaction equation:

$$liquid_{C'} \rightleftharpoons \gamma_{E'} + C_{gr} \tag{8.23}$$

The eutectic microstructure in the stable system is called *graphite-eutectic*.

During further cooling, the carbon dissolving capability of the $\gamma_{E'}$-solid solution corresponding to the composition represented by point E' decreases along the $E'S'$ curve; therefore, graphite precipitates continuously from the austenite. (Since this segregation occurs from the previously solidified γ-phase, this is called secondary graphite and is denoted by II. C_{gr}). Because of this secondary segregation, the carbon content of the $\gamma_{E'}$-solid solution changes to $\gamma_{S'}$ corresponding to the eutectoid composition. (The same changes occur within the graphite eutectic with the component $\gamma_{E'}$ but the graphite segregating from $\gamma_{E'}$ within the graphite eutectic does not appear as an independent phase.)

The $\gamma_{S'}$-solid solution having a composition of S' transforms to *graphite eutectoid* according to the reaction equation:

$$\gamma_{S'} \rightleftharpoons \alpha_{P'} + C_{gr}.$$ (8.24)

Accordingly, the alloy at room temperature consists of *graphite eutectoid*, secondary *graphite* and *graphite eutectic*. (Theoretically, a $P'Q'$ curve can be found in the stable system representing the limited solubility in the stable system similarly to the PQ-curve in the metastable system. However, below the temperature A_1 the segregation of graphite is extremely slow and the quantity of tertiary graphite segregated along this curve is insignificant ($\approx 0.015\,\%$). Therefore, it can be neglected in practical calculations.)

The previously described crystallization process occurs perfectly according to the rules of stable crystallization. In engineering practice, the primary crystallization and transformation of industrial cast irons (the so-called *gray cast irons*) above the temperature A_1 usually follows the rules of the stable system, while the transformation at temperature A_1 in the solid state follows the rules of the metastable system. Accordingly, the microstructure of these cast irons (in the case of optimal composition and appropriate cooling speed) includes pearlite + graphite at room temperature. Therefore, these cast irons are called *pearlite-graphite gray cast irons*. The word gray in the name refers to the fractured surface of the cast iron, which is gray due to the graphite. In contrast, the fractured surface of cast irons containing cementite – the so-called *white cast irons* – is characteristically white due to the cementite in the microstructure.

9. NON-EQUILIBRIUM CRYSTALLIZATION OF IRON-CARBON ALLOYS

9.1. Thermodynamic principles of $\gamma \rightarrow \alpha$ phase transformation of steels

When we analyzed the cooling curve of pure iron, we could see that iron possesses three allotropic modifications. Below the temperature $T = 911\ ^{\circ}C$ the so-called α_{Fe} exists: it has a body-centered cubic crystal lattice. In the temperature range of $T = 911\text{-}1392\ ^{\circ}C$ iron has a face-centered cubic structure called γ_{Fe}. Finally, between $T = 1392\text{ - }1536\ ^{\circ}C$, iron has again a body-centered cubic crystal lattice. To distinguish it from α-iron it is denoted by δ_{Fe} and called δ-iron.

From the analysis of the crystallization of iron-carbon alloys, it is also well known that three allotropic modifications of steels exist. All are interstitial solid solutions of iron and carbon. The so-called α-ferrite is the solid solution of iron and carbon crystallizing according to the body-centered cubic crystal lattice. This can be found below the temperature line A_1 (T = 723 $^{\circ}C$), as well as between the $A_1 - A_3$ temperature lines. The γ-solid solution (called *austenite*) is the solid solution of iron and carbon crystallizing according to the face-centered cubic system. Homogeneous austenite can be found only in the homogeneous γ-zone (above the temperature lines A_3 and A_{cm}). The δ-ferrite – which is also a solid solution of iron and carbon with a body-centered cubic crystal lattice – can be found only in steels having a carbon content of $C \leq 0,16\ \%$ between T = 1392 - 1536 $^{\circ}C$.

The allotropic modifications are the results of allotropic phase transformations. Among them, the $\alpha \rightarrow \gamma$ and $\gamma \rightarrow \alpha$ phase transformations are of outstanding significance, since these two allotropic transformations form the theoretical basis for most of the heat treatment processes of steels. It is also important to know the exact temperature values where these allotropic transformations occur. The theoretical background for their determination is based on the differences between the specific volumes of the allotropic modifications. It is known from crystallography that the atomic packing factor of the body-centered cubic crystal is equal to $T = 0.68$. The atomic packing factor of the face-centered cubic crystal lattice is $T = 0.74$. Consequently – despite the different lattice constants – the lattice of the γ-phase is highly packed resulting in a decrease in the specific volume during the $\alpha \rightarrow \gamma$ transformation. During the $\gamma \rightarrow \alpha$ phase transformation, an increase in the specific volume is experienced. The shrinkage and the expansion occurring during the phase transformation can be determined with a high precision dilatometer. The specific volume changes for a eutectoid steel measured during phase transformation are shown in Figure 9.1. From the diagram, it can be seen that the change in specific volume during the phase transformation is about 0.25 %.

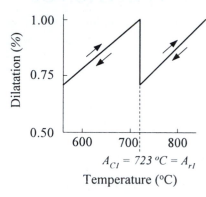

Figure 9.1

Change in specific volume of a eutectoid steel at temperature A_I

The $\gamma \rightarrow \alpha$ phase transformation starts with the formation of crystallization nuclei. Diffusion processes are very important in phase transformations. Accordingly, the three decisive parameters of the process are the *free energy difference (ΔG)*, the *nucleation ability (N)* and the *diffusion coefficient D*. All these parameters are dependent on temperature. Consequently, the time required for the phase transformation is also determined by the temperature dependence of these three parameters.

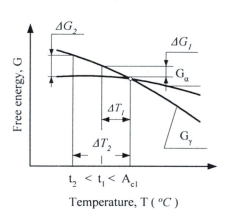

Figure 9.2

Free energy functions of α and γ phases of eutectoid steel at the critical temperature of transformation

Consider first the effect of the free energy differences on the time required to start the process. The driving force of phase transformations is the difference in free energies as in the case of phase transformations during crystallization or recrystallization. The free energies of these two phases taking part in the phase transformation (α-ferrite and γ-austenite) are shown in Figure 9.2. (For the sake of simplicity, the eutectoid steel is selected as an example to analyze the theoretical background of the $\alpha \rightarrow \gamma$ as well as the $\gamma \rightarrow \alpha$ phase transformations, however, the general consequences are valid for all kinds of steels.)

When the temperature is equal to the temperature of phase transformation ($T=T_{crit}=T_{A1}$), the free energies of the two phases are also equal. In this case, there is no driving force to initiate the process and an infinitely long time would be required for the phase transformation. As the temperature decreases, the difference between the free energies of the two phases continuously increases. Thus, increasing the degree of undercooling at the critical temperature – that is the temperature of transformation – an ever-shorter time is required to start the phase transformation. The effect of the difference between the free energies is in close correlation with the degree of undercooling, as shown in Chapter 4 when analyzing the rules of crystallization.

In the previous section, we analyzed the background of the initiation of phase transformation from the point of view of free energies. The other decisive factor is the crystallization aspect of the transformation, which has a significant effect on the microstructure as well. The recrystallization in the solid state (i.e. the phase transformation) also starts with the nucleation of crystallization centers and the mechanism of the growth of these nuclei also plays an important role in the phase transformation process.

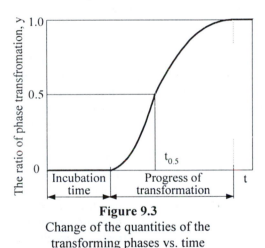

Figure 9.3
Change of the quantities of the transforming phases vs. time

The nucleation ability (N) (see Chapter 4 *"The rules of crystallization"*) is also in close correlation with the ΔT undercooling. If the value of undercooling $\Delta T \to 0$, the size of the critical nucleus follows $r_{crit} \to \infty$. Thus, the amount of time required to start phase transformation tends to $t \to \infty$. This is one of the reasons why recrystallization cannot start at the critical temperature of transformation, but only after a ΔT undercooling. However, even in this case, a certain time is required to start the transformation, as can be seen in Figure 9.3. The time required to start the transformation at a certain temperature is called the *incubation (latent) time* or *incubation period*.

In Figure 9.3, the second stage of transformation (i.e. the growth of the crystallization nuclei) is also shown. The vertical axis of the diagram represents the quantity of the transformed phase, while the horizontal axis represents the time required for the transformation. The quantity of the transformed phase can be expressed by the so-called *Avrami equation*:

$$y = 1 - e^{-kt^n},$$
(9.1)

where y represents the quantity of the transformed phase, k and n are time-dependent reaction constants. It is generally accepted to characterize the rate of transformation by the reciprocal value of the time elapsed when 50 % of the phase transformation is completed. Using the notations of Figure 9.3, it can be given as:

$$r = \frac{1}{t_{0,5}} \,. \tag{9.2}$$

The rate of transformation – as a diffusion process – can be expressed by the following *Arrhenius-type* equation:

$$r = A\,e^{-\frac{Q}{RT}} \tag{9.3}$$

where A is a reaction constant independent of time,
       Q is the activation energy of the reaction,
       R is the universal gas constant,
       T is the absolute temperature.

Processes that can be characterized by Equation (9.3) are called *thermally activated processes*.

It follows from the previously described analysis that the ability of nucleation has similar effects on the time required to start the phase transformation as the differences between the free energies of phases. This means that its effect on time can be also connected to the amount of undercooling. If the undercooling is $\Delta T \to 0$, the size of the critical nucleus is $r_{crit} \to \infty$. This means that the time required to start the transformation also tends to $t \to \infty$. Therefore, the curve illustrating the effect of the ability of nucleation asymptotically approaches the temperature $T = T_{crit}$ (see Figure 9.4). However, the greater the undercooling, the smaller the size of the critical nucleus. This would mean a shorter time required to start the transformation if only this effect were present.

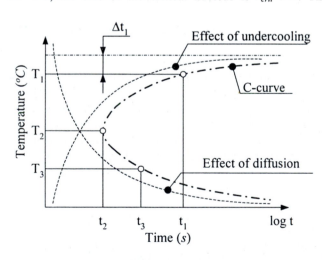

Figure 9.4
The amount of time required to start the $\gamma \to \alpha$ phase transformation vs. temperature

At the same time, the diffusion coefficient determining the rate of nucleation has an opposite effect. The value of the diffusion coefficient is higher at higher temperatures and it decreases exponentially with decreasing temperature. Consequently, at higher temperatures the time requirement of diffusion processes is smaller. As temperature decreases, the time required for diffusion processes increases exponentially. This effect is illustrated in Figure 9.4. (Naturally, this effect of diffusion is valid only in those processes where diffusion exists. This remark is very important from the point of view of diffusionless processes, like the martensitic transformation of steels to be discussed later.)

In reality, the time required to start a phase transformation is determined by the joint effect of these previously analyzed three factors. This is illustrated by the resultant curve drawn in Figure 9.4. Due to its characteristic shape, it is called a *C-curve*. In fact, the time required to complete the phase transformation is also determined by the same factors. This means that a similar *C*-shaped curve is obtained when analyzing the time requirement of completing the transformations.

These types of processes – occurring at a constant temperature (along an isotherm) – are called *isothermal transformations*. The diagrams illustrating these transformations are called *isothermal phase transformation diagrams*. The isothermal phase transformation diagrams provide useful assistance in the heat treatment of steels; therefore, we shall discuss in detail their experimental determination and the main types of these diagrams.

9.2. Non-equilibrium transformations of steels

9.2.1. Isothermal time-temperature transformation diagrams of steels – TTT diagrams

Since the time required for the phase transformation varies in a very wide interval, the scale of the time axis is logarithmic, i.e. the transformation diagrams are drawn in the T - $lg\ t$ coordinate system. This is the reason why these diagrams are also named *time-temperature transformation diagrams* (*TTT-diagrams*). (Keep in mind that near the critical transformation temperature, the curve asymptotically approaches the isotherm denoting the critical temperature of transformation. Thus, the time required for the phase transformation is $t \rightarrow \infty$. However, the shortest transformation time is in the order of 1-2 seconds.)

9.2.1.1. Experimental determination of isothermal transformation diagrams

The significance of isothermal transformations was recognized in the middle of the 19th century, but the theoretical bases of transformations were explored only in the beginning of the 20th century. In this work, Davenport and Bain played an outstanding role in the years 1928-1934.

The substance of the "*classical*" method elaborated for the experimental determination of isothermal phase transformation diagrams is summarized below. About 80-100 specimens are manufactured from the material to be examined. The size of the specimen should be small enough to provide the isothermal conditions. The large number of test specimens is justified by the fact that about 10-12 specimens are required for the accurate determination of the starting and end points of phase transformation at a given temperature. The tests should be carried out at a minimum of 8-10 different temperatures in order to determine the shape of the curve as precisely as possible. In Figure 9.5, the procedure and the results of an experiment are shown for unalloyed steel having an eutectoid composition. The specimens are heated in a tube furnace up to the temperature resulting in homogeneous austenite. Due to the small sizes of the specimens, a few minutes are sufficient to achieve the homogeneous austenite microstructure. When the homogeneous austenite condition is fulfilled, the specimens are dropped into salt baths heated to the desired transformation temperatures and are held at these constant temperatures for various periods. Then, in order to "*freeze in*" the occurred transformations, the specimens

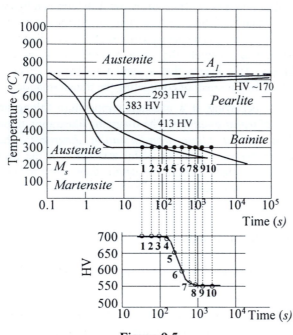

Figure 9.5
Experimental determination of isothermal transformation diagram

are quenched quickly, usually in water. After cooling, the hardness of the specimens is measured and the *hardness curve* is plotted as a function of the holding time at the examined temperature.

The hardness of those specimens in which the decomposition of austenite has not started is equivalent to the hardness of the hardened steel and shows a practically constant value (test specimens denoted by 1-3 in the diagram). If the decomposition of austenite has already started due to the longer holding time in the salt bath, a decreasing hardness can be experienced. The decrease in hardness is proportional to the degree of decomposition of austenite. After a certain holding time, the phase transformation of austenite is completed. Therefore, the hardness of these specimens corresponds to the hardness of the microstructure formed during the phase transformation from the austenite. The hardness of these (fully transformed) specimens has a constant value represented by a horizontal line in the diagram (points 8, 9 and 10). Having a sufficient number of test specimens, the holding time corresponding to the beginning and end of the transformation can be determined with sufficient accuracy. Projecting these points to the isotherm representing the temperature of examination, the starting and finishing points of the phase transformation can be determined.

By repeating this procedure at various temperatures, the isothermal phase transformation diagram can be determined with appropriate accuracy. Usually, simultaneously with the hardness tests, a microstructure analysis is also performed. It can provide further evidence about the progress of phase transformation.

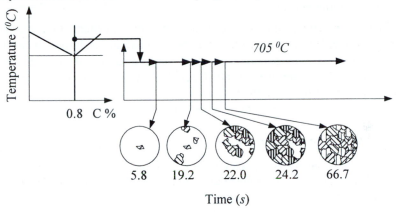

Figure 9.6
Examination of the microstructure to follow the isothermal phase transformations

The microstructural changes in a steel having an eutectoid composition are shown as a function of time for a selected isotherm (T = 705 °C) in Figure 9.6. It can be seen in the Figure that following the holding time of t = 5.8 s, the transformation of austenite to pearlite has already started, but due to the short holding time, only a small quantity of it has been transformed. As the holding time increases, the quantity of the transformed austenite increases as well. The total time required for the complete transformation at the temperature of examination is t = 66.7 s.

9.2.1.2. The types and mechanisms of $\gamma \rightarrow \alpha$ phase transformation in steels

Earlier, we could see that the parameters determining the transformation at various temperatures might result in very different times required for the transformation. The factors influencing phase transformation essentially depend on temperature. Therefore, not only the time required for the transformation, but also the types and mechanisms of the transformations differ largely depending on the deviation from the equilibrium temperature of transformation.

Consider again the phase transformation of a eutectoid steel. The transformation in the metastable system – under quasi-equilibrium conditions – occurs according to the already well known eutectoid reaction:

$$\gamma_S \rightleftharpoons \alpha_P + Fe_3C \ . \tag{9.4}$$

The result of this phase transformation is the so-called *lamellar pearlite*. During this reaction, from the homogeneous single phase austenite a two-phase, heterogeneous microstructure, pearlite is formed containing ferrite and cementite as constitutive phases. Thus, the pearlitic transformation may be considered as a decomposition process in which a heterogeneous two-phase microstructure is formed from a homogeneous, single phase. In this phase transformation, diffusion plays a very important role.

Under non-equilibrium cooling conditions, the transformation mechanism depends on the degree of deviation from the equilibrium conditions. In steels, the $\gamma \rightarrow \alpha$ phase transformation can occur according to three characteristic mechanisms. These are the so-called *pearlitic*, *bainitic*, or *martensitic* phase transformation mechanisms. In the following sections, the general mechanisms and the most important features of these three transformations will be analyzed.

Pearlitic transformation mechanism

The pearlitic mechanism is the typical form of the $\gamma \rightarrow \alpha$ phase transformation occurring at relatively high temperatures. The nucleation capability of iron carbide is high near the critical transformation temperature A_1; therefore, the transformation starts with the nucleation of iron carbide on the grain boundaries of austenite as illustrated in Figure 9.7. The nuclei of iron carbide grow faster longitudinally to form cementite plates from the carbon atoms diffusing from the austenite. The austenite, becoming also lamellar-shaped between the cementite plates, loses its carbon content due to the segregation of carbon atoms, and by a simple lattice rearrangement it transforms into the α-phase, that is ferrite. The pearlite nodes (pearlite islands) that are formed on the austenite grain boundaries consist of parallel plates of cementite and ferrite. The pearlite grains grow towards the center of the austenite until the transformation of the austenite is completed. In eutectoid steels, the phase transformation occurs according to this mechanism in the temperature range of $T = 550\ ^oC - A_1$. The cementite plates of pearlite are embedded in ferrite having the same orientation. Ferrite is a formable (ductile) phase. Since ferrite is the embedding phase (the so-called matrix material), this property is partially inherited by pearlite, although the ductility of pearlite is decreased significantly by the rigid

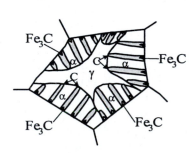

Figure 9.7
Schematic illustration of the mechanism of pearlitic transformation

cementite plates. Conversely, due to the high strength and hardness of cementite, pearlite has significantly higher strength and hardness than ferrite. The tensile strength and hardness of pure pearlite are nearly three times those of ferrite. Thus, pearlite may be considered as a microstructure of high strength with a certain degree of ductility. The process of phase transformation as a function of time is shown in Figure 9.8. In this figure, the changes in the quantities of the transformed phases are shown as a function of the time required for the transformation at a selected constant temperature ($T = 675\ ^{\circ}C$). This is in good agreement with the relevant section of the isothermal transformation diagram.

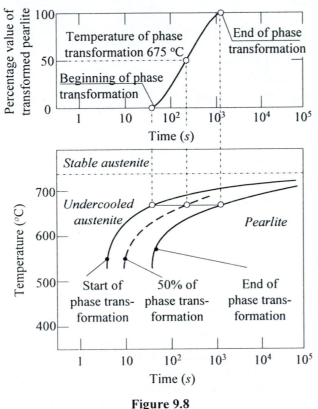

Figure 9.8

The quantities of transformed phases vs. time

From the diagram, it can be seen that the time required to start the transformation is very long close to the critical temperature (A_1) corresponding to the equilibrium transformation, and it decreases with decreasing temperature. It is also true that at temperatures close to A_1, due to the high diffusion coefficient of carbon, the cementite plates form relatively thick layers. During metastable transformation, the thickness of the ferrite/cementite layers can be characterized by a ratio of 8:1. This is called *coarse lamellar pearlite*. With the decrease in temperature, the diffusion coefficient decreases and the plates of pearlite, become thinner. This is referred to as *fine lamellar pearlite*. The coarse lamellar pearlite formed near the A_1 temperature gradually transforms into an ever-finer lamellar pearlite with the increase in cooling speed.

In Figure 9.9, the microstructures of coarse and fine plated pearlite are compared. The pictures were shot under an optical microscope. The lamellar structure of pearlite can be recognized on both photos, but the thickness and the length of the plates – which in turn means that the size of the pearlite islands formed from the austenite grains - are ever smaller. This means that the formed pearlite globally possesses an ever-finer microstructure.

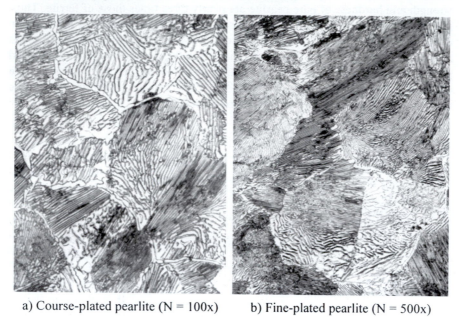

a) Course-plated pearlite (N = 100x)   b) Fine-plated pearlite (N = 500x)

Figure 9.9
Microstructure of pearlite (Etched with: 3 % solution of HNO_3)

Bainitic transformation mechanism

When an eutectoid steel is cooled down from the homogeneous austenite field to a temperature below $T = 550\ ^oC$ and held at a constant (isothermal) temperature, the phase transformation follows the so-called *bainitic mechanism*. In this transformation, certain features are as in the pearlitic mechanism, while others differ from the pearlitic one. The common features of the two different transformations are the following. In both cases, the homogeneous austenite phase transforms into a heterogeneous microstructure consisting of ferrite and cementite constituents. In both cases, the transformation can be regarded as a *decomposition process*. Furthermore, it is common that both transformations are *diffusion-type processes*. However, there are significant differences in the quality of the crystallization nuclei and the morphology of the microstructure. This can be explained essentially by the change in the ability of nucleation and the decrease in the diffusion coefficients with decreasing temperature.

The bainitic transformation may occur according to two different mechanisms depending on the temperature. In the upper temperature range of the bainitic transformation, the so-called *upper bainite* is formed. It starts with the nucleation of ferrite nuclei at the grain boundaries of austenite in the temperature range of

$T = 450$ - $550 \, ^{O}C$, as can be seen in Figure 9.10. Since ferrite is capable of dissolving significantly less carbon than the original austenite phase, the "*excess*" carbon is precipitated in the form of iron carbide along the ferrite nuclei. The nucleation ability of ferrite in this temperature range is much higher than the diffusion rate of carbon. Therefore, iron carbide is unable to grow to form cementite plates as it occurs in the pearlitic transformation. Iron carbide is located along the ferrite nuclei in the form of tiny particles. The small iron carbide particles are embedded in the ferrite matrix as the ferrite needles overgrow them.

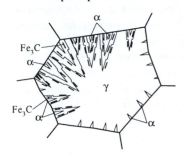

Figure 9.10
The mechanism of upper bainitic transformation

In part a) of Figure 9.11, an optical microscope image of isothermally transformed upper bainite microstructure can be seen. The isothermal transformation occurred at $T = 495$ ^{O}C. On the $N = 750$ times enlarged picture, the ferrite needles in the microstructure of upper bainite can be well distinguished.

The above-described mechanism of upper bainitic transformation is even better shown in part b) of Figure 9.11. This is a transmission electron microscopic image of an upper bainitic microstructure, with a magnification of $N = 10,000$-times.

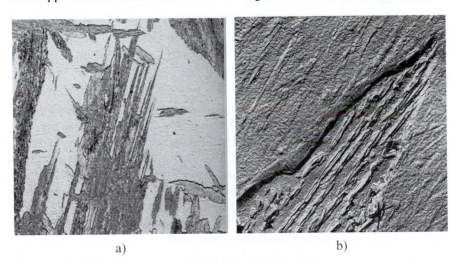

a)   b)

Figure 9.11
a) Optical microscopic image of upper bainite (N = 750x)
b) Electron microscopic image of upper bainite (N = 10,000x)

Figure 9.12

Optical microscopic image of lower bainite (N = 500x, etched with: Picral)

The lower bainite – concerning its final microstructure morphology – is very similar to upper bainite: it is also a two-phase heterogeneous decomposition product of the homogeneous austenite, containing tiny cementite particles embedded in ferrite needles. However, there is an essential difference in the kinetics of formation and, as a result, in the morphology of carbide. The ferrite crystals appearing in lower bainite are formed as martensite needles (see the martensitic transformation later). Diffusion may play a role in the tempering of martensite needles and in the formation of the so-called ε-carbide ($Fe_{2.4}C$ – which is named *Hägg's* carbide after its discoverer). Since this carbide is formed during the tempering of martensite, its shape differs from the disk-shaped carbide of upper bainite. It rather resembles the spherical carbide of spheroidite. In Figure 9.12, a lower bainitic microstructure is shown transformed at the isothermal temperature of $T = 320$ °C. As a consequence of its transformation mechanism, the lower bainite can be well distinguished from upper bainite, but in certain cases, it is difficult to distinguish it from martensite in an optical microscopic image.

Martensitic transformation mechanism

The third mechanism of the $\gamma \rightarrow \alpha$ phase transformation of steels is the so-called *martensitic* transformation, which essentially differs from the described mechanisms of pearlitic and bainitic transformations. We could see that the common feature of pearlitic and bainitic transformations is that both transformations are actually decomposition processes. In both cases, a heterogeneous microstructure (containing ferrite and cementite) is formed from the homogeneous austenite and in these transformations, the diffusion plays an outstanding role. In contrast to these, the martensitic transformation is a diffusionless process and the product of transformation is a single-phase martensite.

Since the major step of the transformation of austenite to martensite is the rearrangement of the γ-lattice into the α-lattice, let us analyze the possible interstitial locations of carbon atoms among the iron atoms in the various lattices. This is illustrated in Figure 9.13, showing all the possible interstitial points of the lattices of austenite, martensite and ferrite. According to Figure 9.13, the *C*-atoms in the γ-lattice are located in the body center and at the edge centers of the lattice. In reality, it can never happen that *C*-atoms are found at every possible point of the lattice, since this would mean a carbon content of approximately 17.7 %. It is well known that the maximum carbon dissolving capability of austenite is represented by point E, corresponding to a carbon content of $C = 2.06$ %! Even in this case (i.e. at a carbon content of $C = 2.06$ %), carbon atoms can be found only at every eighth possible interstitial location.

In the α-lattice, C-atoms may be located at the edge centers of the four parallel sides, as well as at the face centers of the faces perpendicular to the side edges. The possible interstitial locations of C-atoms are at the six apexes of a regular octahedron.

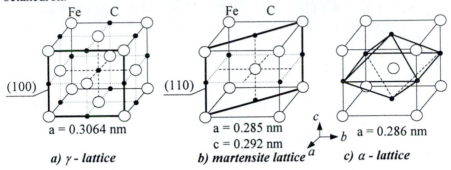

a) γ - lattice   b) martensite lattice   c) α - lattice

Figure 9.13
The possible interstitial locations of C-atoms in various lattices

If we consider only the geometric location of the atoms, we can see that the same type of atomic arrangement exists in the {100} planes of the γ-lattice as in the {110} planes of the α-lattice. However, the sizes of the lattices and, consequently, the distances between the atoms are different. Because of this, the so-called $\gamma \rightarrow \alpha$ lattice rearrangement can occur without changes in the location of atoms with the changes in their distances. If there is sufficient time available for diffusion, the "excess" carbon atoms dissolved in the γ-austenite can diffuse out as Fe_3C iron carbide, as we could see in the pearlitic and upper bainitic transformations. In these cases, the slight changes in lattice parameters necessary for the lattice rearrangement during the $\gamma \rightarrow \alpha$ phase transformation can be realized without significant distortion of the α-lattice.

The possible interstitial positions of C-atoms are the same in the martensite lattice as in the α-lattice. However, considering that the *martensitic transformation* occurs *without diffusion* with extremely rapid cooling, the carbon atoms solved in the γ-austenite can not precipitate, i.e. they remain in the transforming lattice, which results in an oversaturated solid solution. It also leads to significant deformation of the lattice. This deformation is greater in the direction of one of the axes than in the other two directions. This distorts the lattice of martensite to a tetragonal lattice, as illustrated in Figure 9.14. The great lattice deformation caused by the "*excess*" C-atoms in the martensite lattice results in the great hardness of the martensitic microstructure. This lattice distortion leads to high lattice stresses, providing the theoretical background for *martensitic hardening,* which results in the greatest hardness of steels.

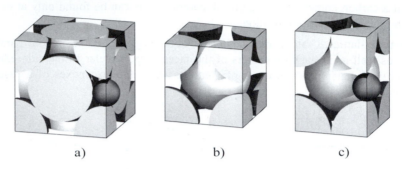

a)   b)   c)

Figure 9.14
Illustration of the lattice deformation effect of the $\gamma \rightarrow \alpha$ lattice rearrangement
a) The face-centered lattice of γ-austenite with interstitial C-atom
b) The body-centered cubic lattice of α-ferrite
c) The distorted tetragonal lattice of α'-martensite

The crystalline planes of martensite are not flat planes because of the interstitial C-atoms, therefore sliding of crystalline planes over each other can be achieved only under very high stresses. Therefore, martensite is an extremely hard and rigid microstructure provided its carbon content is high enough ($C > 0.6$ %). Martensite having a low carbon content ($C < 0.1$ %) is the so called "diluted" martensite. This is not as hard and rigid as those having a high carbon content.

Based on the previous discussion, we can state that martensite is a non-equilibrium phase. It can be considered as an interstitial, supersaturated solid solution of iron and carbon. The temperature at which the transformation of austenite to martensite starts is called the *martensite start* and is denoted by M_S. Similarly, the temperature at which the martensitic transformation finishes is called the *martensite finish* and is denoted by M_f. Since martensitic transformation is a non-equilibrium process, the temperature lines M_S and M_f are not parts of the *Fe-Fe$_3$C* equilibrium diagram. Both temperature lines change as a function of carbon content. As the carbon content increases, the temperature of the martensitic trans-formation decreases, as can be seen in Figure 9.15.

It is important to note that the temperature lines of martensitic transformation (the temperature lines M_S and M_f), opposite to the critical temperatures of diffusion-type transformations, have no hysteresis. This means that for a given steel, martensitic transformation always starts and finishes at the same temperatures if the cooling rate exceeds the critical value at which the martensitic transformation occurs. (The values of the M_S and M_f temperatures depend only on the carbon content for plain-carbon steels). When transformation starts with the martensitic mechanism, the quantity of martensite transformed depends only on the finishing temperature of the transformation and does not depend on the holding time in the temperature range bordered by the M_S and M_f temperatures.

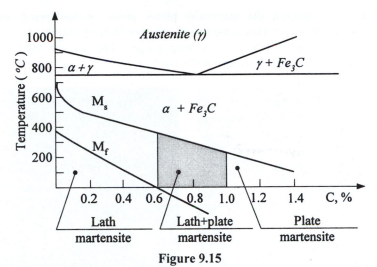

Figure 9.15
Effect of the carbon content on the martensite start (M_S) and martensite finish (M_f) temperatures, as well as on the morphology of martensitic transformation

A further significant feature of martensitic transformation is the fact that it is an irreversible process. The undercooled austenite always transforms to martensite at the same temperature once it is below the M_S line. When it is heated, it does not transform to austenite at the M_S temperature line, but with increasing temperature, it first decomposes into a two-component equilibrium phase called spheroidite, which will be analyzed later. Spheroidite starts to transform to austenite at a temperature A_{c1}, which is much higher than the temperature M_S.

The morphology of martensite transformed from austenite without diffusion varies with carbon content. If the carbon content of steel is lower than 0.6 %, the so called *lath martensite* is formed which consists of martensite needles having different but well-defined orientations, as can be seen in part a) of Figure 9.16.

When the carbon content is above $C = 1$ %, the so-called *plate martensite* is formed. A 500 times enlarged microscopic image of it is shown in part b) of Figure 9.16. In the beginning of the formation of plate martensite, parallel martensite plates form, which grow across the complete austenite grain.

Since martensitic transformation is a diffusionless process, the growth of martensite plates occurs within a fraction of a second at a speed close to that of sound. Between the martensite plates crossing the complete austenite grain, the transformation occurs in defined directions ($\alpha = 57^0$). Between these plates, some untransformed austenite always remains. It is called *residual austenite*.

In the range of carbon content $C = 0.6 - 1$ %, martensite is present with a mixed morphology: *lath* and *plate martensite* are equally present in the microstructure. In the lower third of the composition range, the dominant microstructure is the lath martensite, but as the carbon content increases, the plate martensite dominates. It follows from the transformation mechanism of plate martensite that between the martensite plates some untransformed, residual austenite always remains. This is the reason why the M_f temperature line is always below zero centigrade in this composition range. (The M_f line intersects the temperature axis at a carbon content of $C = 0.6$ %.)

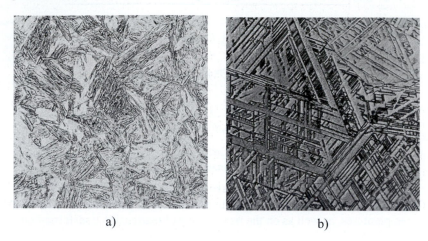

a)   b)

Figure 9.16
The effect of carbon content on the morphology of martensite
a) Lath martensite   b) Plate martensite

9.2.1.3. Types of isothermal time-temperature transformation diagrams

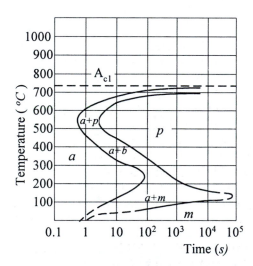

Figure 9.17
"S"-diagram for the isothermal transformation of eutectoid steel (Davenport and Bain, 1930)

It is worth mentioning as an interesting fact of scientific history that the first isothermal transformation diagram was elaborated by Davenport and Bain in 1930 for eutectoid steel. It was named "S"-diagram after its characteristic shape (Figure 9.17).

The isothermal transformation diagram of eutectoid steel applied nowadays is shown in Figure 9.18 indicating the characteristic transformations, too. Concerning the isothermal transformation diagrams, the following has to be noted: the shapes and positions of the diagrams depend mainly on the composition of the steels and on the temperature of austenitization. Therefore, the composition and the temperature of austenitization always have to be shown in the diagrams. It is also important to note that isothermal transformation diagrams may be applied to study the transformations only under isothermal conditions. Additionally, the specimen examined has to be cooled down from the temperature of austenitization to the desired temperature without any transformation until it reaches the temperature of analysis.

In the diagram of Figure 9.18, the *"C"-curves"*, representing the starting and finishing temperatures of pearlitic and bainitic transformations, as well as the lines corresponding to the martensitic transformations of various degrees are indicated. On the right hand side of the diagram, the temperature ranges with the various transformation mechanisms (i.e. the pearlitic, bainitic and martensitic) are also indicated. It can be seen that between the so-called "nose-point" of the C-curve, corresponding to the shortest incubation time, and the temperature $T = 450\ ^\circ C$, the phase transformation starts with the bainitic mode (upper bainite) and finishes with the pearlitic mechanism. This range in the diagram is indicated by the curve drawn as a dotted line between points B_S and B_f.

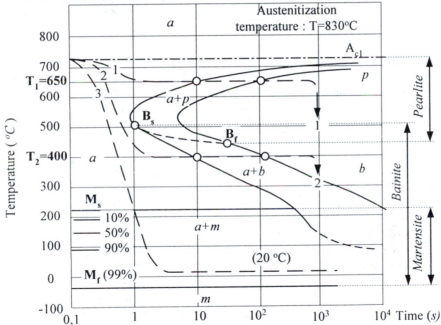

Figure 9.18
The TTT-diagram of eutectoid steel

As already mentioned, the *transformations* in the isothermal diagrams may be studied *only along isotherms*. From these diagrams, we can determine the time between the start and finish of the transformation, as well as what microstructure can be achieved. For example, along the isotherm temperature $T_1 = 650\ ^\circ C$, the transformation starts after 10 s according to the pearlitic mechanism. Before the transformation starts, the austenite is in the so-called *undercooled austenite* state. The complete transformation at this temperature requires 100 s. After completing the phase transformation, the alloy can be further cooled at any rate because the total quantity of austenite has already been transformed. Accordingly, the alloy consists of 100 % pearlite with a lamellar microstructure at room temperature.

The analysis of the transformation can be performed in a similar manner along the isotherm corresponding to the temperature $T_2 = 400$ °C. According to the diagram, the transformation starts after 10 s, too. However, at this temperature the transformation occurs according to the bainitic mechanism and finishes following a holding time of about 110 s. The microstructure of the alloy in this case consists of 100 % bainite at room temperature. It is valid for any transformation above the M_S temperature, and after the phase transformation has been completed, we can find only pearlite or bainite depending on the temperature of the transformation, but neither residual austenite nor martensite can be found in the microstructure.

In contrast, the transformations occurring below the M_S temperature are determined by the finishing temperature of cooling. This is indicated by curve 3 in the diagram. In this case, the alloy is cooled down to room temperature so fast ($T = 20$ °C) that prior to this no transformations could commence. As the degree of martensitic transformation is determined only by the finishing temperature of the transformation, at this temperature about 98% of austenite is transformed to martensite. This follows from the fact that the M_f temperature line of eutectoid steel can be found below zero according to Figure 9.18. Therefore, in the eutectoid steel after the transformation at room temperature there is always some untransformed, so-called *residual austenite*, as well. It is important to note that the residual austenite transforms to bainite after a long period, in some cases after years. This is explained by the fact that the elongation of the C curve section, indicating the bainitic transformation, is intersected below the M_S temperature (see the dotted line of the diagram). Since this transformation occurs with an increase in specific volume, this may cause problems in certain applications (for example, in the case of gauges or high precision tools). Therefore, in such cases, to avoid the presence of residual austenite, the martensitic transformation is performed below room temperature (at a negative temperature selected in correspondence with the M_f temperature according to Figure 9.18) by "*undercooling*".

The *isothermal transformation diagram of hypoeutectoid steels* (Figure 9.19) differs essentially from that of the eutectoid steel. In hypoeutectoid steels, prior to pearlitic transformation, we can find a curve representing the so-called *pro-eutectoid ferrite segregation*. This is in full agreement with the equilibrium diagram since in equilibrium cooling of hypoeutectoid steels, the eutectoid transformation occurring at temperature A_1 is preceded by the precipitation of pro-eutectoid ferrite between the temperatures $A_3 \rightarrow A_1$. Nevertheless, this can be observed in Figure 9.19 indicating the relationship between the equilibrium diagram and the isothermal phase transformation diagram. An infinitely slow cooling rate would correspond to equilibrium cooling. The curves of the isothermal transformation diagrams asymptotically approach the isotherms representing the critical temperatures of transformation in the equilibrium diagram. (The curve of proeutectoid ferrite segregation approaches the isotherm A_3 and the curve representing the pearlitic transformation approaches the A_1 temperature line).

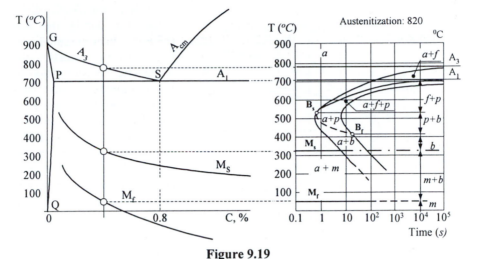

Figure 9.19
The isothermal transformation diagram of hypoeutectoid steel and its relationship to the equilibrium diagram

It can also be seen from the diagram that the curve representing the segregation of proeutectoid ferrite starts from the nose point of the curve representing the start of the pearlitic transformation. Consequently, in isothermal transformation the pearlitic transformation is always preceded by the proeutectoid segregation of ferrite, although its quantity decreases with temperature. Accordingly, the micro-structure of hypoeutectoid steel transformed at a temperature above the nose point always consists of a pearlite + ferrite microstructure at room temperature. At the same time, it is obvious from the diagram that in the range of bainitic transforma-tion and martensitic transformation there are no significant differences concerning the essential processes of transformation compared to the eutectoid steels.

The isothermal transformation diagrams of hypereutectoid steels are similar to the diagrams of hypoeutectoid steels. One of the main differences is that the eutectoid transformation is preceded by the *proeutectoid segregation of cementite* starting at the A_{cm} temperature (see Figure 9.20). This is also in perfect agreement with the equilibrium diagrams, since in hypereutectoid steels the eutectoid transformation in equilibrium cooling is also preceded by the segregation of proeutectoid cementite (secondary cementite) in the temperature range of $A_{cm} \rightarrow A_1$. The curve of proeutectoid cementite segregation also starts from the nose point of the starting curve of pearlitic transformation and approaches the A_{cm} temperature asymptotically. Consequently, the pearlitic transformation in hyper-eutectoid steels is always preceded by the proeutectoid segregation of cementite. The quantity of precipitated cementite continuously decreases with temperature. At room temperature the microstructure of hypereutectoid steel, transformed along the isotherm corresponding to the temperature above the nose point, consists of proeutectoid cementite and pearlite.

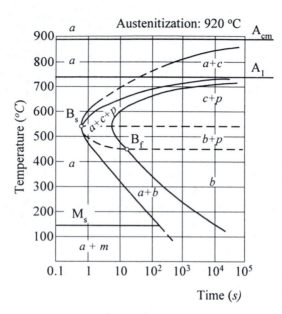

Figure 9.20
Isothermal transformation diagram of hyper-eutectoid steel having a carbon content of C = 1.2 %

It is a remarkable difference that during the isothermal transformation of hypereutectoid steels, the martensitic transformation is never complete at room temperature. Residual austenite always remains, since in these steels the M_f temperature line can always be found below 0 °C. It is a common feature of all the isothermal transformation diagrams that between the temperature corresponding to point B_f indicating the finish of bainitic transformation, and M_S temperature line, the transformation always occurs completely according to the bainitic mechanism. It is also important to note that a *purely bainitic transformation can be achieved only with an isothermal transformation.*

9.2.2. Continuous cooling time-temperature transformation diagrams of steels – CCT diagrams

The isothermal transformation diagrams are important theoretical and practical aids of heat treatments. However, under industrial circumstances, phase transformations achieved by continuous cooling are much more widely applied. In order to understand the conceptual basis of transformation diagrams valid for continuous cooling, it is necessary to overview the effects of the rate of continuous cooling on the transformation and microstructure of steels.

9.2.2.1. The effect of a continuous cooling rate on the transformation of steels

Before analyzing the effect of the cooling rate, we shall briefly summarize the main steps of the $\gamma \rightarrow \alpha$ phase transformation of steels. This can be split into four elementary processes for both the hypo- and the hypereutectoid steels. These steps are conceptually common but they differ somewhat in the realization. Therefore, it is useful to analyze them separately.

The $\gamma \rightarrow \alpha$ phase transformation in hypoeutectoid steels under equilibrium conditions can be theoretically split into the following four processes:

1. At the temperature A_3, the segregation of proeutectoid ferrite starts along the grain boundaries of austenite. Since ferrite is capable of dissolving significantly less carbon than austenite, carbon rediffuses from the ferrite into the austenite. Therefore, the carbon content of austenite continuously increases until the temperature A_1 is reached. At this temperature, the carbon content of austenite is enriched to $C = 0.8$ % corresponding to the eutectoid composition.

2. According to the eutectoid reaction, – from the austenite having a composition enriched to eutectoid – the nucleation of iron carbide starts with the pearlitic mechanism. Due to the continuous diffusion of carbon atoms from the transforming austenite, growth of the iron carbide nuclei to cementite plates occurs.

3. The γ-lattice of austenite, becoming poorer in carbon – due to the nucleation of iron carbide – transforms to the α-lattice of ferrite by a simple lattice rearrangement. This stage of the transformation is diffusionless and occurs even at very high cooling rates. Therefore, the time required for this is insignificant – especially in comparison to diffusion processes.

4. The shape of the cementite plate is not an equilibrium one. Therefore, when sufficient time is available, the cementite plates near the temperature A_1 transform to spheres representing the least surface energy (that is, cementite becomes "spherical"). The lamellar pearlite becomes "granular" corresponding better to the equilibrium state. It contains small spheres of cementite embedded in a ferrite matrix. (The granular pearlite possesses much better toughness properties than the lamellar pearlite. In particular, its resistance against dynamic effects is more favorable.)

Conceptually, the transformation of hypereutectoid steels includes the same four elementary steps with the differences arising from the differences in composition. These steps are described below:

1. At the temperature A_{cm}, the precipitation of *proeutectoid cementite* starts along the grain boundaries of austenite. This results in a decrease in the carbon content of austenite to a value of $C = 0.8$ % corresponding to the eutectoid composition just at the temperature A_1.

Steps 2, 3 and 4 practically occur in the same way as described for hypoeutectic steels. Thus, at first the eutectoid steel decomposes into lamellar pearlite by the pearlitic mechanism obviously including the $\gamma \rightarrow \alpha$ lattice rearrangement described in Step 3. Then lamellar pearlite is transformed into granular pearlite if there is sufficient time for it.

In both hypoeutectoid and hypereutectoid steels, among the elementary steps listed, 1, 2. and 4 are diffusion processes and step 3 (the $\gamma \rightarrow \alpha$ lattice re-arrangement) is diffusionless. Therefore, when increasing the cooling rate, first the fourth step does not occur, i.e. the pearlite is not transformed into granular pearlite. Upon further increasing the cooling rate, hysteresis of the critical transformation temperatures can be observed. This means that the transformations occur at ever-lower temperatures, as illustrated in Figure 9.21.

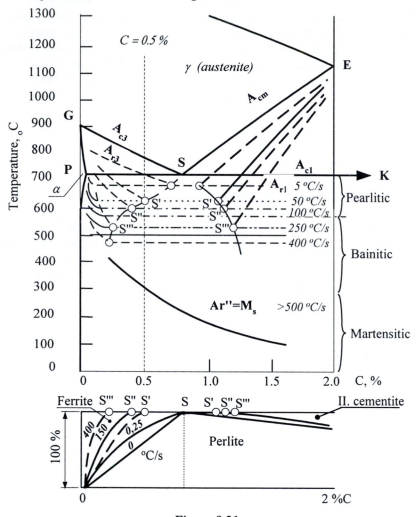

Figure 9.21
The effect of cooling rate on the γ → α transformation and on the critical transformation temperatures of plain carbon steels

With increasing cooling rate, the time available for the precipitation of pro-eutectoid phases (that is the proeutectoid ferrite in hypoeutectoid steels, as well as the proeutectoid cementite in hypereutectoid steel) decreases. Because of this, the quantities of the proeutectoid phases also decrease. An additional consequence is that the quantity of austenite transforming by the pearlitic mechanism also increases. It appears on the equilibrium diagram as if the eutectoid composition would widen to an increasing composition range instead of the defined C = 0.8 % composition corresponding to the equilibrium transformation.

By increasing the cooling rate, changes also occur in the eutectoid transformation. The so-called coarse lamellar pearlite is formed at cooling rates close to the equilibrium value. By increasing the cooling rate, the time available for the diffusion of carbides also decreases. Therefore, the plates of lamellar pearlite become progressively thinner, resulting in a finer pearlitic microstructure. Above a certain cooling rate, the lamellar structure of pearlite can be recognized only with a significant increase of magnification of the microscopic image.

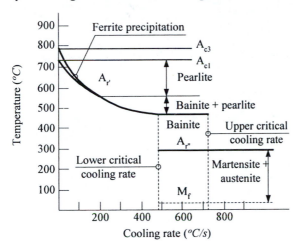

Increasing further the cooling rate, the transformation starts at ever-lower temperatures and besides pearlite, bainite also appears in the microstructure. If the cooling rate is increased above the value of v = 700 °C/s only the $\gamma \rightarrow \alpha$ lattice rearrangement remains from the main steps of $\gamma \rightarrow \alpha$ phase transformation, i.e. the austenite will be transformed to martensite. These changes can be followed in Figure

Figure 9.22

Effect of cooling rate on the microstructure of hypoeutectoid steel

9.22 illustrating the various transformation temperatures and the microstructural elements of a hypoeutectoid steel having a carbon content of C = 0.5 %.

Besides the quality (type) of transformation products, the cooling rate also has a great influence on the size of the products formed. It is known that the grain size of the products of austenitic transformation depends on the initial grain size of the austenite. The grain size of austenite is determined by the temperature of austenitization. The grain size of austenite is the smallest close to the transformation temperature. For eutectoid steel, this is the temperature, which at the same time is the temperature of the smallest homogeneous austenite grain. In hypoeutectoid steels – as discussed in the previous chapter – above the A_1 temperature, austenite is transformed directly from pearlite but besides this, there are the larger grains of ferrite. The hypoeutectoid steels have the smallest homogeneous austenite grain size just above the A_3 temperature. In hypereutectoid steels, homogeneous austenite can be found only above the A_{cm} temperature. With increasing carbon content, it represents such a high temperature that the danger of coarsening the grain size of austenite increases. The grain size of austenite above the A_1 and A_{cm} temperatures, upon heating in the homogeneous γ-zone, is continuously increasing. The largest grain size of austenite can be seen just below the solidus curve. Obviously, the grain size is also influenced by the holding time, as is well known in materials science. The grain size changes of austenite are illustrated in Figure 9.23.

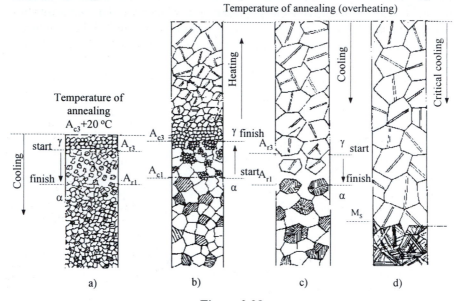

Figure 9.23
The change of the grain size of austenite
a) Cooling from the temperature A_3+20 ^{0}C, b) Heating to a too high temperature,
c)   Slow cooling from the temperature of overheating,
d)   Fast cooling from the temperature of overheating (v > $v_{upper\ critical}$)

It is an important feature of austenite that the grain size obtained upon heating remains unchanged during cooling to the lower limiting temperature of the γ-zone (i.e. to the A_3 or A_{cm} temperature), or up to the M_s temperature of martensitic transformation during diffusionless transformation (see Figure 9.23).

Thus, the products of the austenite transformation of hypoeutectoid steels will be the finest when the steel is heated just above the A_3 temperature (usually to a temperature of A_3+30 ^{0}C). The hypereutectoid steels are heated maximum to point G (900 °C) above the A_{cm} temperature, because when heating steels with higher carbon contents above the A_{cm} temperature, the microstructure becomes rather coarse. Therefore, during hardening, steels with a carbon content higher than 1.2 % are heated only to the temperature A_1+30 ^{0}C.

The grain sizes of the products of austenite transformation also change with respect to the initial grain size of austenite. Thus, the grain size increases with the temperature of austenitization, but the grain size of the products is always smaller than that of the original austenite. In contrast, the grain size of the transformation product formed without diffusion, that is the grain size of martensite, can reach the size of the initial austenite. It means that from coarse grain sized austenite, coarse grain sized martensite is formed (see Section d. of Figure 9.23). Therefore, it is important to prevent the coarsening of the austenite grain size during heating.

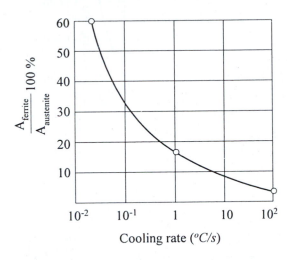

Figure 9.24
The effect of cooling rate on the grain size of ferrite with respect to the grain size of austenite

In steels having a carbon content in the range of C < 0.2 %, we can find mainly ferrite; thus, in these steels the grain size of ferrite is of primary interest. The grain size of ferrite transformed from the austenite is illustrated as the percentage of the initial grain size of austenite in Figure 9.24. The figure shows that the grain size of ferrite is about half the grain size of austenite with slow cooling and it is less than one tenth of the grain size of austenite during rapid cooling.

The effect of cooling rate on the pearlitic transformation can be observed when the transformation of eutectoid steel is analyzed. Increasing the cooling rate also increases the undercooling of austenite and thus the temperature of decomposition decreases. Increasing the degree of undercooling increases the number of cementite nuclei formed in unit time; this means that the formation of pearlite isles starts at several places on the boundary of the austenite crystal. On increasing the cooling rate, the time available for the diffusion of carbon atoms decreases, and thus the distance of diffusion is also decreased. In this way, the thickness of the cementite plates and, at the same time, the thickness of the embedding ferrite matrix decreases. However, the hardness of pearlite also increases as the number of cementite plates increases. The greater the number of cementite plates, the higher the resistance to plastic deformation, leading to an increase in the yield limit and the tensile strength.

These phenomena are shown for eutectoid steel in Figure 9.25. In this figure, the thickness of the pearlite plates, the size of the pearlite isles and the change in the hardness of the pearlite are indicated as a function of cooling rate.

From this illustration, it can be seen that increasing the cooling rate from 10^{-2} °C/s to 10^2 °C/s decreases the thickness of the pearlite plates (which means the total thickness of a cementite and ferrite plate) to 1/6th of the initial value. The size of the pearlite isles also decreases to 1/6th. At the same time, the hardness of pearlite is doubled.

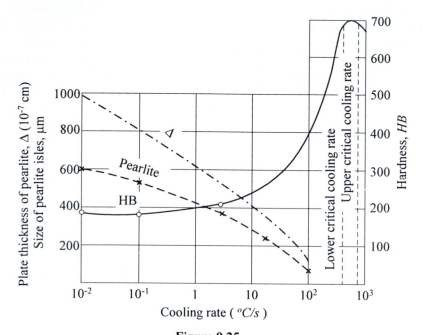

Figure 9.25
The effect of the cooling rate on the microstructure and hardness of pearlite in eutectoid steels

In bainitic and martensitic microstructures, it is more adequate to speak about the increase in dispersity of the products of phase transformation when the cooling rate is increased above the value of 100 °C/s. Together with the increased dispersion, a further significant increase in hardness can be observed. Increasing the cooling rate from 100 °C/s to 500 °C/s increases the hardness of eutectoid steel from $HB = 400$ to $HB = 700$.

Previously, we could see that during continuous cooling austenite transforms into the microstructural elements as during isothermal transformations. However, while the transformation products of the isothermal transformations depend only on the temperature of transformation, in continuous cooling, the quality and number of the transformation products are determined by the cooling rate. A further significant difference is that in the case of continuous cooling the transformations occur at somewhat lower temperatures and after a longer incubation period. This can also be characterized by the fact that the transformation curves are shifted downwards (towards lower temperatures) and to the right (towards a longer incubation period) compared to isothermal transformation. This is illustrated in Figure 9.26, where the lines indicating the pearlitic transformation valid for continuous cooling are plotted over the isothermal transformation diagram of eutectoid steel represented by dotted lines.

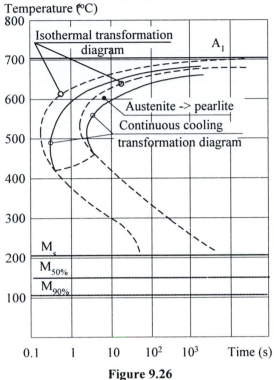

Figure 9.26
Comparison of transformation diagrams valid for the isothermal and continuous cooling

The transformation diagrams valid for continuous cooling are determined by the examination of small specimens cooled at different cooling rates. The specimens are transformed into a homogeneous austenite phase (by a short holding time at a temperature 20-30 °C higher than the temperature A_{c3} for hypoeutectoid steels and at a temperature 20-30 °C higher than the temperature A_{cm}, for hypereutectoid steels). Then the specimens are cooled in a cooling medium having different cooling intensities (calm air, blown air, oil, water) at varying cooling rates continuously to room temperature. Meanwhile, the critical transformation temperatures are recorded. In order to observe the critical transformation temperatures, the *differential dilatometric measurement technique* is usually applied to determine the change in specific volume during phase transformations.

On the polished surface of the cooled specimens, microscopic examinations are also carried out to quantitatively analyze the transformed microstructural elements. The transformation diagrams relating to continuous cooling, similarly to the isothermal transformation diagrams, are drawn in a logarithmic time-temperature coordinate system. On the vertical axis, the temperature values and on the horizontal axis the logarithmic time values are measured.

9.2.2.2. The types of transformation diagrams relating to continuous cooling

In Figure 9.27, a continuous cooling transformation diagram (often regarded as a CCT diagram) can be seen for hypoeutectoid steel with a carbon content of $C = 0.45$ %. In CCT diagrams, the temperature of austenitization and the critical transformation temperatures (A_{c1}, A_{c3}, M_S) have to be indicated. The diagram includes several characteristic cooling curves, representing different cooling rates achieved with cooling media of different cooling intensities.

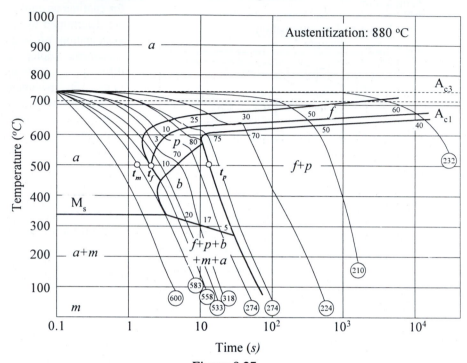

Figure 9.27
Continuous cooling transformation diagram valid for hypoeutectoid steel with a carbon content of C = 0.45 %

The percentage values of transformed microstructural elements are indicated at the intersections of the lines representing the finish of the individual transformations (proeutectoid ferrite, pearlite, and bainite) along the cooling curves. The hardness of the final microstructure is shown in circles at the end of the cooling curves. The three-digit hardness values indicate Vickers (HV), the two-digit hardness values mean Rockwell hardness (HRC). It is important to note that *the analysis of continuous cooling transformation diagrams may be performed only along the cooling curves representing continuous cooling!*

By analyzing the transformation curves of the diagram, we can say that the curve representing the precipitation of proeutectoid ferrite asymptotically approaches the A_{c3} temperature at long transformation times (representing near equilibrium cooling conditions). Those representing the start and finish of pearlitic transformation asymptotically approach the A_{c1} temperature. The bainitic transformation zone appears only above a certain cooling rate. This zone is connected to the field of pearlitic transformation from above and the starting line of martensitic transformation borders it from below. It obviously follows that, contrary to the isothermal transformation, a microstructure containing 100 %

bainite cannot be produced with continuous cooling. In addition to bainite, martensite always arises and, depending on the cooling rate, pearlite or pro-eutectoid ferrite can be found in the microstructure.

It is worth mentioning that the curves of pearlitic and bainitic transformations occurring with the decomposition of austenite are shifted to the right and downwards compared to the isothermal transformation diagram. Conversely, the curves representing martensitic transformation depend only on the composition and can be found at the same temperatures in both diagrams. However, if the transformation of austenite to proeutectoid ferrite, pearlite or bainite has already started prior to the martensitic transformation, the line representing martensitic transformation runs at a temperature proportionally decreasing with the amount of preliminarily decomposed austenite (see the straight section sloping down below the bainite zone). This can be explained by the fact that due to the transformations preceding the martensitic transformation, the composition of austenite changes. Its carbon content gradually increases because of the precipitation of proeutectoid ferrite. The martensitic transformation of austenite having a higher carbon content occurs at a proportionally decreasing temperature in accordance with Figure 9.27.

In the phase transformation diagram (Figure 9.27), some cooling curves are also drawn being of outstanding importance for the technological planning of heat treatments. The cooling rates corresponding to the curves that touch the bainite zone from the left and the right side, as well as the curve bordering the zone of proeutectoid ferrite, are of utmost importance. Cooling rates are characterized by their mean values. It is generally accepted that the cooling rate is calculated by using the elapsed time during the cooling from the temperature of austenitization down to the isotherm temperature $T = 500\ ^{o}C$ using the relationship:

$$v = \frac{T_{aust} - 500}{t_{500}} \tag{9.5}$$

where

v — the mean value of cooling rate,

T_{aust} — the temperature of austenitization,

t_{500} — the time elapsed to reach the temperature $T = 500\ ^{o}C$.

The t_p, t_f and t_m time values, shown at the isotherm $T=500^{o}C$ in the diagram, belong to three critical cooling rates, respectively.

The cooling curve touching the bainite zone from the right represents the so-called *lower critical cooling rate*. It means that above this cooling rate, at any small increment, martensite appears in the microstructure of the steel. The lower critical cooling rate can be calculated from a relationship similar to Equation (9.5):

$$v_{crit}^{lower} = \frac{T_{aust} - 500}{t_p}. \tag{9.6}$$

The so-called upper critical cooling rate, touching the bainite zone from the left can be defined in a similar manner. The *upper critical cooling rate* refers to the cooling rate above which by any increment only martensite is formed during continuous cooling. The upper critical cooling rate can be calculated from a relationship similar to Equation (9.5):

$$v_{crit}^{upper} = \frac{T_{aust} - 500}{t_m} . \qquad (9.7)$$

The diagram also indicates the time t_f at the intersection of the cooling curve touching the zone of proeutectoid ferrite from the left and the isotherm at temperature $T = 500\,^{\circ}C$. Using this t_f value, that critical cooling rate can be calculated at which the formation of proeutectoid ferrite can be avoided, namely:

$$v_{crit}^{ferrite} = \frac{T_{aust} - 500}{t_f} . \qquad (9.8)$$

This is of great practical importance in those cases where steel structure is subjected to a cyclic load. It is well known that the fatigue strength of ferrite is rather low. (This cooling rate is usually applied in patenting of cable wires of various elevators, lifts and conveyors in order to avoid the formation of pro-eutectoid ferrite.)

For the practical application of continuous cooling transformation diagrams, consider the fifth cooling curve from the left in Figure 9.27. (As emphasized earlier, the transformation diagrams valid for continuous cooling should be studied only along the cooling curves!)

When cooling the steel continuously from the temperature of austenitization (from the temperature $T = 880\,^{\circ}C$ for the steel studied), the cooling curve intersects the curve representing the proeutectoid precipitation of ferrite at the temperature $T \approx 615\,^{\circ}C$. At this moment (about 1.5 s after the beginning of cooling) precipitation of proeutectoid ferrite from austenite starts. At the temperature $T \approx 560\,^{\circ}C$, when the cooling curve intersects the curve representing the start of the pearlitic transformation, the decomposition of austenite to pearlite starts. The quantity of proeutectoid ferrite is 3 % according to the value written beside the cooling curve in the figure. As cooling proceeds, when the boundary curve of bainite zone is crossed (at this moment the quantity of pearlite is 70 %), the transformation of austenite continues according to the bainitic mechanism. When reaching the M_S temperature line, representing the start of martensitic transformation, the quantity of transformed bainite is 17 %. Between the M_S and M_f temperature lines, the total remaining austenite is transformed to martensite. Accordingly, at room temperature the microstructure of the steel contains 3 % proeutectoid ferrite, 70 % pearlite, 17 % bainite and 10 % martensite. The average hardness of the steel – according to the figure written at the end of the cooling curve – is $HV_{10} = 318$.

Figure 9.28 shows the transformation diagram for the continuous cooling of a hypereutectoid steel having a carbon content of $C = 1\,\%$. Comparing this with the diagram of the hypoeutectoid steel shown in Figure 9.27, the following differences can be observed. For hypereutectoid steels, the proeutectoid phase is cementite instead of ferrite. During slow cooling, the transformation line representing the segregation of cementite asymptotically approaches the A_{cm} temperature line. The path of the martensitic start line M_S, bordering the bainite field from below, also differs. Contrary to hypoeutectoid steels, where the constant temperature line M_S representing the pure martensitic transformation slopes downward underneath the bainite zone, for hypereutectoid steels, this tendency is opposite. This is due to the decrease in the carbon content of austenite caused by the segregation of pro-eutectoid cementite. This leads to an increased martensite start temperature with an increasing quantity of decomposed austenite before the martensitic transformation.

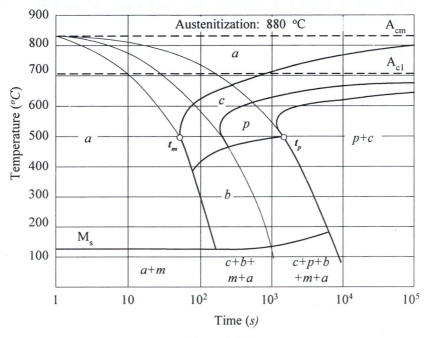

Figure 9.28
Continuous cooling transformation diagram valid for hypereutectoid steel having a carbon content of C = 1 %

A further important difference is that in hypereutectoid steels some residual austenite is always present in the microstructure at room temperature. This is because in this composition range the M_f temperature line can be found at negative temperatures. In all other respects, the rules formulated for the continuous cooling diagrams of hypoeutectoid steels are also valid for the continuous cooling transformation diagrams of hypereutectoid steels.

9.3. The metallographic background of hardening and tempering steels

9.3.1. The term hardened steel

Hardening is regarded as one of the most important heat treatment processes for steels. The primary aim of hardening is to provide high hardness and high strength. During the analysis of the various methods of $\gamma \rightarrow \alpha$ phase transformations of steels, we noticed that the highest hardness could be achieved when the greatest possible amount of austenite was transformed to martensite. The martensitic transformation may be achieved by a cooling rate higher than the upper critical rate. However, besides its hardness, martensite is rigid, its specific elongation and dynamic impact resistance are practically zero. Apart from this, a fully martensitic microstructure – especially in the case of plain carbon steels – can be provided only in small cross sections because of the extremely high value of the cooling rate ($v_{crit}^{upper} = 500 - 700^{\circ}C/s$).

The cross section to be hardened completely can be characterized by the diameter of a bar that contains fully martensitic microstructure throughout. Because of the finite value of convection and thermal conductivity, the cooling rate is often sufficient only for the bainitic and perhaps fine-grained pearlitic transformations in the middle of a greater cross section. Therefore, in engineering practice, the term *hardened steel* means those steels that contain more than 50 % of martensite. In this case, it is the core of bar cooled at a rate falling into the range between the upper and lower critical cooling rates.

9.3.2. The hardness of martensite and hardened steels

In Figure 9.29, the change in the hardness of steels is shown as a function of carbon content. The curve drawn with dotted lines represents the change in the hardness of mild steel for the sake of comparison. In the mild state, the hardness increases linearly with the increase in carbon content. The hardness changes from $HV = 60$ (the hardness of pure ferrite which is practically free of carbon) to about $HV = 250$ for the eutectoid steel having a carbon content of $C = 0.8$ %. The increase in hardness continues in the hypereutectoid range as well, though the degree of increase (the slope of the hardness curve) is smaller than that in the hypoeutectoid steels.

The hardness of martensite is the result of the lattice distortion due to the effect of "*excess*" carbon atoms in the ferrite lattice, as noted during the analysis of the theoretical basis of $\gamma \rightarrow \alpha$ phase transformations. Since the carbon dissolving capability of ferrite at room temperature is equal to $C = 0.006$ %, the hardness of martensite quickly increases even at low carbon contents. The hardness of "pure" martensite exceeds the value of $HV = 400$ at a carbon content of $C = 0.1$ % and it reaches its maximum value of $HV \approx 900$ at around a carbon content $C = 0.6$ %.

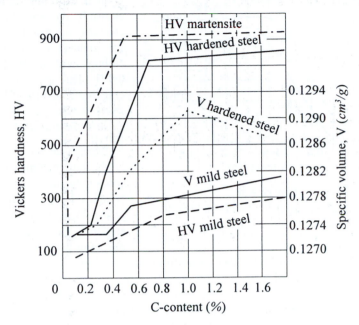

Figure 9.29
The hardness of steels in various states vs. carbon content

The change in the hardness of hardened steel (containing 50 % martensite) shows a character somewhat different from that of the fully martensitic state. The hardness of hardened steels is significantly less than that of pure martensite. The obvious reason for this is that, besides 50 % martensite, hardened steels contain other, relatively softer microstructural elements like bainite, pearlite and, in some cases even proeutectoid ferrite.

Comparing the curves indicating the changes in the hardness of mild and hardened steels, we can see that hardness increases linearly in the whole range of composition for mild steels – although the slope of the increase is smaller in the hypereutectoid range. The hardness of hardened steels increases quickly only up to a carbon content of C = 0.7 %; above this value the further increase is insignificant. This can be explained by the change in the specific volume of steel as a function of carbon content.

In Figure 9.29, the change in the specific volumes (V) of mild and hardened steels is also shown. From these curves, it is evident that the change in hardness follows the change in the specific volume. The correlation between the change in hardness and that of the specific volume can be shown more clearly if the differences between the values of the hardness and specific volume are plotted as a function of carbon content. This is illustrated in Figure 9.30.

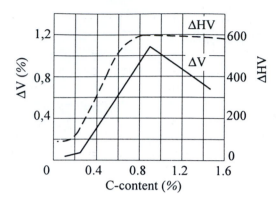

Figure 9.30

The changes in specific volume and hardness
as a function of carbon content

It is especially striking, as shown by the diagram, that the changes in specific volumes are greatest in the range of carbon content of $C = 0.8 - 0.9\,\%$. This explains the high cracking tendency of hardened steels having a composition close to the eutectoid one.

Martensite and consequently, the hardened microstructure containing martensite cannot be considered an equilibrium phase. Therefore, it has a great tendency to change towards the equilibrium state. However, it is well known that any phase transformation requires a driving force. Thus, these changes also need a driving force and activation energy. This is not sufficient at room temperature; therefore, the transformation producing a microstructure closer to equilibrium cannot occur at room temperature.

In the non-equilibrium martensite, besides the lattice stresses providing the required hardness, there are also undesired stresses, caused by the extremely fast cooling. Under extreme circumstances, these undesirable stresses may cause cracking or fracture of the hardened material. To avoid the cracking of hardened steels, a decrease in the undesired stresses can be achieved by a heat treatment called *tempering* following the hardening of steels. This decreases lattice stress and the gradual change of the microstructure towards the equilibrium can take place. The magnitude of these changes depends essentially on the temperature and on the holding time. These effects and the consequent changes are discussed in the forthcoming subsection.

9.3.3. Tempering of hardened steel

Tempering means a heat treatment that follows hardening. During tempering, the lattice stresses generated by hardening are decreased and the microstructure changes towards the equilibrium state. The changes depend essentially on the temperature of tempering, therefore, in the forthcoming paragraphs the effect of the tempering temperature will be analyzed in detail.

The effect of the tempering temperature on the change of the hardness of steels can be studied in Figure 9.31 for steels with various carbon contents. The diagram shows the hardness changes of steels. The curves belong to four different carbon contents as a function of the tempering temperature for a 1-hour tempering time. (It is important to indicate the holding time, since in heat treatments involving diffusion processes, the increase in time at a certain temperature acts as if the temperature had been increased.)

For plain carbon steels, the diagram can be divided into three sections where characteristic changes occur according to the following.

In stage I of tempering (T < 150 °C), a decrease in the lattice distortion of the tetragonal martensite lattice (called α'-lattice) occurs. This lattice distortion is due to the "excess" carbon atoms frozen in the lattice during the γ → α phase transformation. In this stage of tempering, carbon atoms precipitate in the form of ε-*carbide* having the composition $Fe_{2.4}C$ ($Fe_{12}C_5$). As the result of the segregation of ε-carbide having a hexagonal lattice, the carbon content of martensite decreases to *C=0.1-0.3 %* and at the same time the degree of lattice deformation also decreases. Consequently, the hardness of the steel also decreases. This decrease in hardness is insignificant and is compensated by the hardness of the dispersely segregated ε-carbide. Due to the segregation of ε-carbide, the distorted tetragonal lattice of martensite called α'-phase transforms to a regular martensite lattice denoted by α".

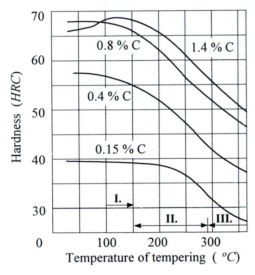

Figure 9.31
The effect of tempering temperature on the change of hardness of steels for various carbon contents

In stage II of tempering (*T=150-280* °C), the residual austenite transforms to bainite (provided there was residual austenite in the steel). Since bainite has a much higher hardness than austenite, this transformation leads to an increased hardness.

This increase in hardness can be experienced only in the steel having a carbon content of *C* = 1.4 % among those shown in Figure 9.31. This is in full agreement with what was learnt earlier, because martensitic transformation of this steel finishes well below 0 °C. When cooling has finished at room temperature, a significant amount of residual austenite remains in the steel. The transformation of this residual austenite contributes to the increase in hardness seen in the diagram. The increase in hardness originating from the transformation of a small quantity of residual austenite is compensated by the fact, that in this range, the decrease in the carbon content of martensite (with the continuous segregation of ε-carbide) has continued. Towards the end of this temperature range, the gradual transformation of ε-carbide to *Fe₃C* iron carbide starts.

In stage III of tempering (*T=280-A_{c1}*), the replacement of ε-carbide by *Fe₃C*-iron carbide further continues. The iron carbide segregated in this way has initially a disk-like shape. As temperature increases, the thickness of the cementite disks grows and they continuously assume spherical shape corresponding better to the equilibrium state.

The martensite in this stage has already lost all its excess carbon content and, by loosing the lattice distortions, it transforms into ferrite having a regular α-lattice. Accordingly, the martensitic microstructure of hardened steel changes to mild ferrite with cementite spheres embedded. This microstructure – showing the best resemblance to the equilibrium state – is called *spheroidite*. In Figure 9.32, the spheroidite microstructure of steel is shown after annealing for 6 hours at the temperature $T = 705\ ^{o}C$-on. The embedding microstructure is ferrite in which the spherical cementite originating from the decomposition of martensite is clearly seen in the form of cementite spheres darkened by the etching agent.

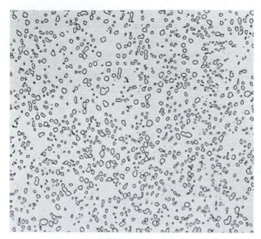

Figure 9.32
Tempered spheroidite microstructure of hardened steel, N = 1000x,
Etching agent: Picral

Although in this section we are dealing with the tempering of unalloyed steels, it is worth mentioning here that there is a fourth step of tempering occurring only in certain alloyed steels. The first three steps of tempering occurs in the same way in both alloyed and unalloyed steels with the difference that the alloying elements may hinder the diffusion processes, thus increasing the tempering endurance of martensite. (It means that the hardness of martensite decreases to a lesser extent, and it keeps its hardness even at higher temperatures).

Further, the most significant difference can be observed when the alloyed steel contains strong carbide-forming alloying elements. The alloyed carbides are more stable than the iron carbide. These carbide-forming elements partially or completely substitute iron in the iron carbide during high temperature tempering ($T > 500\ ^{o}C$) and form carbides of higher hardness and greater stability (for example Cr_2C_3, VC, WC, TiC). These carbides are called secondary carbides because they are formed in the solid state after the formation of iron carbide.

As the result of the formation of secondary carbides, a significant increase in hardness can be observed on the hardness curve corresponding to the formation temperatures of these secondary carbides. This secondary hardness, in certain cases, may be greater than the primary (martensitic) hardness. A further important feature of this hardness is that in alloyed steels it is preserved up to the formation temperatures of the secondary carbides (i.e. up to temperatures $T > 500\ ^{o}C$). This property ensures a wide range of application possibilities of these steels, for example, as high speed steels in tool making where high hardness and wear resistance are important requirements even at elevated temperatures.

Earlier we analyzed the effect of the tempering temperature on the hardness of hardened steels at constant holding times. Holding time affects the processes occurring during tempering in a similar manner as the temperature. Figure 9.33 illustrates the change in the hardness of unalloyed steel (carbon content $C = 0.35$ %) as a function of tempering (holding) time.

It is obvious from the diagram, that in the heat treatment of steels, temperature and holding time are parameters substituting each other within certain limitations. This means that the same degree of hardness can be achieved at a lower temperature with a longer holding time as at a higher temperature with a shorter tempering time.

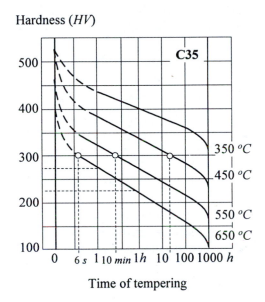

Figure 9.33
The change in hardness of hardened steel vs. tempering time

The simultaneous effects of temperature and time are usually illustrated in a single diagram. For this purpose introduce the following parameter:

$$p = T\left(\lg t + k\right).\tag{9.9}$$

In the Equation (9.9), p is the so-called temperature-time joint parameter, k is a material constant (its value for steels can be taken as $k = 18$). In Figure 9.34, the change in the hardness of hardened steel is shown as a function of the joint p-parameter.

This diagram may be used as a guide to heat treatments as follows. When the required hardness as the result of tempering is known, the value of the joint parameter p corresponding to the expected hardness is determined on the basis of the diagram. Following this, the required tempering time is calculated according to relationship (9.9) or by fixing the value of temperature or vice versa.

Obviously, this diagram can also be used to determine the expected hardness after tempering by selecting the appropriate corresponding value pairs of temperature and tempering time.

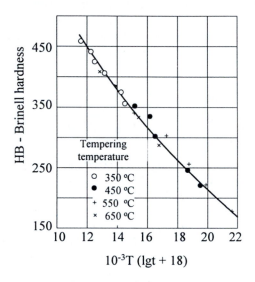

Figure 9.34
Change in hardness of hardened steel as a function of the joint
temperature-time parameter

10. PLAIN CARBON STEELS

10.1. The terminology of plain carbon steel

The unalloyed steels in engineering practice are simply called plain carbon steels. These are not pure *Fe-C* alloys. Certain elements (e.g. *Si, Mn, S* and *P)* are unavoidably included in the alloy during the production of iron and steel. If the content of these elements does not exceed certain limits, they are not considered as alloying elements. The permitted values of these elements are as follows: Si=0.35-0.5 %, Mn=0.5-0.8 %, S=0.01-0.035 %, P=0.01-0.035 %. Besides these elements, steels may contain other elements in very small quantities. The base material of steel production is pig iron, usually containing 3-6 % of carbon. The "excess" carbon is burnt out by oxidization during steel-making. However, in this process, some of the iron oxidizes to iron oxide (*Fe0*) which is dissolved in the liquid steel. It affects the strength parameters of steel disadvantageously; therefore, *Fe0* is deoxidized by *FeMn* and *FeSi*. The excess *Mn* and *Si* remain in the steel as alloying element. *Cr* and *Cu* in steels originate from the scrap iron, similarly *S* and *P* also arise from the pig iron and scrap iron. Thus, in fact, the "unalloyed" steel in everyday practice is an alloyed steel with several components, but the elements entering the steel unintentionally are not taken into consideration below the limits given above. This is quite reasonable, since the listed elements in quantities below the permissible limits do not significantly affect the essential properties of the *Fe-C* two-component system. Therefore, unalloyed steels can be regarded as two-component *Fe-C* alloys with a carbon content of *C*=0-2.06 %. The solidification (primary crystallization) of steels always occurs according to the metastable *Fe-Fe₃C* phase diagram.

Carbon can exist in steels in two various forms: as an interstitial solid solution (α, and/or δ-ferrite, as well as γ-austenite) or a metallic compound (*Fe₃C*, iron carbide). Carbon in steels always exists in bound form. The presence of *C* in steel as free *C* (graphite) always leads to heavy defects in the steel, reducing its strength properties and resulting in so-called black brittleness.

10.2. Mechanical properties of plain carbon steels

The metastable equilibrium condition of steels is best approached by the so-called annealed state. In the annealed state, there is a strong correlation between the microstructure of the steel and its mechanical properties. This is illustrated in Figure 10.1. In part a) of the figure, the Fe-Fe_3C phase diagram referring to steels is shown, whilst in part b) of the figure, the microstructures of three characteristic steels are shown. In part c), the so-called microstructure diagram, whereas in part d), the changes in some mechanical properties are shown.

If we compare parts c) and d) of Figure 10.1, we can see that the curve representing the tensile strength (R_m) follows the boundary line of the pearlite zone of the microstructure diagram. The tensile strength linearly increases from $R_m = 300\ MPa$ (from the tensile strength of the practically pure ferritic mild steel) to $R_m = 900\ MPa$ (i.e. to the tensile strength of the eutectoid steel consisting of 100 % pearlite). The tensile strength of hypoeutectoid steels containing ferrite and pearlite is linearly proportional to the pearlite content. On the one hand, the relatively high tensile strength of pearlite originates from the high tensile strength of iron carbide (which forms about $1/8^{th}$ of the pearlite). On the other hand, it follows from the fact that the cementite and ferrite plates of lamellar pearlite are formed from the same original microstructure of austenite. Thus, the strong atomic bonding forces are inherited from the austenite keeping the same orientation in the decomposed phases as in the initial "mother microstructure".

The tensile strength of hypereutectoid steels decreases linearly in accordance with the pearlite content. In this composition range, the atomic bonding forces existing among the pearlite grains are reduced proportionally to the thickness of the secondary cementite network segregated along the grain boundaries of austenite. At the grain boundaries, randomly ordered Fe atoms together with C atoms form the Fe_3C compound. The orientations of the ferrite plates forming $7/8^{th}$ of the pearlite grains are different in adjacent grains. Therefore, the bonding forces of Fe atoms located in the cementite network along the grain boundaries are significantly weaker than those of the Fe atoms within the cementite plates. This is the reason for the decrease in tensile strength experienced in hypereutectoid C-steels.

In part d) of Figure 10.1, the change in Brinell hardness is also shown. It can be observed in the diagram that the hardness of hypoeutectoid steels also increases linearly with the pearlite content. Its value is equivalent to about one third of the tensile strength. At the same time, the hardness of hypereutectoid steels further increases above the 100 % pearlite content, too. In this composition range, the hardness increases practically linearly, although to a lower degree as in hypoeutectoid steels in accordance with the increasing quantity of cementite plates.

The yield stress ($R_{p0.2}$) changes in a similar manner to that described for the hardness. The reason for it is that the cementite plates of the pearlite grain obstruct the sliding required for plastic deformation similarly to the secondary cementite networks. The specific elongation (A) characterizing the ductility and formability of steel and the specific reduction in area (denoted by Z) decrease with increasing

C-content. (For example, the specific elongation of the pearlite is less than one fifth and the specific reduction in area is about one fourth compared to ferrite.)

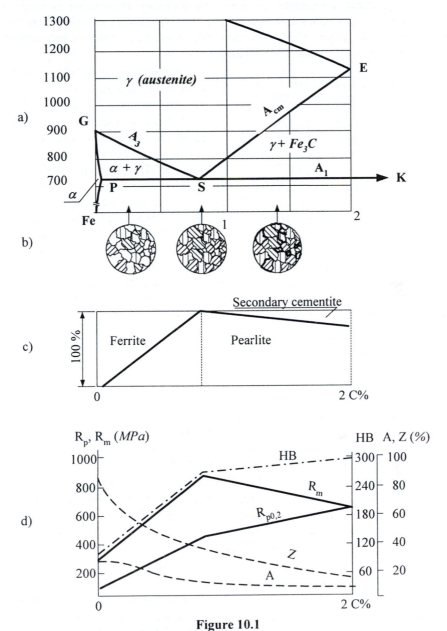

Figure 10.1

The change in mechanical properties of plain carbon steels as a function of carbon content and microstructure

Besides the previously described effects on mechanical properties, the carbon content also significantly influences the transition temperature, which is important in many practical applications of steel. The transition temperature is used in materials testing to characterize the transition from the ductile to the rigid state and is denoted by the letters *TTKV*. (The rigid state means that fracture occurs without significant residual – plastic – deformation.)

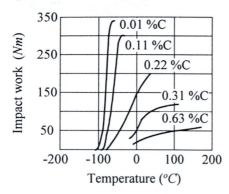

Figure 10.2
The effect of temperature
on the impact work of mild steels
with various *C*-contents

This is particularly dangerous to equipment operating at outdoor-ambient temperatures, and structures subjected to dynamic loads. The resistance against dynamic loads is characterized by the toughness of the material. This is determined by the work exerted during the impact test of a standardized specimen under well-defined conditions. Based on this test, the transient temperature can be determined by various methods. The most objective method is to determine the transient temperature according to a prescribed value of the impact work necessary for fracture. A generally accepted value of impact work for determination of the transient temperature is equal to $KV = 30$ *J*. In Figure 10.2, the change in the impact work is given as a function of temperature for mild steels with various *C*-contents. The transient temperature can be determined corresponding to the inflection point of the curves shown in Figure 10.2.

It can be seen from the diagram that, as carbon content increases, the impact work (the energy required to break the specimen) decreases. On the other hand, it is shifted towards increasing temperatures with increasing carbon content. Both trends are disadvantageous as regards embrittlement.

Steel may be subjected to higher temperatures during processing as well as during normal operation. Therefore, it is important to know the effect of temperature on the mechanical properties. As an example, in Figure 10.3, changes in the properties of a mild steel are shown (the carbon content is about 0.2 %). The tensile strength of this steel at room temperature is $R_m = 450$ MPa.

Several essential conclusions can be drawn from this diagram. At room temperature, representing the normal operating conditions for most structures, besides the tensile strength of $R_m = 450$ MPa, the material tested has a specific elongation of about 25 %. This means a relatively good ductility. As the temperature decreases (at negative temperatures) the strength properties (R_m, $R_{p0.2}$) increase and, at the same time, a sudden fall in ductility properties (*A*, *Z*) can be observed. Besides the impact work, embrittlement can be also characterized by the increase in strength and the decrease in ductility parameters.

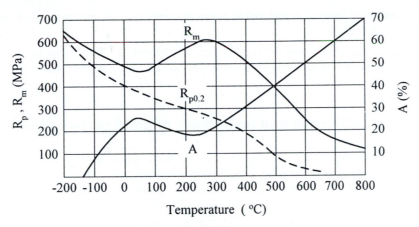

Figure 10.3
The effect of temperature on the mechanical properties of a mild steel

Embrittlement can be considered particularly dangerous at decreasing temperatures, since the increase in the yield strength is more intense than the increase in the tensile strength. Consequently, a drastic decrease in plastic deformation occurs.

When analyzing the effect of temperature on changes in the mechanical properties, it can be stated that at increasing temperatures a decrease in strength and an increase in ductility occur. However, a somewhat deviant behavior from this general trend can be observed in the temperature range of 50-250 $^{\circ}$C. In this range, a local minimum of strength and at the same time a local maximum of ductility parameters can be experienced at around the temperature $T=50$ $^{\circ}$C. Above this temperature, the strength parameters increase and the ductility parameters decrease. There is another local extreme in the temperature range $T=200$-250 $^{\circ}$C. It represents a local maximum of strength and a local minimum of ductility. These changes can also be considered as a kind of embrittlement but this is less dangerous than the previously discussed complete embrittlement typically occurring at negative temperatures. Since at this temperature, steel possesses a characteristic blue color, this embrittlement is called the "*blue brittleness*" of steel.

The blue brittleness of steel is due to the interstitial elements of small atomic diameters (carbon and nitrogen atoms). Obviously, the interstitially dissolved atoms can only be included in the lattice of dissolving metal at the expense of crystal lattice distortion, since the diameters of the atoms are greater than the diameter of the sphere fitting into the gaps of the α-lattice. Therefore, the interstitially located atoms tend to find "more spacious" locations in the lattice. The neighborhood of dislocations can be considered ideal locations for interstitial atoms. Interstitial atoms included in the expanded zones of the dislocations cause smaller lattice distortions. Therefore, these are in more stable positions than those located in dislocation-free interstices of the lattice. When these atoms occupy the expanded zones of dislocations, they obstruct their free movement. Therefore, an extra amount of energy is required to move dislocations. This is experienced as an increase in strength.

Further deviations from the general trend of decreasing strength and increasing ductility with increasing temperature may be observed if allotropic transformations occur in the metal.

10.3. Classification and main types of plain carbon steels

In Chapter 8, one kind of classification of steels was described, based on the equilibrium diagram and taking into consideration only the composition of steels. A more general classification of steels is possible according to their properties and consequently according to their application fields. (Obviously, the classification of steels by their fields of application depends substantially on their compositions as well). Usually, the application fields of the product and the technological process of production (welding, heat treatment, plastic deformation, machining) raise strongly interwoven requirements which are well illustrated by the forthcoming grouping of steels.

Steels can be divided into two large groups based on their fields of application: these are *structural and tool steels*. The *structural steels* fall mainly in the range of the hypoeutectoid composition. Their carbon content is typically in the range of $C = 0.006 - 0.6$ %. These steels – in harmony with their name – are widely used as structural materials in various machines and machine components. The structural steels can be further divided into several subgroups.

An appropriate carbon content ($C > 0.2$ %) is the basic condition for hardening. Therefore, steels containing less carbon than 0.2 % can be hardened only by a special heat treatment process called *case hardening*. (The essence of case hardening can be summarized as follows. The carbon content of a thin surface layer of steel is increased by diffusion occurring in the solid state to a level necessary for hardening. Then a usual hardening and tempering heat treatment is performed. With this procedure, a very hard surface layer and a tough core are obtained, since the core having a low carbon content is not hardened.) Accordingly, steels having a carbon content of $C < 0.2$ % are called *case-hardening steels* and are utilized in areas where the operating stress requires high surface hardness (good resistance against wear), sufficient strength and essentially good toughness in the core.

Another large subgroup of steels is represented by those which are hardenable according to their carbon content (i.e. their carbon content is higher than 0.2 %). These steels possess high strength, high hardness and good toughness after selecting the proper temperature of tempering after hardening. Such steels are called *temper-grade steels*. The main fields of application include structures, machinery components (shafts, gears, etc.) which besides static stresses are subjected to dynamic loads; therefore, toughness is an essential requirement.

Weldable steels form an outstandingly important group of steels. This is due to the fact that a great majority of structures applied in engineering practice (bridges, cranes, pressure vessels, etc.) consist of a large number of structural elements joined by various welding processes. Weldability is a very complex term. Here, only the metallographic aspects of weldability are discussed. The term *weldability* essentially means that neither during welding nor on subsequent cooling disadvantageous changes in the microstructure should occur resulting in undesired worsening of structural properties. Under the term *undesired effects,* we understand changes that can lead to an increased tendency for cracking or to a significant degree of grain size coarsening. These changes are very disadvantageous from the viewpoint of toughness required to withstand dynamic loads. Considering these statements, steels can be further divided into two subgroups according to their weldability. There are steels that can be welded *unconditionally* and others that can be welded if certain conditions are fulfilled. The main danger is the hardening of steel after welding, and as a consequence, the formation of weld cracking. Therefore, steels that cannot be hardened due to their low carbon content (less than $C = 0.2$ %) are considered to be *unconditionally weldable*. Steels having a higher carbon content (and particularly alloyed steels) may only be welded if during welding they are preheated and held at well-defined temperatures depending on the composition of the steel and on the complexity of the structure to be welded. These steels are called *conditionally weldable*.

A large part of structural elements and machine components (or at least their semifinished products) are manufactured by various forming processes. Therefore, ductility is also an important criterion. *Ductility* and *formability* are also very complex terms. Generally, steel is considered well formable when the desired shape and dimensions can be achieved without the danger of cracking with as low energy as possible.

Plain carbon steels with a very low carbon content ($C < 0.1$ %) are very ductile. As the pearlite content increases – due to the rigid cementite phase of pearlite – ductility gradually reduces. Steels having a higher carbon content (thus containing greater quantities of pearlite) are also ductile when cementite is in the form of disperse carbide particles uniformly distributed in the ferrite matrix. (Such pearlite is called *globular pearlite* to differentiate it from spheroidite.) In steels containing more carbon than $C = 0.2$ %, the structure of lamellar pearlite has to be modified into globular pearlite to achieve good formability. This can be achieved by so-called simple annealing below the A_1 temperature. The presence of a certain amount of lamellar pearlite does not significantly decrease the ductility. It is a well-established requirement of industrial practice that the *pearlite-value number* (representing the ratio of the globular pearlite to the total amount of pearlite in the steel) should be above 80 %.

Tool steels form the other main group of steels. The carbon content of tool steels is in the range of $C = 0.6\text{-}2.06$ %. The hardness and yield strength of these steels increases linearly with increasing carbon content (although the slopes of the curves are different for the hypo- and hypereutectoid ranges). Tool steels may be hardened to high strength and hardness; therefore, these are mainly used for manufacturing of various types of tools, as indicated by their name. Tool steels are classified into subgroups depending on the processing technologies. According to the technological processes, we can distinguish, for example, cold forming, hot forming and machining tools. However, it should be noted that tools subject to extremely high loads are made of alloyed steels to be discussed later.

10.4. Characteristic microstructures of plain carbon steels

The main types of microstructural elements formed in the metastable (quasi-equilibrium) crystallization of plain carbon steels or in the non-equilibrium (isothermal or continuous cooling) transformations were already described in the previous sections. In this section the main characteristics of the microstructural elements of steels are summarized.

The microstructural name of the γ-phase is *austenite* (see Figure 10.4). Austenite in the two-component, binary *Fe-Fe₃C* system is stable only at temperatures above A_1=723 °C. This means that at room temperature it is not present in plain carbon steels following the equilibrium cooling. Therefore, it cannot be studied by usual common techniques in plain carbon steels. The microstructure of austenite may be examined on a specially prepared microsection. It is inductively heated in vacuum even above 1000 °C. The micrographs of austenite in plain carbon

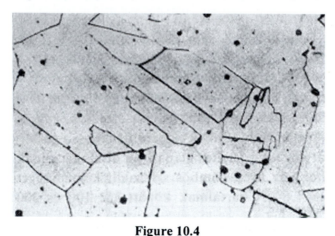

Figure 10.4
The microstructure of the so-called austenitic stainless steel containing 18 % Cr and 8 % Ni (N=500x)

steels are equivalent to the austenite of alloyed steels in which the austenite zone is expanded by alloying elements (*Mn, Ni*) down to room temperature. The microscopic picture of austenitic stainless steel containing 18 % *Cr* and 8 % *Ni* is shown in Figure 10.4. In this picture, polygonal bright grains bordered by straight lines can be seen. It can be distinguished from ferrite by the presence of twin crystals appearing as parallel lines at twin boundaries. Steels having an austenitic microstructure are ductile due to the face-centered cubic structure. An important feature of austenite is that it cannot be magnetized (i.e. the austenite is paramagnetic). This is a useful property to observe, for example, the $\gamma \rightarrow \alpha$ phase transformation. The thermal conductivity of austenite is worse and its thermal expansion coefficient is about 50 % higher than that of ferrite.

Ferrite is the microstructural name of the α-phase crystallizing in the body-centered cubic system. Under an optical microscope, ferrite also has bright polygonal grains with thin curved boundaries. In Figure 10.5, we can see the microstructure of a 99.985 % purity *Armco iron* containing practically pure ferrite. Ferrite is the basic material of spheroidite. Ferrite is the matrix of spheroidite embedding spherical cementite grains. Ferrite is the ductile phase in lamellar pearlite. The carbon content of pure ferrite is less than 0.006 % at room temperature. Thus, it can be considered as carbon-free iron. Ferrite is the softest microstructure in steels. Its Brinell hardness is $HB = 90$; therefore, it can be easily deformed. It is a ferromagnetic phase and is regarded as a soft magnetic material.

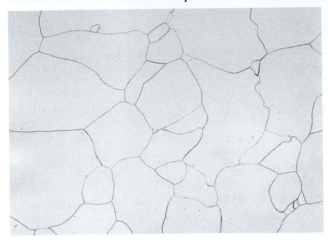

Figure 10.5

Practically pure ferrite microstructure of Armco-iron containing C = 0.015 % carbon
(Etching agent: 3 % of HNO_3, N=500x)

Cementite is the microstructural name of the Fe_3C metallic compound, iron carbide. Cementite can exist in three different forms in $Fe\text{-}Fe_3C$ alloys. These are primary, secondary and tertiary cementite.

Primary cementite crystals can be found only in the microstructure of hypereutectic pig irons in the form of wide cementite plates or in the form of wide needles embedded in ledeburite. *Secondary cementite* can be found in hypereutectoid steels in the form of a cementite network along the grain boundaries of pearlite (see Figure 10.6). *Tertiary cementite* precipitates along the *PQ* curve below 723 °C in hypoeutectoid mild steels. It forms such a thin network along the grain boundaries due to its very low quantities that it often may be neglected in quantitative analyses.

Apart from the previous occurrences, cementite can be found in spheroidite in the form of spheres (Figure 10.11) and in the lamellar pearlite in the form of cementite plates of varying thickness. Cementite is not etched by the etching agents most commonly applied for the preparation of microscopic specimens (for example HNO_3). Therefore, it always remains bright, having a white color and turning dark in polarized light. Cementite, like all metallic compounds is very rigid and hard. This is the hardest microstructure in plain carbon steels with a hardness $HV = 750$.

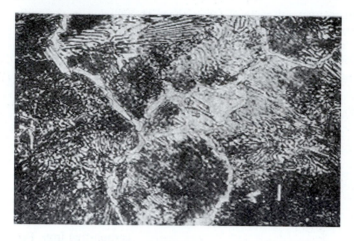

Pearlite is a heterogeneous, two-phase microstructure originating from the decomposition of eutectoid steel (its carbon content is $C = 0.8$ %). It usually consists of cementite and ferrite plates (see Figure 10.7). It is also referred to as *lamellar pearlite* due to the plate-like shape of its components. The thickness of the plates decreases with the temperature of phase transformation.

Figure 10.6
Secondary cementite network in hypereutectoid steel
(C=1.3 %, etching agent: 3 % of HNO_3, N=500×)

The microstructure of the ordinary lamellar pearlite can be recognized at a magnification of 250:1, while the so-called fine-coarse pearlite can be seen at a resolution corresponding to a magnification of $N = 1000×$. The finest pearlite consists of even thinner cementite plates which can be recognized only under the electron microscope with a magnification of $N = 5000×$. In microscopic etching of pearlite, the finest plates etch more quickly to reveal a dark image.

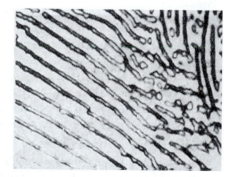

Figure 10.7
Pearlitic microstructure of eutectoid steel containing $C = 0.8$ % carbon
(Etching agent: 3 % HNO_3, magnification: a) N=500× b) N=3000×)

Bainite, depending on the temperature range of its formation, may have different microstructures.

The so-called *upper bainite* is a two-phase microstructure formed by the bainitic mechanism during non-equilibrium transformation of austenite. It contains tiny cementite discs embedded in ferrite, which cannot be seen under an optical microscope. Its microstructure is similar to that of martensite, but it has darker

Figure 10.8
The upper bainitic microstructure of rail steel containing C = 0.5 % of carbon (Etching agent: 2 % of *HNO₃*, N = 500×)

needles. In some cases, the needles of bainite have parallel boundaries: they look like small sticks. Upper bainite has a higher hardness than pearlite, but lower than martensite (see Figure 10.8).

The so-called *lower bainite* is formed near the starting temperature of martensitic transformation (M_S-line). It is, similarly to upper bainite, a heterogeneous two-phase decomposition product of austenite, containing tiny grains of

cementite embedded in ferrite needles. However, there is an essential difference in the kinetics of formation, and consequently, in the morphology of carbide as well. The ferrite crystals appearing in lower bainite are actually formed as martensite needles, while carbide is formed during the tempering of

Figure 10.9
Lower bainitic microstructure of steel with C = 0.5 % carbon content (Etching agent: 2 % of *HNO₃*, N = 500×)

martensite. Therefore, the shape of this carbide differs from the disc-shaped carbide of upper bainite, rather resembling the globular carbide of spheroidite (Figure 10.9).

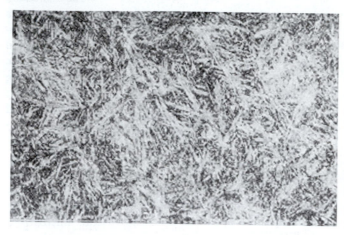

Figure 10.10
Martensitic microstructure of plain carbon steel
(C = 0.8 %; etching agent: 2 % HNO_3, N = 500×)

Martensite is the solid solution of oversaturated α-ferrite. Its microstructure may differ depending on the carbon content, as shown in Figure 9.16. Martensite is formed from austenite when steel is cooled at a rate faster than the critical cooling rate. Its composition is equivalent to that of the austenite from which it is formed. This is the hardest microstructure of plain carbon steel, its hardness may reach the value of HV = 900. In Figure 10.10, the martensitic microstructure of plain carbon steel is shown.

Spheroidite is a decomposition product of martensite formed during tempering. In the lower range of the tempering temperatures (200 –550 °C), the excess carbon atoms within the martensite lattice of hardened steel precipitate in the form of tiny, thin cementite discs. In the meantime, the stressed, distorted tetragonal lattice of α'-martensite gradually changes to a regular α''-lattice having a lower carbon content. At 600°C the cementite discs obtain a globular shape. Thus, in the tempe-

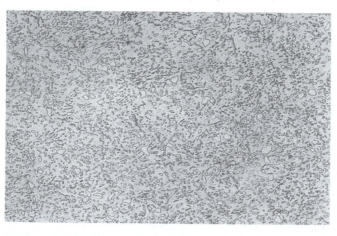

Figure 10.11
Micrograph of spheroidite formed during the tempering of
martensite (Etching agent: 2 % HNO_3; N = 500×)

rature range of 600-700 °C, that is from martensite tempered below the A_1 temperature, spheroidite is formed consisting of a two-phase microstructure of globular cementite embedded in ferrite (see Figure 10.11). When tempering martensite in the lower temperature range, the microscopic picture shows a stronger etching of the microstructure and a blunting of the martensite needles. Later, these martensite needles break up into pieces. The cementite that became globular during tempering in the temperature range of 630-720 °C can be clearly seen in the spheroidite even at a relatively low magnification (N = 250×).

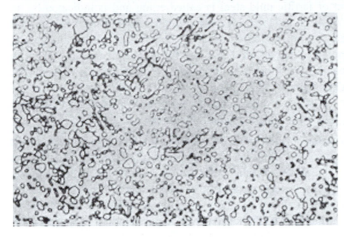

Figure 10.12
Microscopic picture of globular pearlite held for 6 hours at a temperature of T = 650 °C (Etching agent: 2 % HNO_3, Magnification: N = 500×)

Similarly, globular cementite embedded in ferrite is formed from lamellar pearlite when it is held for a longer period below the A_1 temperature, at about 650-700°C. During this annealing, the cementite plates break up due to the surface tension and become globular.

In Figure 10.12, the microstructure of globular pearlite held for 6 hours at T = 650 °C is shown. This microstructure is called *globular pearlite* in order to distinguish it from the spheroidite formed during tempering of hardened steel.

10.5. The contaminants of steel

The contaminants of steel include those undesired elements that enter the steel during the iron and steel making processes or during further processing. These contaminants are mostly *N*, *O*, *H*, *P* and *S*.

10.5.1. Nitrogen in steel

The amount of nitrogen entering steel during the iron and steel making processes is usually small and depends on the processes applied. Its concentration usually varies between 0.001-0.03 %. The amount of nitrogen to be introduced by alloying is in the range of 0.2-0.3 %. Nitrogen is alloyed into steel during surface alloying processes known as nitriding in order to achieve excellent surface properties of the steel. High hardness, good resistance to wear and advantageous sliding properties of the surface layer may be produced by nitriding. Thus, nitrogen may have two opposing effects: it may be a detrimental contaminant as well as a useful alloying element.

a) Nitrogen as a detrimental contaminant

The detrimental effect of nitrogen is manifested in so-called *aging* and *alkaline embrittlement* of steel. These are usually caused together with oxygen.

The embrittlement of ferritic mild steel is called aging and is indicated by an increased difference between the lower and upper yield points of the steel, as well as by a decrease in elongation. The aging of mild steels is attributed to the *atomic clouds* of nitrogen moving relatively freely in the ferrite lattice and accumulating in the expanded zones of dislocations. During plastic deformation, the N-atoms are unable to leave the dislocations and in this way they block them and obstruct the movements necessary to produce plastic deformation. Steels having a N-content higher than 0.004 % have a significant tendency for aging.

The tendency for aging in mild steels is tested by the so-called aging test. It may be summarized as follows: after a 10 % plastic deformation, the specimen is boiled for one hour at 250 °C. Plastic deformation is applied to increase the number of dislocations and heating is used to accelerate the diffusion processes. The value of impact work obtained for the artificially aged steel is compared to that measured in the original state. The decrease in impact work is proportional to the aging tendency.

The *alkaline embrittlement* is a type of grain boundary corrosion. It appears particularly in boiler sheets made of steel contaminated by N and O with a tendency for aging. When these types of mild steel are exposed under tensile stress to the etching effect of warm alkaline acids and salt solutions they become brittle. Due to the effect of tensile stresses, cracks can arise along the grain boundaries and often propagate across the whole cross section. Such phenomena frequently occur in the chemical industry in caustic evaporator boilers as well as in steam boilers supplied with water softened by soda.

In order to avoid the two detrimental phenomena, the tendency for aging and the alkaline embrittlement, steel is treated with Al during steel making. The result is the so-called *killed steel*. Al is put into the molten steel just before it is cast to ingots. Aluminum has a high affinity to both O and N, so they can be bonded in the form of Al_2O_3 and AlN, which are stable compounds. With this treatment, both the tendency for aging and the alkaline embrittlement of mild steel can be decreased, since alkaline embrittlement has the same reasons as aging.

b) Nitrogen as a useful alloying element

Nitrogen as a useful alloying element plays a role in several fields of application. One of these applications is the surface hardening process called *nitriding*. During this process, natural hardness of the surface is achieved by diffusion of nitrogen. Nitrogen can be applied as an alloying element to expand the γ-zone. For this purpose, 0.2 % N can replace 2-4 % Ni in austenitic $CrNi$ steels. Nitrogen forms stable nitrides with V and Nb in normalized structural steels. In this way nitrogen as a microalloying element increases the strength of steels and prevents grain coarsening.

10.5.2. Oxygen in steel

Oxygen can exist in two forms in steels: either in the dissolved state in ferrite or in bonded form as oxide inclusions. In the former state, together with nitrogen it contributes to the aging and alkaline embrittlement of steel. The oxide inclusions are undesired impurities that are particularly dangerous when their melting points are lower than the temperature of forging. The oxide inclusions with low melting points may cause *red brittleness* during forging in the same manner as the *Fe-FeS* eutectic.

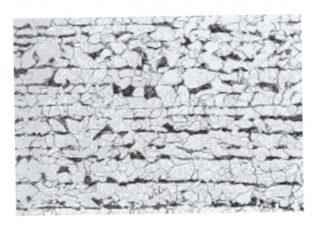

Oxide inclusions with high melting points are also harmful, since they can hinder diffusion and contribute to a coarse or fibrous structure. The oxides with high melting points are positioned as an elongated chain in the rolling direction. Following rolling during recrystallization, the proeutectoid ferrite crystallizes on these slag threads consisting partially of these oxides as nuclei. Thus, pearlite also has an ordered position as shown in Figure 10.13. This

Figure 10.13
The coursed structure of steel containing C = 0.15 % carbon, Etching agent: 3 % solution of HNO_3, Magnification: $N = 100\times$

fibrous structure is the reason for the lower strength and elongation properties of rolled sheets in the transverse direction compared to the longitudinal direction.

10.5.3. Hydrogen in steel

Hydrogen in steel can be present both in a dissolved and in an absorbed gaseous form (H_2). It is always considered as a detrimental contaminant. The H_2 absorbing capacity of steel decreases with decreasing temperature and therefore about half of the absorbed hydrogen is "pushed out" upon solidification. The removal of hydrogen may be promoted by heating the steel to a temperature of 200 °C. The absorbed H_2 makes the steel rigid and significantly decreases its elongation. Steel may pick up H_2 during cinder removing by pickling or acidic etching applied after the recrystallization heating. If steel is cold-rolled immediately after etching, without allowing it to rest for a few days, or heating it up to 200 °C or the wire is drawn, it can easily break. Therefore, etched products have to be stored or have to be boiled at a temperature of 200°C prior to further processing. During heating to 200°C, the H_2 content of steel leaves the steel so fast that the stress caused by this event can even break it. This is the so-called *pickle embrittlement*. In such cases, the surfaces of sheets get bulgy at the places containing gas bulbs.

Flakiness or *lustrous taint* is also a detrimental problem of steels, particularly in stainless steels. It is formed by H_2 gathered in cavities and intending to leave under great tension. However, the blocked hydrogen has a cracking effect, which is indicated by shiny patches on the fracture surface. Another detrimental effect of H_2 is its strong C reduction effect. H_2 is particularly harmful if it is present in the atmospheres of annealing furnaces. Hydrogen has further unfavorable effects in the welding of steels as well. Hydrogen increases the danger of cracking near the welds. H_2 can enter the welds from the electrodes, particularly if they are damp.

10.5.4. Phosphorus in steel

Generally, phosphorus is considered a contaminant of steel because it reduces the impact work and, thus, the toughness of steels even at concentration as low as 0.1%. In structural steels, the upper limit of phosphorus is given as 0.05 %. Phosphorus is applied in the so-called automatic steels to make the chips fragile. It is also added to cast iron to produce thin walled segmented castings.

10.5.5. Sulfur in steel

The presence of sulfur in steel causes the danger of so-called "*red brittleness*" and "*hot brittleness*". Red brittleness may occur during hot forging or hot rolling of steels having a high sulfur content (sometimes a sulfur content of 0.03 % is considered high enough). Thus, it may occur in the temperature range of 900-1000°C. The reason is that the *Fe-FeS* eutectic melts at 985°C. Its melting point is decreased in the presence of iron oxide. Iron sulfide and nickel sulfide solidify last from the liquid, as a network along grain boundaries. Therefore, cracks may form during forging.

Hot brittleness may occur at around 1200°C due to the iron sulfide network solidifying along grain boundaries at a lower temperature than steel. Both red and hot brittleness can be prevented by adding manganese to steel. Since the affinity of *Mn* to *S* is higher than that of *Fe*, manganese sulfide (*MnS)* is formed instead of iron sulfide. It has a much higher melting point (1620°C) than iron. Furthermore, manganese sulfide is present in the form of small, dispersed inclusions, which do not cause fracture.

Sulfur is a useful alloying element in the so-called automatic steels. When it is alloyed in a quantity of 0.15-0.30 %, the chips become fragile and thus it improves the machinability of mild steels on automatic machining equipment.

11. ALLOYED STEELS

Steel is considered *alloyed* when in addition to the basic components (*Fe* and *C*), *other alloying elements* are *added intentionally* to assure certain properties. It is important to emphasize that these elements are added intentionally, since during the analysis of plain carbon steels, we could see that certain elements (*Si, Mn, S, P*, etc.) unavoidably entered the steel during the iron- and steel-making processes. However, these are not considered alloying elements unless their content exceeds the amount corresponding to the production process.

11.1. The main alloying elements of steels and their characteristics

The purpose of alloying steel – as is generally the purpose of alloying – is to provide such physical, chemical, mechanical or other properties that cannot be assured without alloying. The most important and most frequently applied alloying elements of steel are manganese (*Mn*), nickel (*Ni*), chromium (*Cr*), tungsten (*W*), vanadium (*V*), molybdenum (*Mo*), titanium (*Ti*), niobium (*Nb*) and boron (*B*).

The main aim of alloying steels is to improve mechanical properties (for example, strength, ductility, toughness), to increase resistance to corrosion (chemical resistance), to improve certain physical properties (for example, magnetic, electrical properties) and in many cases to improve complex properties of technological workability (for example, formability, weldability, machinability). These objectives can be achieved in various ways depending on the relationship of the alloying elements to the base material (*Fe*) and to the fundamental alloying element of steels, that is carbon.

The alloying elements can form *solid solutions* and *metallic compounds* with iron. The solid solutions formed with iron can be substitutional and interstitial solid solutions. Only alloying elements (e.g. *Cr, Mn, Ni, Co, V*) with atomic diameters similar to that of iron can form *substitutional solid solutions* with iron (note the atomic diameter of iron, $d_{Fe}=2.47\times10^{-10}$ m). *Interstitial solid solutions* are formed in steel by elements with small atomic diameters (for example, *C*, *N*, *B*). Many elements form *metallic compounds* with iron – among others - nitrogen (*Fe₂N*, *Fe₄N*), aluminum (*FeAl₂*), silicon (*FeSi*), phosphorus (*Fe₃P*, *Fe₂P*), titanium (*Fe₂Ti*), vanadium (*FeV*), chromium (*FeCr*), molybdenum (*FeMo*) and tungsten (*Fe₂W*, *Fe₃W*). It can be seen from this list that certain alloying elements (e.g. *N, P, W*) can form several types of metallic compounds with iron as opposed to others (as, for example, *Mn* and *Ni*) which never form metallic compounds with iron in steels. The greater the differences between the atomic radii of elements forming the metallic compounds, the more stable the metallic compounds of steels. These differences are greatest in the case of *C* and *N* forming metallic compounds in steels. This is the reason why carbides and nitrides are usually more stable in steels than other metallic compounds. (However, it should be noted that the stability of compounds also depends on several other factors, among them the type of chemical bond.)

The relationship between the alloying elements and carbon has a decisive role as regards several properties of alloyed steels (especially hardness and resistance to wear). Some alloying elements never form carbides with carbon (for example *Ni* and *Cu*). Some other elements generally form carbides in alloys, but never in steels. These are *Al* and *Si*: the latter forms *SiC*, which never can be found in steels.

Obviously, those alloying elements that also form carbides in steels are of utmost importance in alloyed steels. These are called *carbide-forming alloying elements*. The essential significance of carbides lies in the fact that they maintain their hardness (called *natural hardness*) up to a much higher temperature than martensite. (We have seen in earlier chapters that *martensitic hardness is an artificial one* existing only in the hardened, non-equilibrium state and quickly decreasing with increasing temperature.) As a rule, it can be stated that the higher the melting point and the hardness of the carbides, the more stable they are. The most important carbides existing in alloyed steels are summarized in Table 11.1 in the order of their stability.

According to Table 11.1, the least stable carbide in the absolute sense is iron carbide (*Fe₃C*) and the most stable one is *TiC*. It is worth mentioning that the three least stable carbides in the series crystallize in the rhombic (or orthorhombic) crystal system, which are relatively rare in metallic systems. The most stable carbides crystallize in the cubic system. Carbides crystallizing in the hexagonal system can also be characterized by great hardness and stability.

It is also worth mentioning that in alloyed steels, besides the iron carbide that is also present in plain carbon steels, the carbides of other alloying elements denoted by Me_xC_y can usually be found. *Me* in the formula stands for the symbol of a metallic element forming the carbide. These carbides (the iron carbide and the carbides of the alloying elements) can mutually dissolve each other. Therefore, besides "simple" carbides there are so-called complex carbides – especially in highly alloyed, multi-component systems – which are symbolized by the formula *(Fe, Me)ₓCᵧ*.

Table 11.1 Characteristic properties of carbides in alloyed steels

Carbides			
Symbol of carbides	Lattice structure	Melting point T, [°C]	Hardness HV
Fe_3C	Rhombic	1250	900
Mn_3C	Rhombic	1050	1100
Cr_3C_2	Rhombic	1890	1300
Mo_2C	Hexagonal	2690	1500
Cr_7C_3	Hexagonal	1665	2100
WC	Hexagonal	2850	2400
W_2C	Hexagonal	2860	3000
$Cr_{23}C_6$	Cubic	1550	1650
VC	Cubic	2810	2800
TiC	Cubic	3140	3200

Since the carbon dissolving capability of austenite is high, the carbides of the alloying elements also dissolve in austenite although at much higher temperatures than iron carbide. According to the decrease in carbon dissolving capability, during cooling, the more stable alloyed carbides precipitate first, which is indicated by the carbide precipitation line in the phase transformation diagrams (see, for example, the upper dotted line in Figure 11.3). The properties of these carbides, and thus, the properties of alloyed steels may be changed within a wide range by controlling carbide precipitation. This forms an important theoretical basis for the heat treatment of alloyed steels.

Summarizing these features, it may be stated that the alloying elements influence the mechanical properties of steels through both solid solutions and metallic compounds formed with the fundamental element (iron) as well as through the mechanisms of carbide formation. These effects are manifested in the influences of alloying elements on the equilibrium phase diagrams, as well as on the non-equilibrium phase transformation diagrams valid for isothermal and continuous cooling. These effects will be analyzed in detail in the next sections.

11.1.1. The effects of alloying elements on the phase diagrams

Consider first the effect of alloying elements on the *Fe-Me* binary phase diagrams (where *Me* denotes the alloying element). From this point of view, the alloying elements are classified according to their effects on the positions of A_3

and A_4 temperature lines representing the critical phase transformations of steels (see Figure 11.1).

1. The first group includes those alloying elements that decrease the A_3 and increase the A_4 temperature line proportionally with the increase in alloying content. (These alloying elements are called *austenite forming* since they widen the austenitic zone). This group of elements can be divided into two further subgroups:

    *   The most characteristic austenite forming elements are those that shift the A_3 temperature line down to room temperature creating a so-called *open γ-zone*. This means that austenite can be found in these alloys even at room temperature. *Mn* and *Ni* are two typical alloying elements belonging to this subgroup. The characteristic binary (*Fe-Me*) phase diagram of austenite forming elements can be seen in part a) of Figure 11.1.

    *   The second subgroup of austenite forming alloying elements leads to the widening of the γ-zone in a certain composition range, but austenite cannot be found at room temperature in the alloy system (similarly to plain carbon steels). These alloying elements are called *γ-zone expanding elements*. The fundamental alloying element of steels, i.e. carbon itself, belongs to this group (see the changes in the critical transformation temperatures A_3 and A_4 with the increase in carbon content). Furthermore, *Cu* and *N* also belong to this group of alloying elements. The characteristic binary (*Fe-Me*) phase diagram with the γ-field expanding, austenite-forming elements is shown in part b) of Figure 11.1.

2. In the second main group, the so-called *ferrite forming alloying elements* are included. The effects of these alloying elements are opposed to those of the austenite-forming elements. They mostly shift the critical transformation temperature line A_3 upward and the A_4 temperature line downward. This leads to the narrowing of the γ-zone and, simultaneously, to the expansion of the α-zone (in certain cases creating an open α-zone). There are three subgroups within this group of alloying elements:

    *   Alloying elements, which create a closed γ-zone and simultaneously open the δ and α-zones into a joint zone, belong to the most characteristic subgroup of ferrite-forming elements. In this way, a continuous *open ferrite zone* is created as shown in part c) of Figure 11.1. *Cr* and *V* are the main alloying elements in this subgroup.

    *   The second subgroup of ferrite-forming elements includes those alloying elements which open the δ and α-ferrite zones only in a narrow range rather than in the whole alloy system. These are the alloying elements forming the so-called *partially open ferrite zone*. *Al, Si, Ti, Mo* and *W* are the characteristic members of this subgroup. The equilibrium phase diagram characteristic of the partially open ferrite zone is shown in part e) of Figure 11.1.

    *   Alloying elements, which expand the α-zone when their concentration increases but do not open the δ and α-ferrite zones into one joint zone,

belong to the third subgroup of ferrite forming elements. Due to this, they also narrow the γ-zone with increasing concentration. This subgroup includes *B, Zr, Nb, Ta*. The characteristic binary phase diagram of these alloys is shown in part d) of Figure 11.1.

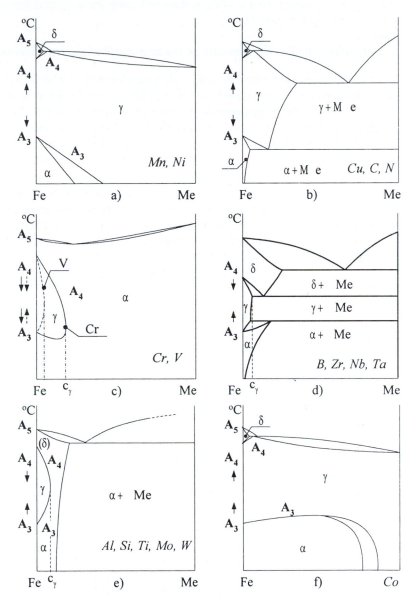

Figure 11.1
The effect of alloying elements on the *Fe-Me* binary phase diagrams

3. Concerning the effects of alloying elements on the equilibrium phase diagram, *Co* forms a unique group of alloying elements. *Co,* besides expanding the α-zone, forms austenite that remains stable even at room temperature at high *Co*-concentrations, i.e. it also creates an open γ-zone. The equilibrium phase diagram of the binary *Fe-Co* alloy system reflecting these special features is shown in part f) of Figure 11.1.

11.1.2. The effect of alloying elements on the non-equilibrium transformations

Besides the equilibrium phase diagrams, alloying elements of steel significantly influence the transformations referring to non-equilibrium conditions as well as the shape and position of the transformation diagrams. Summarizing, these effects can be described as follows: increasing the quantities of alloying elements increases the shortest latent (incubation) times, which is manifested by the definite shift of the transformation diagrams to the right along the time axis. On the other hand, an increase in the alloying element content results in a continuous separation of pearlite and bainite zones, which, in strongly alloyed steels, leads to the complete separation of pearlite and bainite zones. The longer latent (incubation) period results in lower critical cooling rates, thus decreasing the danger of cracking during or after hardening as well as increasing the cross section to be fully hardened.

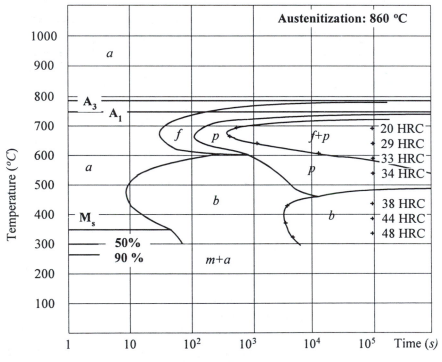

Figure 11.2
Isothermal transformation diagram of low-alloyed chromium steel, *Cr* 135
($C = 0.4$ %, $Cr = 1.5$ %)

A significant increase in the shortest latent times can be observed in Figure 11.2, showing the isothermal transformation diagram of a low-alloyed chromium-steel (*Cr 135*) containing $C = 0.4$ % carbon and $Cr = 1.5$ % chromium. The isothermal transformation diagram of steel *C 45* ($C = 0.45$ %) is shown in Figure 9.19. In this diagram, the shortest latent time is $t = 0.8$ s, which is about one tenth of the shortest latent time found in the isothermal transformation diagram of steel *Cr 135* having the same carbon content ($C = 0.45$ %). In the latter diagram, the shortest latent time ($t = 8$ s) can be found at the nose point of the bainite zone.

Furthermore, an initial separation of pearlite and bainite zones can be observed in this diagram, manifested by the definite separation of pearlitic and bainitic transformation as the content of the alloying elements increases. This is illustrated by Figure 11.3 for the isothermal transformation diagram of the so-called ledeburitic chromium steel ($C = 2$ %, $Cr = 12$ %). There is an extremely long – almost 1000 hours – latent time between the two completely separated pearlite and bainite zones. A similar long – close to half an hour – latent time can be seen at the nose point of the bainitic transformation, which ensures the hardening of these steels without the occurrence of detrimental thermal stresses.

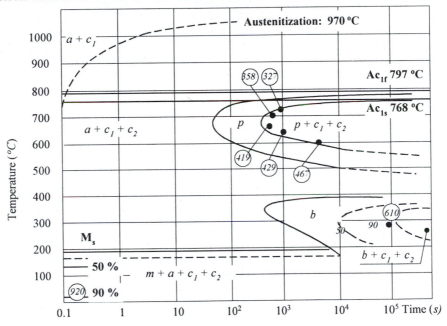

Figure 11.3
Isothermal transformation diagram of ledeburitic chromium steel
($C = 2$ %, $Cr = 12$ %)

The main alloying element of the strongly alloyed ledeburitic chromium steel is the strong carbide-forming element of chromium. As we have seen in Table 11.1, chromium forms several carbides of differing stability in steels. Part of these carbides – formed during the solidification of steel – do not dissolve even at the temperature of austenite formation (they would dissolve only on melting). It is indicated by the mark c_I referring to the primary carbides in the austenite zone of the diagram. The carbides present at the temperature of austenite formation prevent coarsening of austenite grains, which is also very favorable from the point of view of a hardened microstructure. The dotted line in the diagram indicates the borderline of the precipitation of secondary carbides from austenite.

The most important effects of the alloying elements on the isothermal transformation diagrams can be summarized as follows. Increasing the alloying content increases the shortest latent times and separation of pearlite and bainite zones can be observed. These effects can be seen more clearly in the phase transformation diagrams valid for continuous cooling. This is illustrated in Figure 11.4 showing the continuous cooling time-temperature transformation diagram of the ledeburitic chromium steel.

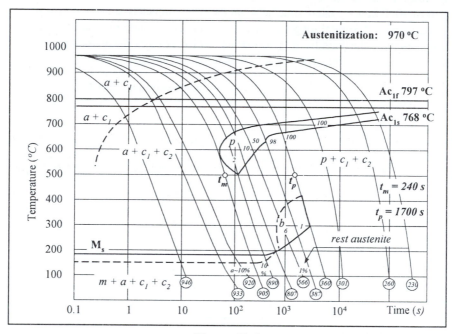

Figure 11.4
Continuous cooling time-temperature transformation diagram of
ledeburitic chromium steel

In the diagram, A_{cls} indicates the starting and A_{clf} the finishing temperature of pearlite to austenite transformation upon heating. The intersection of the isotherm $T=500^oC$ with the curve representing the cooling rate required for fully martensitic transformation can be found at $t_m = 75$ s. From this value, the *upper critical cooling rate* can be calculated from the expression

$$v_{crit}^{upper} = \frac{970-500}{75} = 6.3^{o}C/s.$$

(11.1)

This is about 1/120 of the critical cooling rate of plain carbon steel having the same carbon content. This represents such a low cooling rate that it can be ensured by air cooling for small cross sections and by oil cooling for larger cross sections to achieve martensitic microstructure. As we have seen, in plain carbon steels this could be done only by intense water cooling.

11.2. Main types of alloyed steels

Previously we analyzed the effects of alloying elements on certain properties of steels as well as on the equilibrium and non-equilibrium phase transformations. In this section, the most important properties and the theoretical bases of these properties will be discussed for some essential groups most generally applied in engineering practice. Neither the volume of this book nor the subject makes it possible to deal with an overall analysis of all the alloyed steels. Concerning the most important physical, chemical and mechanical properties and the main application fields of these alloys, we refer to the relevant national standards.

11.2.1. Manganese steels

Manganese is one of the *most powerful austenite-forming* alloying elements. It reduces very strongly the critical transformation temperatures of steels. In steels containing more than 10 % Mn, austenite can be found at room temperature even

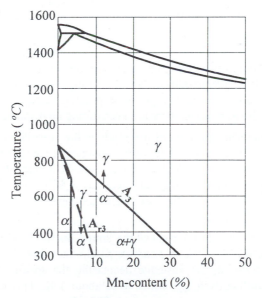

Figure 11.5
Part of *Fe-Mn* phase diagram

after application of high cooling rates. This is illustrated in the binary *Fe-Mn* phase diagram (Figure 11.5). In this figure, the hysteresis of A_3 critical transformation temperature can be clearly seen. At near equilibrium cooling, the homogeneous austenite microstructure can be found at room temperature above 32 % manganese content. (However, austenite can appear even below 10 % *Mn* content on rapid cooling (see line A_{r3} in the diagram). It follows from the previously described relationships of the critical transformation temperatures that the manganese steels can be hardened only up to 10-12 % manganese content. The $\gamma \rightarrow \alpha$ phase transformation, representing the condition for martensitic hardening, exists only up to this Mn-content. (In Figure 11.5, the *Fe-Mn* phase diagram is drawn only up to 50 % *Mn*.)

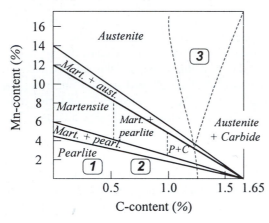

Figure 11.6
Guillet-microstructure diagram of *Mn*-steels

The microstructure of *Mn*-steels is essentially influenced by the carbon content. However, instead of the complicated *Fe-C-Mn* ternary phase diagram, the so-called *Guillet diagram* is used to study their microstructure (see Figure 11.6). In the diagram, the microstructure of small-size, air-cooled test specimens is shown as a function of *Mn* and *C*-contents. (Note that the carbon content $C = 1.65$ % in the diagram basically corresponds to point *E* of the *Fe-Fe₃C* phase diagram since this was thought to be its value at the time when *Guillet* constructed this diagram.) The manganese steels in the middle composition range – resulting in a martensitic microstructure – are not used in engineering practice because they have a martensitic microstructure at room temperature even during air-cooling.

The most characteristic type of *Mn*-steels is the so-called *austenitic Mn-steel* containing 1.2-1.4 % carbon and 12-14 % manganese (see the range marked by 3 in the Figure 11.6). It is called *Hadfield steel* in honor of the scientist who elaborated it. After rapid cooling, this steel has a fully austenitic microstructure characterized by good ductility, excellent toughness and a high degree of hardenability.

In the alloys located to the right of the dotted line in the austenite zone of the Guillet diagram, *(FeMn)₃C* double carbide precipitates at a low cooling rate along the grain boundaries of the austenite decreasing the specific elongation of the otherwise extraordinarily tough austenite to about 1 %. This carbide completely dissolves in austenite at temperatures above 900 °C.

If the steel having a homogeneous austenite microstructure at $T = 950$ °C is cooled in water, there is not enough time for the diffusion of carbides to segregate. Therefore, this steel has a pure austenite microstructure with medium hardness ($HB \approx 250$), high strength (R_m=900-1200 MPa) and high elongation (A=30-60%). Consequently, the alloy has excellent toughness properties as well. The hardening capability of austenitic manganese steels is one of the highest among the alloys. Therefore, these steels are mainly used to manufacture frogs of rail crossings and turnouts, jaws of clamp cheek grippers and stone crushers subjected to dynamic loading. During operation, local plastic deformations occur resulting in intense strain hardening and in increased wear resistance.

Mn-steels are also used as structural and tool steels. In both cases, their *Mn*-content is in the range of 1.5-2 %, while the carbon content is equal to C=0.3-0.5 % for structural steels (zone 1 in Figure 11.6) and C=0.8-0.9 % for tool steels (zone 2 in Figure 11.6). The manganese steel with a carbon content corresponding to the eutectoid composition can be characterized by good shape and dimensional stability. This may be explained by the low M_s-temperature, the low critical cooling rate and minimal thermal and shrinkage stresses during non-equilibrium cooling. Therefore, these tool steels are used to make gauges and deburring tools.

11.2.2. Nickel steels

Nickel as an alloying element leads to similar results to manganese in many respects. It creates an *open γ-zone*, being an austenite-forming alloying element. *Ni* also reduces the critical cooling rate. From the Guillet diagram, it can be seen that, compared to *Mn*-steels a double quantity of *Ni* is required to achieve the same effect (see Figure 11.7).

Ni in steels acts as the alloying element of "*physical properties*". This is because many of the physical properties of steels are significantly improved by alloying with *Ni*. For example, *Ni* increases the magnetic permeability by two orders of magnitude compared to that of pure iron. Due to its

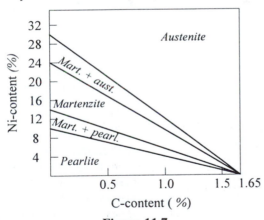

Figure 11.7
Guillet microstructure diagram of Ni-steels

effect on magnetic properties, *Ni* is one of the most popular alloying elements in permanent magnets. It provides constant modulus of elasticity across a wide range of temperature. Therefore, *Ni*-steels are used in the production of torsion springs of various measuring instruments (for example the *Eötvös torsion pendulum*).

It also changes the thermal expansion coefficient of steels in a wide range. The change of the thermal expansion coefficient as a function of *Ni*-content is shown in

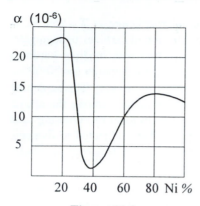

Figure 11.8
Thermal expansion coefficient vs. Ni-content

Figure 11.8. This provides the theoretical bases for various applications of *Ni*-steels. The thermal expansion coefficient of steel has its lowest value at a *Ni* content of 36 % (the linear thermal expansion coefficient is $\alpha = 1.5 \times 10^{-6}$ 1/°C).

This steel is called *invar steel*. Since the parts produced from invar steel hardly change their dimensions, it is widely utilized in the manufacturing of high precision instruments and in the precision mechanics industry. The greatest value of thermal expansion coefficient ($\alpha = 22.5 \times 10^{-6}$ 1/°C) is experienced at a *Ni*-content of 25 %. If a strip made of this alloy is welded together with another made of invar steel, we obtain a so-called *bimetal* used in thermostats and temperature switches. The thermal expansion coefficient of the alloy containing 41 % *Ni* is similar to that of glass and therefore, it is used as the current leads of light bulbs.

11.2.3. Chromium steels

Chromium is the most widely used alloying element in steels. Chromium is a ferrite forming alloying element that creates a closed γ-zone and "*joins*" the δ and α-ferrite zones creating a so-called open ferrite zone as can be seen in the binary phase diagram of *Fe-Cr* (Figure 11.9). It can also be seen from the diagram that the carbon content significantly expands the closed γ-zone. A further characteristic feature of the *Fe-Cr* phase diagram is the occurrence of the so-called σ-phase. It is formed at $T=820^{\circ}C$ and $Cr=45 \%$. In fact, it is an ordered solid solution. However, due to its extreme rigidity it was thought to be a

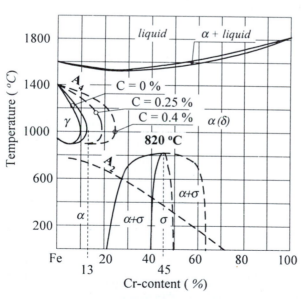

Figure 11.9
The *Fe-Cr* binary phase diagram

metallic compound for a long time. During prolonged annealing, the σ-phase can already appear in the concentration range of 21-61 % *Cr* forming an α+σ mixture. Alloys containing more than 30 % of *Cr* are so rigid due to the significant amount of σ-phase that these are not used in practice. The chromium content of steels in engineering practice does not usually exceed 18 % and in most cases it is in the range of 12-14 % *Cr*.

If nickel is called the alloying element of physical properties then *chromium* can rightly be considered as the *alloying element of chemical properties*. This is because chromium can form a whole series of corrosion- and acid-resistant steels. Its perfect protective effect is due to its passivating nature. In an oxidizing environment or in the presence of etching agents, a thin oxide film (0.1-0.2 μm) is formed on the surface of *Cr*-steels thus protecting the rest of the alloy from further oxidation. The main types of corrosion resistant *Cr*-steels are distinguished according to their microstructure. Based on this, ferritic, pearlite-martensitic and austenitic *Cr-Ni* steels are distinguished.

The composition of *ferritic corrosion-resistant Cr-steels* is characterized by a carbon content in the range of 0.05-0.11 % and a chromium content in the range of 12-18 %. Owing to the low carbon and high chromium contents, there is no $\gamma \rightarrow \alpha$ transformation in these steels. Therefore, they cannot be hardened and the grain size cannot be modified by heat treatment. These are applied in the chemical and food industry to produce various appliances resistant to weak acids.

The *pearlite-martensitic chromium-steels* have a carbon content in the range of 0.2-0.8 % and a *Cr* content in the range of 12-14 %. These steels can be hardened but they have poor weldability. They are used to manufacture various machine parts, turbine blades and medical scalpels.

The *austenitic corrosion resistant steels* are the most popular types of chromium-nickel steels. The most characteristic representative is the so-called 18/8 austenitic chromium-nickel steel that contains 18 % *Cr* and 8 % *Ni*. The carbon content of these steels can be found in the range of $C = 0.05$-0.11 %. These are the best corrosion- and acid-resistant steels. The corrosion resistance is enhanced by the homogeneous austenitic microstructure besides the passivating effect of chromium. This is one of the most frequently utilized structural materials in the construction of stainless steel vessels. The wide range of application of this alloy is also due to its excellent weldability.

The corrosion resistance of austenitic *Cr-Ni* steels may be deteriorated by the segregation of chromium carbide ($Cr_{23}C_6$) along the grain boundaries of austenite. It may occur during either welding or heat treatment if the alloy is heated up to a temperature of $T = 700$-800 °C and cooled down slowly. The decrease in corrosion resistance is strongly related to, on the one hand with disintegration of the homogeneous austenitic microstructure and, on the other hand, to the reduced chromium content in the neighborhood of grain boundaries. It is well known that austenite containing less chromium has a reduced degree of corrosion resistance. This phenomenon is called *grain boundary corrosion*. It may be avoided by preventing the segregation of chromium carbide along the grain boundaries of austenite. This may be achieved in several ways.

One of the most widely applied methods to avoid corrosion along the grain boundaries is the application of the so-called *austenitic cooling*. If, following the thermal effect initiating carbide segregation, the steel is heated up to 1000 °C, the precipitated carbides dissolve and a homogeneous austenitic microstructure is obtained. By applying rapid cooling from this temperature, there is no time for the precipitation of carbides. Thus, a homogeneous austenite with uniform *Cr*-content remains without grain boundary corrosion aptitude.

Another method to prevent carbide segregation is limitation of the carbon content to a value as low as possible. If the carbon content in austenitic *Cr-Ni* steels is below $C = 0.05$ %, there is insufficient carbon to form carbides that lead to grain boundary corrosion.

Additionally, the segregation of carbides causing grain boundary corrosion may be avoided by adding alloying elements to the steel with a higher chemical affinity for carbon than chromium. *Ti* and *Nb* are the most powerful carbide-forming elements to prevent grain boundary corrosion in austenitic steels. For this purpose, the titanium content should be five times and the niobium content ten times the carbon content to form *TiC* or *NbC* to prevent the formation of $Cr_{23}C_6$ which causes the grain boundary corrosion. Both *TiC* and *NbC* segregate inside the grains in the form of small carbide particles with a disperse distribution. Steels alloyed with the necessary amount of titanium or niobium adequate to prevent the grain boundary corrosion are called *stabilized steels*.

Besides corrosion-resistant steels, chromium is also an important alloying element of various structural and tool steels. The outstanding role of chromium as an alloying element is due to the following features:

* The high affinity of *Cr* to *C*, *N* and *O* promotes the formation of various – uniformly distributed – carbides, nitrides and oxides which, serving as crystallization nuclei, refine the grain size of steels. This results in excellent mechanical properties (the increased toughness due to the finer grain size should be emphasized).

* Among the advantageous effects of chromium, its strong carbide-forming ability should be emphasized. Chromium forms a series of stable carbides in steels (from the rhombic Cr_3C_2, dissolving at the lowest temperature, through the hexagonal Cr_7C_3 up to the cubic $Cr_{23}C_6$, showing the highest degree of stability). Besides these, chromium also forms $(Cr, Me)_7C_3$-type complex carbides in multi-component alloy systems. The previously listed carbides are stable to elevated temperatures in chromium steels providing high natural hardness and outstanding resistance to the wear of these steels.

* Chromium is an important alloying element from the point of view of increasing the cross section that can be fully hardened and reducing the danger of cracking on hardening. This follows from the effects of *Cr* on the phase transformation diagrams valid for isothermal and continuous cooling. As the

chromium content increases, the critical cooling rate required for hardening decreases. This is summarized in Table 11.2, which shows the change in the upper critical cooling rate with respect to increasing chromium content. (The data in the Table refer to the same $C = 0.45$ % carbon content.)

Table 11.2 Changes in the critical cooling rate with the chromium content

Chromium content	Time characteristic of the martensitic transformation	Critical cooling rate	
Cr, %	t_m, [s]	v_{crit}, [$^{\circ}$C/s]	ratio
0.15	1.5	270	1
1.02	8.0	50	1/5
1.70	50.0	8	1/30
13.00	600.0	0.67	1/400

Due to these advantageous effects, chromium steels are widely used in engineering practice. A whole series of case-hardening and temper-grade steels is known. Among the many application fields of chromium steels, the components of heat engines requiring high thermal strength are worth mentioning. Another important field of application of chromium steels is the production of ball bearings. A typical representative is the ball bearing steel having 1 % C and 1.5 % Cr. The chromium alloys having higher carbon contents are used to manufacture tools characterized by high strength and great resistance to wear. The best-known chromium alloyed tool steel is the so-called *ledeburite chromium steel*. It contains 2 % C and 12 % Cr. A unique feature of this steel is that it has ledeburite in its microstructure, which is unusual in steels. The presence of the hard and rigid ledeburite is an advantage in toolmaking.

Besides the corrosion resistance of chromium steels, they exhibit remarkable heat resistance, which increases with increasing chromium content. The heat resistance of chromium steels increases in so-called resistance steps, as shown in Figure 11.10.

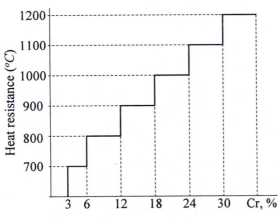

Figure 11.10
The heat resistance steps of *Cr*-steels

11.2.4. Tungsten steels

Tungsten as an alloying element of steel possesses in many respects properties similar to those of chromium. It narrows the γ-zone and creates a partially open α-zone as illustrated by the binary *Fe-W* phase diagram in Figure 11.11. In *W*-steels, the γ → α phase transformation can be observed only below a tungsten content of *W*=8 %. Consequently, steels with a higher tungsten content cannot be hardened. (Similarly to chromium steels, the γ-zone can be significantly expanded by increasing the tungsten content. Accordingly, the limit of hardenability is also shifted toward higher tungsten contents.) Similarly to chromium, tungsten also increases the capability of steels to withstand tempering and raises the thermal strength.

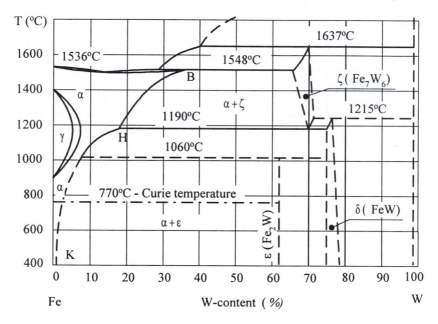

Figure 11.11
The *Fe-W* binary phase diagram

Tungsten steels existing in the region bordered by the line *BHK* (shown in Figure 11.11), and representing limited solid solubility are of utmost practical importance. Along this line, the Fe_7W_6 metallic compound segregates (marked by ζ-phase in the Figure). In the temperature range of 1060 to 1190 °C, it first transforms to the δ-phase (*FeW*), then below the temperature 1060 °C to the ε-phase (Fe_2W metallic compound). The hardness of the tough α-solid solution is significantly raised by the hard Fe_2W metallic compound. Its quantity increases with decreasing temperature. The hardening effect of the Fe_2W metallic compound leads to a much more favorable effect if the steel is rapidly cooled from the homogeneous zone of the α-solid solution (from the zone above the line representing the limited solubility). In this case, segregation of the coarse grained solid solution is prevented (as indicated by the dotted line representing the limited solubility in Figure 11.11). Then on heating the steel to the temperature range of 600-700 ^{o}C, finely dispersed metallic compounds will segregate along the grain boundaries resulting in a very high hardness of the steel.

Tungsten is the most important alloying element of the typical multi-component *high-speed steels*. Besides tungsten, high-speed steels contain chromium, vanadium and in certain cases molybdenum and cobalt as an alloying element. The most characteristic representative of high-speed steels is the one containing 0.9 % C, 18 % W, 4 % Cr and 1 % V. The high hardness of these steels – even at elevated temperatures – is based on the natural hardness of the carbides due to the precipitation hardening briefly described above.

Precipitation hardening is the typical heat treatment of high-speed steels. It consists of hardening performed from a high temperature (from the temperature $T = 1250\text{-}1290$ ^{o}C) and a subsequent high temperature tempering (around $T \geq 550$ ^{o}C). The essence of this combined heat treatment can be summarized as follows. Due to the poor thermal conductivity of high-speed steels, the increase in heat up to the temperature of hardening is performed in three steps. The first two heating steps are performed in a furnace heating the specimen slowly, followed by the final heating in a salt bath. The aim of this gradual heating is to avoid cracking during heating. The optimal temperature of hardening is about 100 ^{o}C lower than the solidus temperature, assuring perfect dissolution of most of the carbides. The high hardening temperature does not cause grain coarsening due to the presence of primary carbides (c_1) formed during the primary crystallization of high-speed steels. At this high temperature, a very short holding time is sufficient (only a few minutes are necessary for austenitization). Based on the isothermal transformation diagrams of high-speed steels, a cooling rate is selected to avoid the lines representing the starting of pearlitic and bainitic transformations. Through this method, the decomposition of austenite is hindered and, because of the definite shift of the transformation diagrams of high-speed steels to the right, cooling in oil or by air is sufficient. This hardening part of the process results in a microstructure containing martensite and residual austenite.

This is followed by high-temperature tempering (at $T \geq 550$ ^{o}C). During the tempering of high-speed steels, basically the same processes occur as in the tempering of plain carbon steels. The only significant difference is that in the fourth stage of tempering a much higher degree of hardening occurs due to the disperse precipitation of secondary carbides. This can be observed in the temperature range of $T = 540 - 570$ ^{o}C, where normally a significant decrease in martensitic hardness occurs, but in high-speed steels it is overcompensated by the precipitation hardening. The effect of secondary hardening on the hardness of high-speed steels can be seen in Figure 11.12. In this figure, the hardness changes of high-speed steels are shown as a function of the tempering temperature. It can be also seen from the diagram that the degree of secondary hardening depends also on the temperature of austenitization. Within certain limits, an increase in the austenitization temperature leads to an increase in precipitation hardening. This is because the dissolution of alloyed carbides (c_2) is greater in austenite at higher

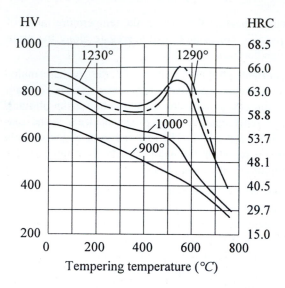

Figure 11.12

The effect of the austenitization temperature on the secondary hardness of high-speed steels

temperatures. During tempering, the secondary carbides precipitate in a dispersed form, providing higher secondary hardness. The tempering of high-speed steels is generally performed repeatedly (usually 2×1 hours, or 3×0.5 hours long).

High-speed steels keep their secondary hardness up to the precipitation temperatures of secondary carbides. Therefore, tools made of high-speed steels played an important role in the spreading of high cutting speed machining technologies. Certain high-speed steels are also widely used to produce hot-forming tools. Since these tools generally have large sizes with large cross sections, these high-speed steels are always alloyed with chromium to assure full hardening.

Tungsten is present in high-speed steels in a high percentage. Since tungsten is a very expensive alloying element, it is often substituted by molybdenum in the so-called *economical* high-speed steels. Additionally, the *Mo* increases the tempering resistance; its further advantage that it is required in a much smaller quantity compared to tungsten. In the so-called *super high-speed steels,* the tungsten content is decreased to 1.5 % and the molybdenum content is in the range of 8.5-9.5 %, having roughly the same machining lifetime.

11.2.5. Miscellaneous alloying elements of steel

11.2.5.1. Molybdenum

Iron-molybdenum alloys crystallize according to a phase diagram similar to the phase diagram characteristic of tungsten steels. Molybdenum can form a solid solution with iron and *Mo*-carbides, as well as double carbides. Molybdenum increases the tempering resistance and the thermal strength of steels. Therefore, the so-called heat-resistant steels always contain 0.5 % *Mo* besides 14-15 % of *Cr*. Molybdenum is not used alone for alloying but as an additional alloying element it has various applications. Molybdenum is added to *CrNi* steels to increase the tempering resistance. It is used for the alloying of 18/8 austenitic *CrNi* steels to increase their resistance against sulfuric acid and chlorinated lime, as well as to increase the tempering resistance of high-speed steels.

11.2.5.2. Vanadium

The binary phase diagram of *Fe-V* alloys is identical to that of the *Fe-Cr* system. Thus, vanadium is closing the γ-zone. Vanadium forms the *FeV* metallic compound with iron and the V_4C_3 carbide with carbon. The main effects of vanadium as an alloying element can be summarized as strong grain refining, powerful deoxidization and nitride formation. All these effects are exerted by vanadium as an additional alloying element leading to a fine grain microstructure and thus good toughness properties of steels. This is why vanadium is also a favored alloying element in tool steels. Due to its strong affinity to oxygen and nitrogen, it is used in the metallurgical processes as a killing and microalloying component. Due to its capability for nitride forming, vanadium is one of the most important alloying elements in nitriding.

11.2.5.3. Titanium and niobium

Ti possesses a very high affinity to oxygen and nitrogen. Therefore, *Ti* is the strongest deoxidizing and denitrating alloying element in steels. Its oxide easily becomes slag. Therefore, steels killed with titanium are free of slag. Because of its high affinity to nitrogen, it is used for the microalloying of structural steels to increase their strength. The affinity of *Ti* to carbon is much stronger than that of chromium. This is utilized in preventing the grain boundary corrosion of 18/8 *CrNi* austenitic stainless steels after welding and heat treatment. In this case, the carbon is bonded in the form of *TiC*, thus preventing the precipitation of high *Cr* content carbides along the grain boundaries of austenite. For this purpose, the *Ti* content should be five times higher than the carbon content of the alloy. Niobium has a similar effect to titanium in preventing the grain boundary corrosion of austenitic *CrNi* steels. However, the niobium content should be ten times higher than the carbon content. Both titanium and niobium are important microalloying elements in steels.

11.2.5.4. Boron

The application of *boron* as an alloying element has a short history. The unique feature of boron is that it is a very powerful alloying element even in very small quantities. It significantly increases the fully hardened cross section. Even with a concentration as small as 0.0005 % boron, it may be effective and its regular concentration in steels is about 0.0025 %. The main disadvantage of alloying with boron is that it frequently causes a significant scattering of mechanical properties and that its effect decreases with increasing the carbon content.

11.2.5.5. Silicon

The affinity of silicon to oxygen is higher than that of iron. Therefore, it is the most important element for producing *"killed steels"*. Silicon creates a closed γ-zone in the *Fe-Si* phase diagram at about 2 % *Si*. However, this limit is expanded to 8 % *Si* by adding 0.3 % carbon to the alloy. The so-called silicon-ferrite zone expands to a content of 8-15 % *Si*. Above 15 % *Si* content, the *Fe$_3$Si$_2$* chemical compound appears. It is very brittle but has good acid resistance. Silicon is used for alloying structural steels in the concentration range of *Si*=1-3 %.

One of the most frequent application fields of silicon is the production of steels for springs. The normal silicon content of transformer core sheets is 3.6-4.4 %. *Si* effectively decreases the power loss. The *Fe$_3$Si$_2$* compound formed above 15 % *Si* is resistant to hot sulfuric and nitric acid. This property is utilized in the production of evaporators for nitric and sulfuric acid. In these cases, the silicon content is in the range of 15-18 %. This type of steel is very hard, rigid and brittle; therefore, it can be machined only by grinding or with hard metals. Silicon has a strong graphitization effect in cast irons. Therefore, silicon is an important alloying element of gray cast irons. However, this property of silicon is disadvantageous in steels causing the danger of "*black fracture*" of structural steels. This may occur during heat treatment of steels if a steel with high silicon content is held at the A_1 temperature for a sufficient time to start graphite segregation. Therefore, the silicon structural steels require very careful and precise heat treatment. Heating should be performed at the lower limiting value of the required temperature for the shortest possible period.

11.2.5.6. Aluminum

Aluminum can be dissolved in α-iron up to a concentration of 15 %. It reduces the γ-zone. The great affinity of aluminum to oxygen and nitrogen is utilized in steel making processes. Due to its high affinity to oxygen, aluminum is one of the most effective deoxidizing elements. The aluminum oxides formed during deoxidation act as tiny foreign crystallization nuclei resulting in significant grain refinement. The affinity of aluminum to nitrogen leads to the formation of the aluminum nitride (*AlN*) metallic compound. Thus, aluminum bonding of the free nitrogen effectively contributes to preventing of aging and embrittlement of mild steels. These effects of aluminum are utilized in sophisticated steel making using aluminum as microalloying element. Aluminum is used for surface hardening of steels. Adding 1-1.5 % *Al* to steel and diffusing nitrogen, a very hard surface layer of *AlN* is created providing good resistance to wear.

Adding 1-3% *Al* to heat resistant *Cr*-steels increases the thermal resistance of these steels. 8-15% *Al is* alloyed to produce very strong *Al-Ni* and *Al-Ni-Co* permanent magnets.

11.3. Recent development trends in steels

In the past few decades, steel making has had to face particularly great challenges. This was partly connected with heavy economic changes worldwide and with the development of various new structural materials. All these increased the need to elaborate new grades of steels fulfilling simultaneously more and more complex requirements: higher strength, good workability (formability, weldability,

machinability, etc.), toughness, resistance to brittle fracture, etc. in an ever cheaper and more economical manner. In the following section, these requirements will briefly be overviewed.

11.3.1. Requirements to meet new challenges in the production of steels

Increasing the strength of steels has been one of the central problems of materials science since steel making commenced. This problem was emphatically put in the foreground in the past decades and especially following the oil price boom in the 1970's. Tremendous efforts were exerted all over the world (and are exerted even today) to manufacture steels of higher strength at a lower cost. Applying structural materials with higher strength properties and higher load bearing capacity results in less material consumption. This in turn increases the economy of application. The best examples can be found in the automotive industry, one of the biggest "consumers" of the products of steel making. In order to justify this, in Table 11.3, the technical data of the *Volkswagen Beetle*, the most sold car in the world in all times and those of *Volkswagen Golf*, elected as the car of the year in 1992, are compared. From this table, it can be seen that the specific mass was halved due to the application of higher strength materials (as well as of other automotive developments). At the same time, the power is doubled and the final speed is 50 % higher with 35 % less consumption.

Table 11.3 Comparison of technical data of two VW models

Type	Power [kW/HP]	Mass [kg]	Consumption [l/100 km]	Final speed [km/h]	Specific mass [kg/kW]
VW Beetle	25/34	760	10.0	114	30.4
VW Golf	51/70	780	6.5	170	15.2

Similar examples can also be found in other application fields. A spectacular development can be introduced by comparing a few parameters of the old and the new Elisabeth bridge crossing the River Danube in Budapest (see Table 11.4). From the data of this table, it can be seen that 11,700 tons of steel was used to construct the old Elisabeth bridge between 1898-1903 and only 6,300 tons of steel was used to build the new one in 1961-1964. At the same time, the road width was increased by one half due to the higher loadability of the structure. (The strength column in the table indicates the permitted strength of the cables applied.)

Table 11.4 Comparison of technical data of old and new Elisabeth bridges

Name	Construction time	Strength [N/mm^2]	Built-in mass [tons]
"Old" Elisabeth bridge	1898-1903	110	11,700
"New" Elisabeth bridge	1961-1964	480	6,300

Beside these requirements (high strength, good formability, weldability), the brittle fracture resistance is a frequent demand in the case of structures operating under outdoor conditions. (The danger of brittle fracture was brought to the foreground of materials science research following some great catastrophes. The embrittlement of structures operating at low ambient – sometimes negative – temperatures were first experienced several decades ago. The sinking of the huge *Liberty* tanker during the World War II or the collapse of the steel structure of the railway station in Berlin can both be explained by brittle fracture.)

However, these requirements sometimes contradict each other. It is well known that with increasing strength, the degree of ductility decreases. In many cases alloying, aimed at increasing the strength results in an increased transition temperature and, consequently this increases the tendency for brittle fracture. Therefore, in the forthcoming sections, we shall briefly overview the materials science background explaining how these requirements are met and we shall analyze the latest efforts in the development of materials based on these concepts.

11.3.2. Methods to increase the strength of steels

Several methods are known for increasing the strength. It is well known that the mechanical properties of steels are essentially determined by the chemical composition, dislocation structure and microstructure. Consequently, an increase in strength can be achieved by modification of these parameters. Some possibilities to increase strength are similar for both unalloyed and alloyed steels, while some are different. (For example, martensitic transformation, grain refinement or plastic deformation can equally be applied as mechanisms of increasing the strength of unalloyed and alloyed steels, however, dissolution and segregation as strength increasing mechanisms basically apply only to alloyed steels).

The change in the chemical composition of plain carbon steels is quite limited. This means only a change in carbon content. In Figure 11.13, the effect of carbon content on the mechanical properties is shown. It can be seen in the diagram that as the strength properties ($R_{p0.2}$, R_m, HB) improve, the ductility parameters (A, Z) decrease. This verifies the previously mentioned contradiction among requirements. It is also a well-known fact that the number of dislocations can be increased with cold plastic deformation.

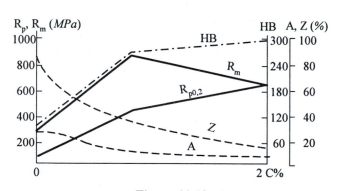

Figure 11.13
The change in strength and ductility parameters of plain carbon steels vs. carbon content

This results in an increased energy level of the deformed crystals; consequently, the strength parameters of the metal compared to the mild state can be significantly improved. Usually, the strength improving effect of plastic deformation can be characterized by the so-called hardening law describing the relationship between the true strain (φ) and the true stress (σ'). One of the most generally applied approximate relationships is the so-called power law of hardening established by Nádai:

$$\sigma' = K\varphi^n, \tag{11.2}$$

where K is the hardening coefficient (its value is equal to the true stress that results in a unit true strain), n is the hardening exponent that can be taken equal to the value of the true uniform elongation – i.e. $n = \varphi_m$ – which can be determined by a uniaxial tensile test). Usually, the increase of strength achieved by plastic deformation is expressed as a function of the dislocation density, namely:

$$\Delta\sigma = k\sqrt{\rho}, \tag{11.3}$$

where ρ stands for the dislocation density and k is a material constant depending on the type of dislocation. It is obvious from earlier studies that the hardening effect of plastic deformation may be applied to improve the strength parameters only in those cases where the operating temperature never exceeds the temperature of recrystallization.

An even greater possibility to increase the strength of plain carbon steels is provided by modification of the microstructure. Analyzing the non-equilibrium transformations in Chapter 9, we could see that the mechanical properties of plain carbon steels might be modified in a wide range by applying different cooling rates. Increasing the cooling rate results first in the refinement of pearlite plates, thus increasing the strength of the pearlitic microstructure. Increasing the cooling rate further, bainitic and finally martensitic microstructure can be achieved. It is known that the highest strength can be obtained by the martensitic transformation.

In martensitic transformations, several strength-improving mechanisms play a role. Due to the difference between the specific volumes of austenite and martensite, a high degree of deformation occurs, which leads to the increase of dislocation density from $\rho = 10^6$ cm^{-2} to a value of $\rho = 10^{12}$ cm^{-2}. The former value corresponds to the value in mild austenite, while the latter corresponds to that of the hardened martensite. Taking into consideration the relationship (11.3), it can be seen that this increase in the dislocation density results in such a high increase in strength as that achieved by plastic deformation with the true strain $\varphi = 0.6$-0.8.

The significant increase in strength obtained by martensitic transformation is caused by the distortion of the crystal lattice, originating from the excess carbon that was originally dissolved in austenite and remained in the martensite lattice due to the diffusionless transformation. The increase in strength resulting from this can be estimated by the so-called *solution strengthening mechanism*, which will be described later when discussing the strengthening mechanisms of alloyed steels.

In alloyed steels, there are further strengthening mechanisms besides the previously described ones. The strengthening effect of alloying elements may vary depending on whether they can form solid solutions with the fundamental metal or whether they create separate phases (discrete particles). Two different types of separate phases can be distinguished: segregation or dispersed particles.

If the alloying element forms solid solutions with the fundamental metal, the strengthening effect is exerted through the so-called *solution hardening mechanism*. In solid solutions, part of the alloying atoms is located in the elastically distorted region of dislocations. Alloying atoms with an atomic radius smaller than that of the fundamental metal are typically located in the *compressed zone*, while alloying atoms with a larger atomic radius are typically located in the *stretched zone* of dislocations. These are shown in Figure 11.14. This neighborhood of the dislocations, rich in foreign atoms, is called the *Cottrell atmosphere*. These foreign atoms obstruct the movement of dislocations. This is one of the fundamental reasons for the strengthening effect of solved atoms. The solute atoms – regardless of being smaller or larger than the atoms of the fundamental metal – always distort the lattice. Consequently, they increase the energy level of the lattice and thus the strength. If the content of the alloying element is not more than a few percent, their strengthening effect can be determined from the relationship:

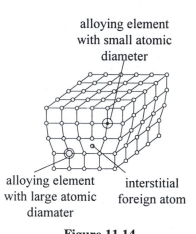

alloying element
with small atomic
diameter

alloying element   interstitial
with large atomic   foreign atom
diamater

Figure 11.14
Locations of dissolved alloying atoms
in the distorted range of edge
dislocation

$$\Delta\sigma = G\varepsilon^2 c \qquad (11.4)$$

where $\Delta\sigma$ is the yield strength increment achieved by alloying, G is the shear elastic modulus of the material, c is the concentration of the dissolved alloying element and ε is the so-called *specific difference between the ion diameters*. It may be calculated from the expression:

$$\varepsilon = \frac{d_o - d}{d_o}. \qquad (11.5)$$

In the above expression, d_o is the diameter of the ions of the fundamental metal, while d represents the same value for the alloying ions. According to relationship (11.4), the higher the concentration of the alloying element and the greater the specific difference between the diameters of the ions, the greater the strengthening effect. According to the *Hume-Rothery* law, the condition of forming an unlimited solid solution is that the specific difference between the diameters of ions does not

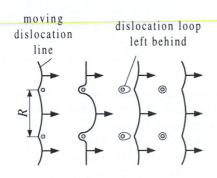

moving dislocation line

dislocation loop left behind

R

Figure 11.15

Dislocation loops formed around segregation

exceed the value of $\varepsilon \leq 0.15$. (This value is called the dimensional factor of the formation of a solid solution.)

Another strengthening mechanism of alloying elements is the so-called *precipitation hardening mechanism*. The alloying elements usually generate much higher strengthening if they form a separate phase, for example, a metallic compound with the fundamental metal or with each other. (Examples of metallic compounds formed by C and N with the fundamental metal are Fe_3C, as well as Fe_2N, Fe_4N. Carbides may be also mentioned as examples for metallic compounds formed by the alloying elements, such as Cr_2C_3, AlN, TiN, and the type $(Fe,Me)_xC_y$ so-called double or complex carbides.)

The increase in strength obtained by precipitation hardening can be calculated by the expression:

$$\Delta\sigma = 2G\,\varepsilon\,c \qquad\qquad (11.6)$$

provided that the precipitate and the fundamental crystal form a coherent phase boundary. It follows from this expression that the precipitation hardening mechanism is more powerful than the solution mechanism. First, this is due to the factor 2 in the expression (11.6). Secondly, the specific difference of ion diameters (ε) that always has a value less than one is raised only to the power of 1 in expression (11.6), as opposed to the expression (11.4) where it is raised to the exponent 2. The strength increase caused by precipitation hardening can be explained as follows. The coherent or semi-coherent phase boundaries hinder the movements of dislocations. The coherent phase boundaries cause distortions of both the fundamental crystals and the segregated grains. The dislocations in the distorted regions can move only under higher stresses. The last step in the precipitation process is the formation of the equilibrium phase, which in most cases is accompanied by the cessation of coherent phase boundaries. The increase in strength caused by the non-dissolving precipitation having incoherent phase boundaries is called *dispersion hardening*. The moving dislocations cannot pass through the material containing a dispersed secondary phase and they therefore move around this secondary phase. This is illustrated by Figure 11.15 showing the dislocation loops around the segregation originated by the movement of a dislocation. The first theoretical interpretation of this mechanism was given by *Orowan*.

The closer the dispersed segregated particles, the higher the stress required for the motion of dislocations. Their effect on the strength increase can be characterized by the relationship:

$$\Delta\sigma = \frac{Gb}{R} \tag{11.7}$$

where b is the absolute value of Burger's vector of dislocations and R is the distance between the dispersed particles.

The magnitude of strength increase obtained by dispersion hardening is comparable to the values achieved by solution hardening, but it remains below the value of strength increase obtained by precipitation hardening. The stability of strength increase achieved by dispersion hardening depends on the change of the ratio of dispersed particles as a function of temperature. The quantity of metallic compounds in the alloys (Fe_3C, and/or Fe_2N, Fe_4N) decreases with increasing temperature. Proportionally to these changes, the strength will also decrease due to the dissolution of particles, and their ratio is also decreased as the average distance between the particles is increased. In many materials, this phenomenon is connected to the term *overaging* resulting in a decrease in strength. Therefore, the strength increase caused by dispersion hardening is much more stable if the ratio of dispersed particles does not change with temperature. Such phases are SiO_2, Al_2O_3, which are insoluble in certain metals and maintain their strength increasing effect even with increasing temperature.

In many cases, the alloying elements cause an increase in strength through their *grain refining effects*. This may be explained by the fact that the dislocations are unable to pass through the grain boundaries. It is obvious that the finer the grain structure, the greater the ratio of grain boundaries and, consequently, the higher the strength increase due to the grain size. The strength increasing effect of grain sizes was already discussed in Chapter 4 (see the *Hall-Petch* relationship).

11.3.3. Dynamic load and the ductile-to-brittle transition temperature

Some of the huge catastrophes of the past decades drew attention to the fact that higher strength properties are not the only criterion of the loadability of structural steels. We could see, for example, that on increasing the strength the ductility parameters may decrease significantly. This could lead to fracture of dynamically loaded parts or structural elements even in cases when the load remains justifiably well below the classical design criteria (i.e. the yield strength). In most of these damages, the low operating temperature also played an important role.

It is well known that the strength properties of steels increase and the ductility parameters decrease with decreasing temperature. This relationship is shown in Figure 10.3 for plain carbon steels. It is very important that metals having a face-centered crystalline structure (for example *Cu, Ni, Al*) keep their ductility to some extent even close to the absolute zero temperature ($0~K$), while metals having a body-centered crystalline structure completely loose their ductility with decreasing temperature. This phenomenon is called *embrittlement*. Embrittlement depends not only on temperature, but also on the loading rate and stress conditions.

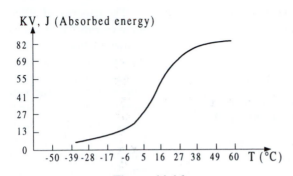

KV, J (Absorbed energy)

Figure 11.16

The impact work necessary to
fracture vs. temperature

Resistance to dynamic loads or the tendency to embrittlement can be analyzed in the simplest way by the so-called *impact test*. During this test, a notched specimen having a cross section of 10x10 mm is broken by an appropriate impact device (the so-called *Charpy pendulum*) by a single impact. The amount of work necessary to fracture the specimen and the fractured surface obtained from this test are suitable for characterization of the resistance of a material to dynamic load. The impact test is the most suitable simple method to determine the tendency for embrittlement. Figure 11.16 shows the results of impact tests performed at different temperatures on the same type of material. From this figure, it can be seen that as the temperature decreases, the amount of work necessary to fracture radically decreases below a certain temperature. The tendency to embrittlement can be characterized by the so-called transition temperature. The *transition temperature* represents the temperature at which the material passes from the ductile to the brittle state. The inflection point of the curve – describing the amount of impact work necessary to fracture vs. temperature – is accepted as the transition temperature. For accurate determination of the transition temperature, several methods have been proposed. In one of these, the transition temperature is assigned to a prescribed impact work (for example, the fracture work is equal to $KV = 40$ J). This transition temperature for a "V"-notched test specimen is denoted by the sign *TTKV* according to convention.

The transition temperature of steels is significantly influenced by the chemical composition. However, the effect of composition is not as decisive as that of the temperature or the stress conditions. The effect of carbon as the main alloying element of steel is shown in Figure 10.2. It can be seen from this figure, that increasing the C-content by 0.6 % (from the value of $C = 0.01$ % to 0.63 %), the transition temperature is increased by about 180°C. This clearly indicates that carbon has an intense embrittlement effect.

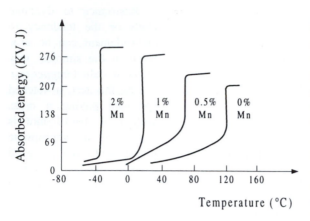

The majority of alloying elements has an effect similar to that of carbon on the transition temperature of steels. Manganese is one of the rear elements showing completely opposite effects. Figure 11.17 clearly shows that an increase in *Mn*-content by 2 % decreases the transition temperature from + 120 °C to - 40 °C. This means that manganese significantly reduces the tendency to embrittlement.

Figure 11.17
The effect of Mn alloying on the transition temperature

11.3.4. Formability, weldability

The grades of steels in common use today have to fulfill both functional requirements as well as demands for relatively simple, easy and cheap workability and further processing. It is well known that all changes that increase the strength simultaneously reduce *formability*. An increase in carbon content decreases formability due to the increasing pearlite content. By analyzing the effects of other alloying elements, it can be stated that those forming solid solutions with the fundamental metal reduce formability to a lesser extent than those increasing the strength by precipitation hardening.

A not too low martensitic start temperature (M_S) is an essential condition of good *weldability*. A low value of M_S and a high hardness of martensite decrease weldability. The reduced weldability of steels with increasing carbon content can be reasoned by the fact that as carbon content increases, the value of the M_S temperature decreases (see Figure 9.15). Obviously, increasing hardness and brittleness with a higher carbon content reduces weldability.

11.3.5. Some recent developments in steel making practice

Previously, we briefly summarized the results achieved by material sciences concerning the improvement of steels in the past few decades. One of the best examples is represented by the **High Strength Low Alloyed Steels** (*HSLA-steels*), whilst the so-called *dual-phase steels* are regarded as another good example.

11.3.5.1. High strength low alloyed steels - the HSLA steels

The most important characteristics of *HSLA* steels are as follows: high strength (R_m = 400-1000 N/mm²), relatively good formability and weldability, low transition temperature and high resistance to brittle fracture. These excellent properties of *HSLA*-steels are ensured by the addition of small quantities of low cost alloying elements. The *HSLA*-steels have a low carbon content (C = 0.1-0.2 %) and contain only manganese as a significant alloying element (the *Mn* content usually varies in the range 1.0-1.7 %). These steels may also contain *V*, *Nb*, *Ti*, *Al* and *N* as microalloying elements. The total quantity of these elements does not usually exceed 0.12 %. (We can speak about m*icroalloying* when the total quantity of the additional alloying elements does not exceed 0.15 %.) Typical microalloying elements are *V*, *Nb*, *Ti*, *Al*, *N*, *Zr*, and *B*.

The *low carbon content* in *HSLA*-steels is of great importance for good formability, weldability and a low transition temperature (in order to have good resistance to brittle fracture). The low carbon content can be ensured with the *LD*-converter processing without significant extra costs.

Mn as an *alloying* element plays an important role from many aspects. *Mn* exerts its strengthening effect through the solution hardening mechanism. However, a too high manganese content is not beneficial because it decreases the starting temperature of martensitic transformation and thus decreases both formability and weldability. Nevertheless, the most advantageous effect of *Mn* is that it significantly reduces the transition temperature and thus, the tendency of steels to brittle fracture.

Microalloying elements exert their beneficial effects in many ways. To increase the strength, they utilize both the solution and the precipitation hardening mechanisms. For example, *Mn* and *Al* are added to increase the strength of ferrite. Precipitation hardening is achieved by those elements (for example vanadium, niobium and titanium) that form a small quantity of hardly soluble dispersed carbides and nitrides. Vanadium, niobium, titanium and aluminum contribute to the increase in strength through grain refinement.

Various types of high strength low alloyed steels are known which typically contain these alloying elements. In Figure 11.18, the relationship between the yield strength and the transition temperature is shown for a group of norma-

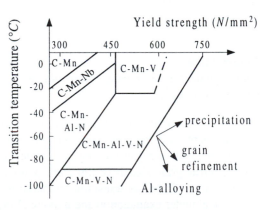

Figure 11.18

Yield strength vs. transition temperature for various *HSLA* steels

lized *HSLA* alloys. In the respective fields, the chemical symbols of the alloying elements characteristic of the given alloy group are shown. Arrows in the figure indicate the direction in which the grain refinement and segregation influence the yield strength and the transition temperature. From this figure, it can be seen that a heat treatment increasing the strength through precipitation hardening will also increase the transition temperature and thus the tendency for brittle fracture. In contrast, microalloying elements increasing the strength through grain refinement also contribute to a decrease in transition temperature.

A more illustrative presentation of these effects is shown in the so-called *Irvine* diagram (Figure 11.19). In this figure, the effects of vanadium and niobium on the yield strength and on the transition temperature of *HSLA*-steels are shown. Since both alloying elements play an outstanding role in precipitation hardening as well as in grain refinement, their effects on the increase of strength and transition temperature may only be understood in its complexity.

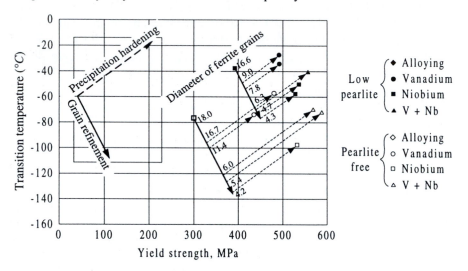

Figure 11.19
The effects of vanadium and niobium on the transition temperature and yield
strength of HSLA-steels

The diagram can be used as follows. The yield strength and the transition temperature of steel not containing the mentioned two microalloying elements determine a point in the coordinate system with a corresponding size of ferrite grains. (Select the point belonging to the ferrite grain size d=18 μm corresponding to R_p = 300 N/mm^2 as yield point and *TTKV* = -75 °C as transition temperature.) By adding vanadium and niobium to the steel under examination and moving along the arrow plotted with a continuous line, the size of the ferrite grains decreases as the alloying content increases. This is indicated by the decreasing size of the ferrite grains in the range of 16.7...4.2 μm in the diagram. If only grain refinement occurred, the yield point would increase and the transition temperature would decrease according to the arrow plotted with a continuous line. However, both alloying elements increase the strength through the precipitation hardening mechanism as well, which increases the transition temperature. The change can be followed along the arrow drawn with the dotted line. This means that starting from an initial grain size (18 μm), the addition of V and Nb alloying elements will decrease the grain size to 5.4 µm. As a result of the simultaneous grain refinement and precipitation hardening, the yield strength of steel will be almost doubled to a value of $R_p \approx 580$ N/mm^2. However, the transition temperature will remain at about the same value ($TTKV \approx -75$ °C) as earlier.

Similarly, we can also study the separate effects of vanadium and niobium. Since the lengths of the vectors drawn in the diagram are proportional to the magnitude of their effects, it may be concluded from the Irvine diagram that vanadium promotes the precipitation hardening mechanism and niobium increases the strength mainly through its grain refinement effect.

We have considered the effects of the parameters influencing the properties of *HSLA*-steels. These steels are widely used in engineering practice with special emphasis on the automotive industry. Generally, the properties of steels are significantly influenced by the production and manufacturing processes. This is particularly valid for *HSLA*-steels. Therefore, in the forthcoming section, the processes occurring during the technological procedures will be discussed.

The excellent properties of *HSLA*-steels are mainly ensured by plastic deformation performed at an optimal temperature followed by a controlled cooling. Therefore, in the production of *HSLA*-steels, the realization of the optimal thermo-mechanical system is of outstanding importance.

Grain refinement also plays an important role in providing the prescribed properties of *HSLA*-steels. This can be achieved primarily by microalloying with *Nb, V, Ti* and *Al* elements. From the point of view of grain refinement, it is more beneficial if the solubility of the alloying element in austenite is low. In this case, the alloying elements precipitating on the grain boundaries hinder the growth of crystallites. This is also the most favorable solution from an economical point of view, since the same effect can be achieved with a smaller quantity of alloying element. (The decrease in the austenite grain size is of utmost importance, since the sizes of the phase transformation products are smaller if the initial grain size of the austenite is smaller.)

The size of ferrite grains is influenced by the initial grain size of the austenite, and whether it is formed from the plastically deformed austenite or from the austenite recrystallized after hot forming. When we analyzed the rules of crystallization, we could see that the critical size of a crystallization nucleus capable of growing is given by the relationship:

$$r_{crit} = \frac{2\sigma}{\Delta G_v} \, , \tag{11.8}$$

where σ is the surface energy of the nucleus and ΔG_v represents the difference between the free energies of austenite and ferrite. In Figure 11.20, the free energy changes of the mild austenite and that of the ferrite are plotted as a function of temperature. It can be seen from the figure that the free energy differences are greater for the deformed austenite than for the undeformed (mild) austenite with the same degree of undercooling. Consequently, the critical size of the nucleus for the deformed austenite as an initial phase is smaller. This means that it is more favorable for the formation of a finer grain structure. The degree of undercooling has an additional effect. The greater the degree of undercooling, the higher the probability of the formation of new ferrite nuclei that are capable of growth, i.e. the higher the

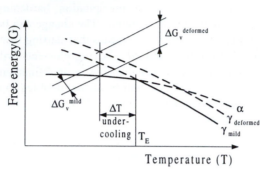

Figure 11.20
The free energy changes of ferrite, mild and deformed austenite vs. temperature

crystallization rate of ferrite. The degree of undercooling increases with the cooling rate and, proportionally, the number of ferrite nuclei formed in a unit time increases resulting in a finer microstructure. At the same time, it is obvious that the cooling rate cannot be increased infinitely, since above a certain cooling rate it would result in the formation of bainite or even martensite. The formation of new ferrite nuclei on the boundaries of austenite grains is much more probable than inside the grains.

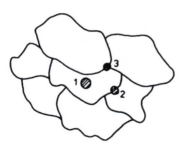

Figure 11.21
The possible locations of new ferrite nuclei

Figure 11.21 indicates the possible positions of formation of new ferrite nuclei. New ferrite nuclei are formed with the greatest probability in position 3 because three austenite grains meet at that point. The surface energy requirement of the new nuclei is the lowest in this location compared to locations marked by 2 and 1. At the same time, it follows from this that a greater number and thus finer ferrite nuclei originate from austenite having a finer grain structure.

To ensure the formation of ferrite grains from the finer, deformed (and not recrystallized) austenite, the addition of microalloying elements to retard recrystallization of austenite is required. In *HSLA*-steels, this is provided by *Nb, Ti* and *V* microalloying elements.

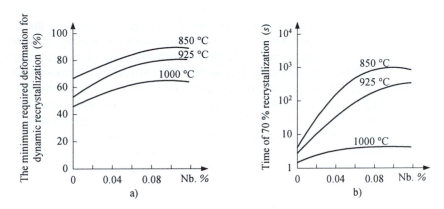

Figure 11.22
The effect of *Nb*-content on the parameters of recrystallization

Microalloying elements can delay recrystallization of austenite because, at the beginning of hot forming, the microalloying elements, as well as *C* and *N*, are in solution. As the temperature decreases, segregation starts. The presence of precipitates hinders recrystallization. The higher the degree of precipitation, the later the recrystallization starts. This is indicated in Figure 11.22 for the *Nb* micro-alloying element. An increase in *Nb*-content increases both the minimum degree of deformation required to start recrystallization (the so-called *threshold value* of recrystallization) and the amount of time required for recrystallization.

The other microalloying elements have similar effects. Only slight differences can be found between the temperatures at which the segregation of carbides and nitrides of these microalloying elements start. This is illustrated in Figure 11.23 for the *Nb*, *Ti*, *V* and *Al* microalloying elements. On the vertical axes, the concentration of the microalloying element, and on the horizontal axes the concentration of the metalloid forming the relevant compound are shown.

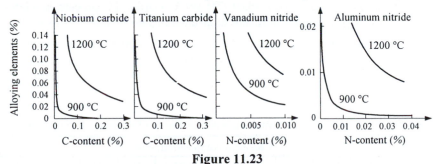

Figure 11.23
The solubility of the most frequent compounds of
microalloying elements in austenite

11.3.5.2. Dual-phase steels

HSLA-steels have many excellent properties. Beside high strength, they possess significant plasticity compared to conventional steels. However, in many cases this relatively good formability is not sufficient for manufacturing of complex parts that require high strength and good formability. This is one of the reasons why researchers' attention has turned to the elaboration of a new type of steels possessing high strength similar to that of *HSLA*-steels with good formability, resembling the unalloyed low *C*-content ($C < 0.1$ %) steels. The first experiments roughly coincided with the first wave of the oil crisis, which forced the users – first, the automotive industry – to achieve energy saving by mass reduction but without increasing prices. Since that time, industrial practice has already proved that the *dual-phase steels (DP-steels)* match the above intentions.

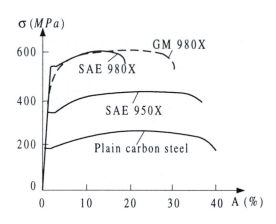

Figure 11.24
The effect of heat treatment of dual-phase steels on the properties of HSLA-steels

In Figure 11.24, the tensile diagrams of three different types of steels are compared. Among them, the properties of the steel having the highest strength (*SAE 980X*) corresponding to *HSLA*-steel have been further improved by the technology characteristic of *DP*-steels. This included the heating of *SAE 980X* steel after the usual production process to a temperature of 787 °C, then holding it for 4 minutes and quenching. The properties of the steel that has been heat treated in this way are indicated by the curve marked *GM 980X*. During this process, a microstructure consisting of martensite islands embedded in a ferrite matrix is formed.

The chemical compositions of steels shown in the figure are summarized in Table 11.5. By comparing the properties of the three different steels, it can be stated that the applied heat treatment significantly improves the properties of commercially available steels.

Table 11.5 Comparison of chemical compositions of various steel grades

Type of steel	Chemical composition, mass %									
	C	Mn	Si	P	S	Mo	V	Al	N	O
Unalloyed	0.05	0.32	0.02	-	-	-	-	-	-	-
HSLA	0.09	0.83	0.12	-	-	0.011	0.05	-	0.011	0.004
DP-steel	0.12	1.46	0.51	0.008	0.009	0.008	0.11	0.11	0.019	0.002

The *DP*-steels are generally continuously heat-treated (in a traction furnace or in a salt bath) or can be heat-treated by controlled cooling following hot rolling. The essence of the two kinds of heat treatments can be summarized as follows. At the so-called intercritical temperature, the desired ferrite/austenite ratio is set in the α+γ zone. In this example, the amount of usually 10-20 % austenite crystallites is surrounded by a greater quantity of deformable ferrite crystallites. The carbon content of the austenite crystallites far exceeds the average value corresponding to this temperature and provides better hardenability. When the steel is hardened by starting from the intercritical temperature, martensite is formed from austenite during the phase transformation. Thus, a ferrite-martensitic microstructure typical of *DP*-steels is obtained in this way. The mechanical properties of these steels depend on the quantity, size and distribution of martensite. In Figure 11.25, the microstructure of an unalloyed plain carbon steel containing 0.1 % carbon is shown after the usual *DP*-type heat treatment.

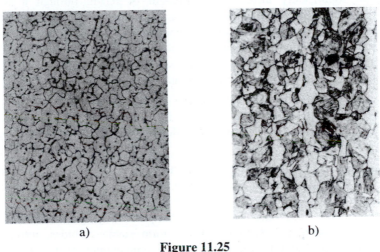

a)   b)

Figure 11.25
Microstructure of an unalloyed steel with 0.1 % of C-content after
a) conventional heat treatment b) DP-type heat treatment

The main characteristics of steels manufactured in the previously described manner are the following:
- They have a mild microstructure consisting of dispersed (10-20 %) martensite islands embedded in a formable ferrite.
- Besides their high tensile strength, they have unusually good plasticity.
- There is no definite yield point on their tensile test diagrams.
- Their deformation hardening exponents are quite high, indicating large uniform elongation.
- Their mechanical properties show slight anisotropy.

With regard to the chemical compositions of dual-phase steels, the following should be taken into consideration. The carbon content of dual-phase steels is usually around 0.1 %. The strengthening capability of carbon in *DP*-steels is of special importance because its strength also depends on the carbon content besides the definite volumetric ratio of martensite. Its drawback is that it decreases weldability and impact work and, consequently, it increases the transition temperature. Therefore, it is advisable to keep carbon content as low as possible.

Besides *C, Mn* is the most significant component of *DP*-steels. Usual quantities are in the concentration range of 1.0-2.2 % (generally between 1.4-1.6 %). It significantly increases the steel strength and decreases the transition temperature. In the previous discussion, we found that the transition temperature is the most characteristic parameter of steels indicating embrittlement. The most important role of *Mn*, besides decreasing the transition temperature, is manifested in the significant decrease in the phase transformation temperature. Thus, the volumetric ratio of ferrite/martensite can be easily controlled. This is extremely important from the viewpoint of determining the properties of *DP*-steels.

Silicon significantly increases the strength of ferrite through a solution hardening mechanism. However, it also increases the transformation temperatures. The increase in the strength of ferrite is accompanied by a reduction in plasticity. Therefore, when plasticity is important, the silicon content in the solid solution has to be kept below certain limits. Generally, steels contain 0.30 % *Si*. In a higher concentration, it increases notch-sensitivity and reduces weldability.

Cr does not increase strength despite the fact that its lattice distorting capability is close to that of *Mn*. The most likely reason is that it forms a more stable carbide with carbon than iron and, therefore, it is present in the form of carbides.

The *Al*-content of *DP*-steels is low. Its strength increasing capacity is half of that of *Mn*. It has strong grain refining effect. *Al* forms a stable compound with *N*. It significantly decreases the transition temperature up to a concentration of 0.2 %. *DP*-steels contain *Nb*, *Ti* and *V* as the most frequent alloying elements with high precipitation hardening capabilities. These form stable carbides, nitrides and carbo-nitrides that start to precipitate at the temperature of hot forming of austenite. Their effects are very beneficial since they delay recrystallization. As the temperature decreases, first the compounds of *Al* then those of *Nb* and *Ti,* and finally the *V* compounds start to crystallize in sequence.

Table 11.6 Chemical composition of typical dual-phase steels

Alloying elements, mass %						
C	Mn	Si	Al	P	S	N
0.005	1.51	0.24	0.010	0.009	0.010	0.006
0.060	1.47	0.23	0.009	0.009	0.010	0.007
0.120	1.47	0.24	0.008	0.009	0.010	0.007
0.160	1.53	0.24	0.024	0.009	0.009	0.005
0.200	1.53	0.25	0.023	0.009	0.010	0.006
0.290	1.51	0.26	0.021	0.009	0.010	0.007
0.400	1.53	0.25	0.026	0.009	0.010	0.005

The most important technological steps in the production of dual-phase steels are the proper selection of the so-called *intercritical temperature*, the *holding time* at this temperature and the subsequent *cooling*. Steel has to be heated up to the intercritical temperature, T_i, providing the formation of the α/γ ratio corresponding to the predetermined ratio of ferrite and martensite. The intercritical temperature can be determined from the *Fe-Fe₃C* phase diagram based on the required volumetric ratio of austenite by applying the second lever rule. However, this cannot be applied directly to multi-component steels. In order to accurately determine the intercritical temperatures of steels, the corresponding multi-component phase diagrams should be known. However, with a good approximation, the binary phase diagrams may be used in most cases.

During heating, the transformation of austenite occurs in three steps. Above A_1, the formation of austenite starts immediately on the ferrite-cementite boundaries because at that point the carbon content is appropriate for phase transformation. These locations are both the pearlite islands and the boundaries of tertiary cementite particles located on the borders of the ferrite. This is then followed by the growth of austenite until all the carbides dissolve. The maximum concentration of carbon in the austenite is determined by the maximum carbon dissolving capacity of austenite at the given temperature. It is represented by the *ES*-curve in the two-component system. Such a concentration exists on the $\gamma\text{-}Fe_3C$ boundary. The minimum concentration in austenite can be found at the $\alpha\text{–}\gamma$ border determined by the *GOS* curve. In step 2, the ferrite $\rightarrow$ austenite transformation continues at a lower rate. This is due to the limiting effect of *C*-diffusion in austenite. (The reason for this is that the diffusion coefficient is one order of magnitude higher in ferrite than in austenite.) The third step of the transformation occurs completely in the austenitic structure until it reaches the equilibrium state after redistribution of carbon and manganese.

The formation of martensite is determined by the almost uniform carbon content and the inhomogeneous distribution of substitutional alloying elements during cooling. This is a very complicated process. Where the *C*- and *Mn*-content is higher, the value of the M_S temperature is lower. At these locations, the formation of martensite occurs more easily. However, in the zones containing less *C* and *Mn*, martensitic transformation starts at a higher temperature if the cooling rate is fast enough. Thus, martensite formation may be different in low and high carbon-concentration zones. Accordingly, the characteristics of *austenite to martensite* transformation may vary within a single part.

This transformation differs from the usual *austenite* $\rightarrow$ martensite transformation. The main difference is that it occurs in the $\alpha+\gamma$ zone in these steels. Consequently, in comparison to the normal hardening of steels, the greatest difference originates from the fact that the austenite crystallites are surrounded by ferrite. The increase in specific volume accompanying the formation of martensite causes plastic deformation in the ferrite, especially in the ranges adjacent to the martensite islands. Accordingly, the density of dislocations in ferrite changes significantly. It is greatest near the α-martensite phase boundary and decreases farther away.

The composition of ferrite changes depending on the alloying elements and on the cooling rate in the cooling process. This unavoidably decreases the plasticity of ferrite. In steels containing *Nb*, *Ti* and *V*, these alloying elements form stable carbides, nitrides and carbo-nitrides. The ferrite/austenite phase ratio in thin sheets is established within 2-10 minutes in a traction furnace. The heating in box-type furnaces requires a much longer time. It may take 1-3 hours depending on the charge weight. It has the great disadvantage of grain coarsening of both phases in the α+γ zone. Though the structure of *DP*-steels is ferrite-martensitic in nature, the properties are reduced compared to the fine-grained structure of continuously heat-treated materials.

The most appropriate solution is the controlled cooling after hot rolling. In this case, if steels contain elements (e.g. *Nb*, *Ti*, *V*, and *Al*) delaying recrystallization during or after hot forming, the hot forming can be performed in the pure γ-zone, above the A_3 temperature. Consequently, the γ→α and/or the γ→martensite transformations occur in the whole volume and the transformation products become very fine. This also has positive effects on the increase in strength, toughness and formability properties.

Several versions of intercritical heat treatments are applied in practice. In Figure 11.26, temperature vs. time diagrams are shown ensuring the characteristic microstructure of *DP*-steels. These examples refer to simple binary *Fe-C* steels. (However, the *DP*-steels are usually multi-component alloys, though the most important features may also be characterized by the two-component steels.)

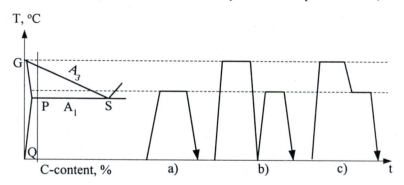

Figure 11.26
Characteristic T-t diagram of dual-phase steels

The simplest solution is shown in part a) of Figure 11.26. It consists of heating the steel to the prescribed T_i intercritical temperature and holding it until the required volumetric ratio of austenite is transformed, and then followed by hardening. This corresponds to the heat treatment of finished parts. The simplicity of the process is advantageous; however, it has the drawback that the ferrite is not transformed and its grain size can grow during heating. Therefore, there are relatively few austenite crystallites at a given volumetric ratio of austenite, leading to coarse martensite islands. This results in a decrease in both the strength and plasticity properties.

The temperature-time program illustrated in part b) of Figure 11.26 eliminates the negative effect of the previous heat treatment. In this case, the steel is heated above the A_3 temperature (above the *GOS* line) into the homogeneous γ-zone. Consequently, all the crystallites are transformed. The microstructure and properties of steel are determined by the cooling from the austenitic zone. This is then followed by the rapid thermal cycle as shown in part b) of Figure 11.26.

In part c) of Figure 11.26, a heat treatment cycle is shown that is indicative of a technological production line, though this method can be utilized only for strip production. Following the last pass of hot rolling in the austenitic zone, a short holding follows at the T_i intercritical temperature. During this holding, the quantity of ferrite corresponding to the equilibrium is transformed from austenite. It has a very fine grain structure since it is formed from the deformed austenite. Obviously, the higher the degree of plastic deformation, the greater the number of nuclei formed in a unit volume. Consequently, the ferrite crystallites have a smaller size. During further cooling, the plastically deformed austenite transforms into martensite. This martensite becomes much finer than the one transformed from undeformed austenite under identical conditions.

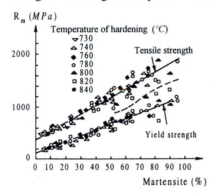

Figure 11.27
The strength of *DP*-steels vs. the martensite content

The most important properties of *DP*-steels are the yield strength, the tensile strength and the elongation. The yield strength and the tensile strength for a given composition are functions of the martensite and/or bainite content.

Figure 11.27 illustrates that the yield strength and the tensile strength change linearly with the martensite content. The volumetric ratio of martensite was changed in two different ways. First, the hardening of steels with different carbon contents was performed from the same intercritical temperature. Then, steels having the same carbon content were hardened from different intercritical temperatures. In the first case, the carbon content of martensite changed due to the difference between the compositions of steels. In the second case, the composition of ferrite and martensite remained unchanged, but the quantity of martensite changed.

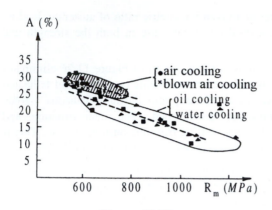

Figure 11.28

Total elongation of DP-steels alloyed with *Mn* and *Si* as a function of tensile strength

The *DP*-steels were developed to provide good formability. In Figure 11.28, the total elongation of steels containing 1.5 % *Mn*, 1.0 % *Si* and 0.037-0.2 % *C* is shown as a function of tensile strength. The groups of steels cooled in air, oil or water can be easily identified in the figure.

The decrease in the plasticity of steels cooled at a higher rate is due to the decrease in the plasticity of ferrite. This may be explained by the larger quantity of dissolved interstitial elements above the solubility limit. This postulation has been verified by microhardness tests on ferrite.

This is illustrated in Figure 11.29 showing that the higher the microhardness of ferrite, the smaller is the elongation of *DP*-steels. In the diagram, the numbers at each measured point represent the quantity of residual austenite.

Obviously, as the microhardness of ferrite increases, the strength of the steel is also increased as shown in Figure 11.30 The numbers at each measured point represent the volumetric percentage of ferrite. Naturally, it may be argued that the cooling rate actually determines the quantities of constituents remaining in the ferrite as a supersaturated solution. Therefore, strength should not be directly considered as the sole consequence of the amount of martensite.

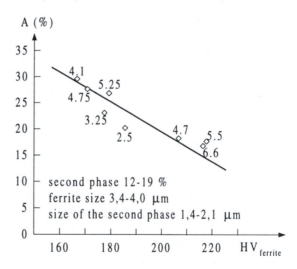

Figure 11.29

The effect of the microhardness of ferrite on total elongation

The plasticity is mainly due to the ferrite. However, according to many experiments, this is further improved by the residual austenite. The significant effect of residual austenite is surprising because this deformable metastable phase

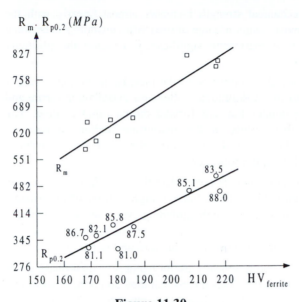

Figure 11.30

The effect of the microhardness of ferrite on the yield
point and tensile strength

at room temperature may only be found enclosed in the islands of martensite. This is because the compression stress that is capable of suppressing the $\gamma \rightarrow$ martensite transformation is high enough only at these locations. It is not the feature of the effect, but rather its magnitude is remarkable.

Formability is closely related to strain hardening characterized by the hardening exponent. The hardening exponent is an important parameter of the well-known relationship attributed to Nádai. In the expression

$$\sigma' = K\varphi^n , \qquad (11.9)$$

σ' is the true stress related to the momentary cross section, φ stands for the so-called true strain that can be calculated from the expression $\varphi = ln\,(l/l_o)$. In this latter expression, l is the current length and, l_o is the original length of the tensile test specimen. K and n are values characteristic of the material that in fact may be considered as constants.

The above function is most generally used to describe the relationship of stress and plastic strain. In Nádai's relationship, the n hardening exponent is equal to the true uniform elongation. It follows from this that the higher the value of the n hardening exponent, the greater the true uniform elongation. Accordingly, the high value of n refers to good formability (actually good stretchability). A high value of n is characteristic of *DP*-steels. Obviously, as the martensite content increases, the degree of formability decreases.

Based on these considerations, the most important features of dual-phase steels can be summarized as follows. *DP*-steels have a dual-phase microstructure, they contain inexpensive and easily accessible alloying elements (usually *Mn, Si, Cr*, less frequently *Mo, V*). Excellent properties can be obtained by starting from the intercritical temperature and rapidly cooling them from the $\alpha+\gamma$ two-phase zone. The production technologies of these steels are relatively simple and inexpensive. Their properties are primarily determined by the martensite islands embedded in ferrite. The tensile strength of these steels is in the range of 700-1000 MPa. At a certain composition, the mechanical strength increases almost linearly with the quantity of martensite. The most common range of martensite quantities is between 10-20 %. Greater quantities of martensite significantly decrease the plasticity characteristics of these steels.

The strength of ferrite may also contribute to the increase in strength. This is essentially due to the – primarily substitutional – elements dissolved in ferrite and to those probable microprecipitates that are formed during rapid cooling. The interstitial elements exert their effect in the martensite and/or through the martensite. Apart from the above two effects, grain refinement also contributes significantly to the increased strength properties.

The formability of dual-phase steels is better than that of *HSLA*-steels having the same strength. This is ensured by the relatively high portion of deformable ferrite (α-ratio > 80 %). *DP*-steels have a high hardening exponent (n) equal to the value of the true uniform elongation. Residual austenite remaining in the microstructure due to rapid cooling also improves the formability parameters.

One of the most remarkable and practically useful features of dual-phase steels is that their tensile curve is continuous. This means that they do not have a definite yield point. (This often means relatively low yield strength.) The change in specific volume accompanying martensitic transformation causes plastic deformation of the ferrite. Firstly, this provides a great number of movable dislocations in ferrite initiating plastic deformation at a small external load, and secondly, due to the large number of dislocation reactions, even a small extent of plastic deformation results in significant hardening.

High strength, low yield point, high plasticity and high hardening exponent, n (referring to a great degree of elongation and hardening capacity) make the *DP*-steels excellent material for parts manufactured by forming processes. These properties are very advantageous in the production of automotive parts.

Earlier we discussed some recent developments based on the complex application of materials science knowledge. These developments are good examples of how the most suitable and most economic products can be achieved by proper selection of the composition of steels, the technological parameters matching the goals and composition (temperature, rate of plastic deformation, degree of deformation), and the setting of the optimal thermo-mechanical system of production.

12. CAST IRONS

12.1. The term cast iron

When we classified iron-carbon alloys, we could see that point E of the equilibrium diagram separates iron-carbon alloys into two large groups, i.e. steels and pig irons. In the foregoing, the first main group, i.e. steels having a carbon content below point E ($C < 2.06$ %), have been discussed. In this chapter, we deal with the other main group, i.e. with pig iron. However, in industrial practice, a rather limited composition range of pig irons (the so-called cast iron) is of great practical importance; therefore, in the following the *industrial cast irons* will be introduced in detail. Cast irons belong to the hypoeutectic group of pig irons. (The carbon content of the eutectic point is $C = 4.3$ % in the *Fe-Fe₃C* metastable system and $C = 4.26$ % in the *Fe-C* stable system.) The hypereutectic pig irons represented by the composition range above the eutectic point do not play any significant role in industrial practice. Therefore, we are not going to discuss this composition range in detail.

12.1.1. Main types and characteristics of cast irons

The properties of cast irons are essentially determined by the form of carbon. In cast irons, carbon may exist in chemically bonded form, i.e. cementite or in a free form of graphite. Thus, two large groups of cast irons are distinguished. Cast irons containing carbon only in a chemically bonded form (i.e. in the form of Fe_3C - cementite) are called *white cast irons*. White cast iron is named after its white, lustrous fracture surface. The properties of white cast irons are determined by the properties of their characteristic microstructure element called ledeburite. *Ledeburite* is a very hard ($HV = 550$ - 600) and rigid microstructure. Due to its great hardness, it has good resistance to wear but, because of its rigidity, it can only be used to manufacture parts and components not subjected to dynamic loads.

Cast iron containing carbon in the free form of graphite is called *gray cast iron* since, due to the graphite in the microstructure, its fractured surface has a characteristic gray color. Gray cast iron has a much wider industrial application range than the white cast iron. Its properties are essentially determined by the shape, size and distribution of the graphite.

When producing various kinds of cast irons, the main task is to control the crystallization of carbon, since it determines the microstructure and properties of cast iron. The mechanism of carbon crystallization can be influenced primarily by modifying the composition and controlling the cooling rate.

12.1.2. Effect of composition on the microstructure of cast iron

The effect of composition on carbon segregation has been well known for a long time. There are alloying elements that promote graphite formation. These elements are called *graphitizing* alloying elements. Naturally, carbon itself belongs to this group of elements together with *Si, Al, Ni, Co* and *Cu*, listed in the reverse sequence of their graphitizing effect. There are alloying elements that stabilize cementite and carbide. These elements are named *whitening elements*, referring to the fracture surface of white cast iron. This group includes *S, V, Cr, Sn, Mo* and *Mn*, listed in the sequence of their *whitening* effect.

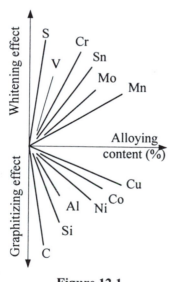

Figure 12.1
Effect of alloying elements on the microstructure of cast iron

The previously mentioned effects of alloying elements are illustrated in the so-called principal diagram of Bunyin (Figure 12.1). The horizontal axis represents the percentage quantities of alloying elements and the vertical axis their *whitening* and the graphitizing effects in the respective directions. Among the alloying elements, carbon plays an outstanding role as regards the microstructure and properties of cast iron. It is well known that the carbon content of the eutectic composition in the two-component *Fe-C* alloy system corresponds to $C = 4.26$ %. The change in the composition of the eutectic point requires special attention since the hypoeutectic composition is more favorable with regard to the strength properties. However, the castability is much better close to the eutectic composition. The alloying elements significantly influence the composition of the eutectic point, even in small quantities. This effect can be expressed by the so-called saturation factor (S). This is the ratio of the actual carbon content (C) and the calculated equivalent eutectic carbon content (C_{eut}) representing the effect of the alloying elements:

$$S = \frac{C}{C_{eut}}. \tag{12.1}$$

The value of C_{eut} may be calculated from the percentage quantities of the alloying elements using the expression:

$$C_{eut} = 4.26 - 0.317Si - 0.33P + 0.027Mn. \tag{12.2}$$

From the relationship (12.2), it can be seen that carbide-forming elements increase the value of the equivalent eutectic composition while graphitizing elements decrease it. The saturation factor of hypoeutectic cast irons is $S < 1$. The eutectic composition corresponds to a value of $S = 1$, while hypereutectic cast irons have saturation factors in the range of $S > 1$.

Concerning the application fields, various malleable and high strength cast irons belong to the group of hypoeutectoid cast iron. These are used as materials for various machine components, machine frames, etc. Cast irons of general quality usually have a eutectic composition and the various pig irons belong to the hypereutectic composition region.

Besides carbon, *silicon* has the strongest effect on the microstructure of cast irons. *Silicon* is a powerful graphitizing and ferrite-forming element that forms metallic compounds with iron. Depending on the composition, complex compounds of $(Fe_3C)_x$ $(Fe_2Si)_y$ are formed in cast iron. Multiplying the silicon content of the complex compound with the quantity of eutectic carbide, the graphitizing tendency of cast iron can be estimated from the following relationship:

$$K = \frac{4\,Si}{3}(1 - \frac{5}{3C + Si}). \tag{12.3}$$

Plotting the values of the K factor calculated for different C and Si contents, the so-called Laplanche-diagram can be obtained, as shown in Figure 12.2. The regions bordered by the curves belonging to different but constant K values represent the characteristic microstructures of cast irons. In Table 12.1, the typical regions of the diagram indicating the corresponding values of the K factor and the fracture surfaces of the characteristic microstructure are summarized.

Table 12.1 Summary of characteristic microstructures of cast iron

The value of the K factor	Fractured surface of cast iron	Characteristic microstructure
< 0.65	white	Ledeburite + pearlite
$0.65 - 0.85$	white + gray	Cementite + pearlite + graphite
$0.85 - 2.05$	gray	Pearlite + graphite
$2.05 - 3.10$	gray	Pearlite + ferrite + graphite
$3.10 <$	gray	Ferrite + graphite

The *effect of manganese* is opposite to that of silicon. Increasing the *Mn*-content decreases graphitization, therefore, the *Mn*-content of gray cast irons must be kept in the range of 0.4 –1.0 %. A higher *Mn*-content stabilizes the iron carbide. At the same time, alloying with *Mn* is important in bonding sulfur in the form of the chemical compound *MnS*. The amount of manganese required to bond sulfur can be calculated from the relationship:

$$Mn = 1.72 \, S + 0.3 \, (\%).$$ (12.4)

If the *Mn*-content is higher than the value calculated from the above relationship, it promotes the formation of pearlitic microstructure. If the *Mn*-content exceeds 1 %, it results in the formation of carbide, i.e. in metastable crystallization. Therefore, tough cast iron is characterized by a low sulfur and low manganese content.

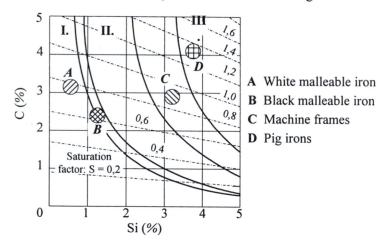

Figure 12.2
The graphitization readiness of cast iron according to the Laplanche-diagram

Sulfur in low-manganese cast irons forms iron sulfide (*FeS*), which damages the mechanical properties of cast iron by reducing strength and increasing brittleness. In addition, sulfur is the most powerful "*whitening*" element. In larger quantities, it prevents graphitization, but even in small quantities, it disadvantageously increases the coarsening of graphite lamellae. When manganese is present in cast iron in quantities corresponding to relationship (12.4), sulfur forms *MnS* with manganese. Since *MnS* has a high melting point ($T = 1650°C$), it partially goes into slag during casting and forms disperse, nonmetallic inclusions. Due to their small size, *MnS* act as crystallization nuclei and consequently, has a positive effect on graphitization as crystallization centers. *Phosphorus* in cast irons forms Fe_3P-iron phosphate, which together with iron and cementite forms a three-component eutectic of Fe-Fe_3C-Fe_3P. This is the so-called *steadite* having a low melting temperature ($T = 953°C$). This microstructure has characteristic concave grain boundaries. Steadite is similar to ledeburite as regards both its microstructure (Figure 12.3) and its properties. It shows a high hardness ($HV \approx 600$) and brittle-

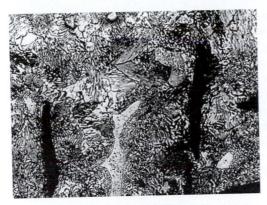

Figure 12.3

Fe-Fe₃C-Fe₃P three-component eutectic microstructure (*steadite*) in gray cast iron
Etching agent: 3 % HNO_3, N=300×

ness, which decreases the properties of cast iron. In view of this, phosphorus is usually an undesired alloying element. However, it has a beneficial effect on castability when producing thin-walled, complex shaped, segmented castings.

Due to its simplicity, in industrial practice the Maurer–diagram (Figure 12.4) is more widely used than the Laplanche-diagram to indicate the possible microstructures of cast iron. In this diagram, the microstructure of cast iron is given as a function of *C* and *Si*. The dotted line drawn for construction purposes at the carbon content $C = 1.7$ % actually refers to the composition corresponding to point *E*, thought to be the separating composition between steels and pig irons when the diagram was first created.

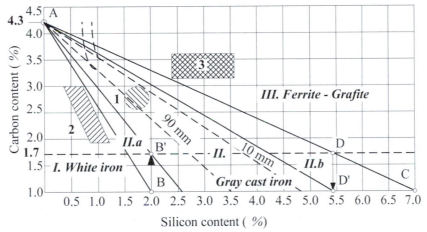

Silicon content (%)

Figure 12.4 – The Maurer-diagram
I. White iron (ledeburite), II/a) Transition between white and gray cast irons (cementite + pearlite + graphite), II.) the ideal gray cast iron (pearlite + graphite), II/b.) pearlite-ferrite-graphite gray cast iron, III.) ferrite + graphite cast iron

12.1.3. The effect of cooling rate on the microstructure of cast iron

Besides composition, the cooling rate plays the most important role in controlling the crystallization of carbon and the microstructure of cast iron. Analyzing the crystallization of iron-carbon alloys, we could see that in equilibrium cooling (i.e. at an infinitely low cooling rate) alloys crystallize in the stable *Fe-C* system. This means that these alloys can contain carbon only in the form of free graphite. At the same time, it was also referred to the theoretical significance of this fact, since an infinitely slow, equilibrium cooling cannot practically be achieved. It is known that the time required for diffusion processes increases exponentially with decreasing temperature. Therefore, in practice it is accepted that high temperature processes (mainly primary crystallization from the liquid) occur according to the stable system and recrystallization in the solid state according to the metastable system. The characteristic microstructure of cast iron solidifying under the above circumstances contains graphite embedded in pearlite as the most favorable microstructure of gray cast irons. At higher cooling rates, the iron-carbon alloys crystallize in the metastable $Fe\text{-}Fe_3C$ system. At higher cooling rates, causing metastable crystallization, carbon can be found in a chemically bonded form, i.e. white cast iron is the product of metastable crystallization.

Since cast irons have rather poor thermal conductivity, the wall thickness strongly influences the cooling rate and, consequently, the microstructure of cast irons. This effect is well illustrated by the Maurer-diagram referring originally to ϕ 30-mm cylindrical rods cast in sand moulds. In the pearlite-graphite zone of Figure 12.4, the dotted lines represent the narrower composition range where castings having a wall thickness of 10 - 90 mm will definitely possess a pearlite-graphite microstructure. These lines also reflect the fact that, for higher cooling rates, due to the smaller wall thickness, higher carbon as well as silicon contents are required to achieve the desired pearlite-graphite microstructure compared to castings with a greater wall thickness.

12.2. Gray cast iron

The ideal microstructure for industrial cast irons is pearlite + graphite. Since carbon and silicon are the most powerful graphitizing elements, their contents are the most decisive factors affecting the pearlite-graphite structure of cast irons. It can be observed in the Maurer-diagram (Figure 12.4) that, to a certain extent, carbon and silicon can mutually substitute each other. This means that the desired pearlite-graphite microstructure may be achieved with a higher silicon content and a lower carbon content and vice versa. The above correlation between carbon and silicon – based on the Maurer-diagram and successfully applied in practice – is described by the following relationship:

$$C + Si = 4,6 \div 6\% .\tag{12.5}$$

(However, it is obvious from the regions marked in the diagram that the most favorable carbon content resulting in an optimal microstructure is in the range of $C = 2.0\text{-}3.5$ %.) Furthermore, it is worth mentioning that higher $C+Si$ values contribute to the formation of many large graphite crystals and to the segregation of ferrite. Therefore, these values cause lower strength properties, while lower $C+Si$ values result in a purely pearlite-graphite microstructure thus providing more favorable strength properties.

Figure 12.5
Microstructure of pearlite-graphite gray cast iron (Etching agent: 3 % HNO_3, N = 300×)

The micrograph of a characteristic pearlite-graphite gray cast iron is shown in Figure 12.5. The matrix is pearlite in which irregularly shaped graphite flakes can be seen. This microstructure is called flake-graphite gray cast iron. The properties of practical cast irons are significantly influenced by the quality of the matrix and the morphology of the graphite flakes.

The matrix of gray cast irons can be pearlite, pearlite + ferrite or pure ferrite. The strength of gray cast iron is strongly influenced by the properties of the matrix. Naturally, a high-strength pearlite matrix results in higher strength than the ferritic matrix.

In Figure 12.6, the micrograph of a cast iron having a pearlite + ferrite matrix with flake graphite is shown. The strength of cast iron possessing a pearlite + ferrite matrix – assuming similar graphite morphology – increases proportionally with the amount of pearlite. This is shown in Table 12.2 summarizing the mechanical properties of the most common cast irons with flake graphite. The numbers adjacent to each material grade refer to the tensile strength in MPa. In the Table, R_m means the tensile strength, HB is the Brinell hardness, R_b stands for bending and R_c for compressive strength.

Figure 12.6
Micrograph of gray cast iron with pearlite-ferrite-graphite microstructure
(Etching agent: 3 % solution of HNO_3, Magnification N = 300×)

The pearlite content of cast irons shown in the table increases in the sequence listed. From the table, it can be seen that all the strength properties increase with increasing pearlite content. It can also be observed that gray cast irons having flake graphite in the microstructure possess low tensile strength, somewhat better bending strength and significantly higher compressive strength. The relatively poor mechanical properties can be attributed to the notch effect of the graphite flakes. One the one hand, great graphite flakes disrupt the metallic contact among the grains of the matrix causing relatively large discontinuities. On the other hand, graphite flakes as sharp notches decrease the strength parameters. These unfavorable effects can be reduced by various methods to produce cast irons having strength properties comparable with those of steels. These methods will be analyzed in detail in the forthcoming section.

Table 12.2 Mechanical properties of typical gray cast irons

ASTM Specification	R_m [MPa]	R_b [MPa]	R_c [MPa]	HB	Matrix of cast iron
A 278 Class 100	100	200-310	500-600	100-150	ferrite
A 278 Class 150	150	230-370	550-700	140-190	ferrite
A 278 Class 200	200	290-430	600-830	170-210	ferrite + pearlite
A 278 Class 250	250	350-490	700-1000	180-240	ferrite + pearlite
A 278 Class 300	300	410-550	820-1200	200-260	ferrite + pearlite
A 278 Class 350	350	470-610	950-1300	210-280	pearlite
A 278 Class 400	400	530-670	1100-1400	230-300	pearlite

12.2.1. Possibilities to improve the properties of cast iron

The strength of cast iron is closely related to the properties of the embedding materials, i.e. the metallic matrix and the graphite morphology. The quality of the matrix primarily depends on the composition of cast iron and the cooling rate, as we could see in the previous sections. In the case of the same matrix material, the properties of gray cast iron can be improved by modification of the graphite morphology – i.e. modification of the shape, size and distribution of graphite.

In Figure 12.7, an unetched micrograph of gray cast iron is shown. In this figure, the coarse graphite flakes can be clearly seen.

The most common and simplest method of refining the graphite structure is over-heating of the molten gray cast iron prior to casting. By heating the liquid of cast iron 100 °C above its melting point, even the crystallization nuclei still present in the melt can be dissolved completely. Due to the significant over-

Figure 12.7
Coarse graphite flakes on an unetched micrograph of gray cast iron (N=100×)

heating, crystallization starts at a temperature significantly lower than the liquidus point determined by the composition of the alloy. This results in a high degree of undercooling and leads to an increased nucleation ability. Therefore, a large number of small nuclei are formed resulting in a finer graphite structure. By overheating cast iron prior to casting, the strength properties may be improved by an average of 20-25 %.

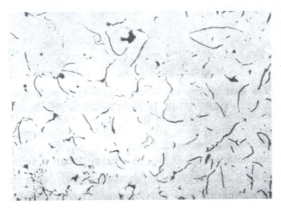

Figure 12.8
Modified fine graphite flakes in an unetched micrograph of gray cast iron (N=100×)

A more progressive method of refining the graphite structure is the so-called *modification.* In this process, such alloying elements (so-called *ladle additives*) are added to the molten cast iron prior to casting that act as foreign crystallization nuclei during solidification. This results in an even finer graphite structure, as shown in Figure 12.8. Ferro-silicon, silicon-calcium, or a mixture of ferro-silicon and aluminum in quantities of 0.2-0.3 % are used as modifiers

By analyzing the mechanism of modification, it was proved that the modifiers remove gas and deoxidize the melt, and by creating fine inclusions, they form crystallization nuclei.

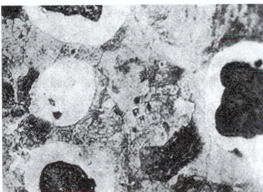

Figure 12.9
The microstructure of cast iron with nodular (spherulitic) graphite
(Etching agent: 3 % HNO_3, N=100×)

These two methods influence only the size of the graphite flakes by refining them. The most efficient strengthening effect as well as a significant improvement in other mechanical properties (elongation, toughness) of cast irons can be achieved by modifying the shape of graphite, provided that spherical graphite is formed instead of the flake graphite. To produce nodular graphite in cast iron, 0.3 – 1.2 % magnesium is added to the melt. The higher the C and Si content of cast iron or the greater the wall thickness of the casting the greater is the amount of magnesium required. 0.3 % Mg relates to a wall thickness of $s = 15$ mm, while 1.2 % Mg is added to $s = 300$ mm. Since Mg is an extremely flammable element, special care should be taken when adding it to the casting ladle. In certain cases – mainly due to its flammability – Mg is substituted by a cerium (Ce) based alloy of rare earth metals although Mg is much more common in the production of nodular cast irons. Mg forms with the Si-content of cast iron a metallic compound $MgSi$. It is insoluble in iron and serves as a crystallization nucleus resulting in spherulitic crystallization. The microstructure of nodular graphite cast iron produced with Mg additions is shown in Figure 12.9. The matrix of nodular graphite cast iron is pearlite, the spherulitic graphite is surrounded by ferrite. However, ferrite-free nodular graphite cast irons also exist.

Table 12.3 Mechanical properties of nodular (ductile) cast irons

ASTM Specification	R_m [MPa]	HB	A %	Matrix
A 897 Gr 350/175/22	350	115-160	22	ferrite
A 897 Gr 400/225/15	400	125-200	15	ferrite
A 897 Gr 500/275/07	500	150-240	7	ferrite + pearlite
A 897 Gr 600/325/03	600	175-290	3	ferrite + pearlite
A 897 Gr 700/400/02	700	210-320	2	pearlite
A 897 Gr 800/500/02	800	230-350	2	pearlite

In Table 12.3, the mechanical properties of standard, nodular graphite cast irons are summarized. The mechanical properties of cast irons possessing nodular graphite compete with steels because there is no notch effect due to the spherical shape of graphite. Beside the considerable increase in tensile strength (it may even reach the value of $R_m = 800 - 900$ MPa), in certain cases the percentage elongation is also remarkable (it is in the range of $A = 8 - 17$ %).

The previously described methods of increasing the strength of cast irons were based on modification of the graphite morphology. The strength of cast irons can also be increased by adding alloying elements that increase the strength of the metallic matrix. The strength of cast irons having a ferrite matrix is lower than that of those having a pearlite matrix. Usually, such alloying elements are added that hinder ferrite segregation and, simultaneously, the strength of pearlite is increased through the refinement of pearlite lamellae. The alloying elements performing this task are Cr, Mo and W. The addition of Sn in a low concentration (≈ 0.1 %) can be effective in preventing ferrite segregation.

From the preceding observations, it is obvious that pearlite-graphite gray cast irons have the widest practical application. The ferrite-graphite cast irons corresponding to region III in the Maurer-diagram are used for castings with increased magnetic permeability (for example, casings of electric equipment). Non-magnetizable cast irons are produced by alloying them with 10 % Ni and 2 % Cu.

The thermal resistance of cast irons is increased by alloying them with 18-20 % *Ni* and 2-3 % *Cr*. Cast irons resistant to hot sulfuric and nitric acid should contain about 14-16 % *Si* and 0.5-0.8 % *C*. As regards its carbon content, this alloy could have been considered a type of steel. However, due to its high silicon content, this is a cast iron with the so-called silicon-ferrite microstructure. This is due to the fact that the composition of the eutectic point is decreased to 0.7 % *C* by adding 11 % *Si* to the alloy. Besides silicon ferrite, this cast iron contains extremely hard and rigid carbides.

12.3. White cast iron

White cast iron is named after its characteristic fracture surface, which shows a metallic, white luster. Carbon exists in white cast iron in the form of chemically bonded Fe_3C (cementite). The crystallization of white cast iron may be ensured by controlling the composition and the cooling rate to provide metastable crystallization. According to Figure 12.1, the crystallization of white cast iron is enhanced by the elements *Mn, Mo, Sn, Cr, V, S*, listed in the sequence of their whitening effect. Simultaneously, it is desirable to limit the silicon content to the range of $Si = 0.5$- 1.2 % to avoid graphite formation. White cast iron is very hard and rigid owing to the ledeburite present in its microstructure. The practical applications of white cast iron are also determined by these properties. It is used to produce various items of equipment for mills, grinding plates and crushing balls, as well as in the production of case-chilled cast iron rolls. In this latter case, the working surfaces of mills subjected to wear are made of ledeburite having high hardness due to the presence of cementite as a result of the metastable crystallization process. In addition, the wear resistance is often increased by alloying 1.5 % Cr. Simultaneously, the bearing surfaces of these rolls crystallize corresponding to gray cast iron, the cooling rate being controlled by *"thermal insulation"* of these surfaces in sand moulds during cooling. Due to the great wall thickness, the core of the rolls (below the hard surface layer) crystallizes according to the stable system. Thus, with this process a case-chilled cast iron roll with a very hard, wear-resistant outer shell and with a tough core can be obtained. The big rolls for paper mills are also manufactured in this way. However, the most important application field of white cast iron is the production of various malleable cast irons.

12.4. Malleable cast irons

The starting material for malleable cast irons is white cast iron with a hypoeutectic composition and a low silicon content. By applying a special heat treatment process called *tempering*, two types of malleable cast irons are produced. These are the black and the white malleable cast irons. The typical compositions of white cast irons used for producing malleable cast irons are listed in Table 12.4.

Table 12.4 Composition of base materials for black and white tempering

Type of malleable cast iron	Composition, [weight %]				
	C	Si	Mn	P	S
White	3.0-3.4	0.5-0.8	0.2-0.5	0.12	0.1-0.3
Black	2.4-2.8	0.8-1.1	0.4-0.7	0.12	0.1

12.4.1. Black tempering

The aim of black tempering is to decompose the cementite, which causes the brittleness of white cast iron into its constituents according to the reaction equation:

$$Fe_3C \rightarrow 3Fe + C.$$ \hfill (12.6)

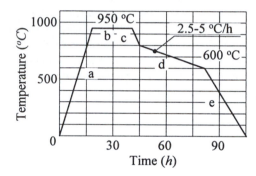

Figure 12.10

The temperature-time diagram of black tempering

The metallographic background of this reaction is that Fe_3C is a non-equilibrium, metastable phase that decomposes into its constituents corresponding to the equilibrium state during prolonged heating at a high temperature. Black tempering is carried out according to the time-temperature diagram shown in Figure 12.10, in an inert gas atmosphere. As a result of the high temperature of annealing ($T = 940$-$960°C$) and the long holding time ($t = 16$-20 h), first the cementite of ledeburite decomposes according to equation (12.6). During subsequent slow cooling, (carried out in the furnace), the secondary cementite – segregating during the cooling – is also decomposed. Decomposition of the cementite within the pearlite, formed at the A_1 temperature during the eutectoid transformation, requires a significantly longer time owing to the lower temperature. This can be realized by a very slow cooling rate ($v = 2$-$5°C/h$) when passing through the temperature range of $A_1 + \Delta T$ to $A_1 - \Delta T$. The total time required for the complete heat treating cycle may reach 50-110 hours depending on the technological conditions and on the composition of the white cast iron. Due to the long heat treatment cycle, the decomposed graphite is accumulated in the ferrite matrix (through atomic diffusion of carbon). The result is the so-called black-tempered malleable cast iron containing *graphite* in the form of *temper carbon* embedded in the ferrite.

The microstructure resulting from this black tempering process is shown in Figure 12.11. Although temper graphite collects in nodes, it can be easily distinguished from the spherulitic graphite of nodular graphite cast iron. Therefore,

Figure 12.11

Microstructure of black-tempered cast iron

Etching agent: 3 % HNO_3, $N = 300\times$

this decomposed graphite in the black tempered cast iron is called *temper carbon* or *flaky graphite*. Though, temper carbon has a nearly spherical shape, its surface is always irregular having a torn and broken outline. If slow cooling is neglected close to the A_1 temperature, pearlite-graphite malleable cast iron can be produced.

The characteristic mechanical properties of typical black-tempered malleable cast irons are summarized in Table 12.5. Besides relatively good strength properties, they possess remarkable elongation compared to that of the gray cast iron with graphite flakes. Therefore, malleable cast irons are also called ductile cast irons. The properties of malleable cast iron may further be improved by quenching prior to tempering. Due to quenching, the diffusion lengths during tempering become smaller. Consequently, the specific elongation may be increased to $A = 10\text{-}15$ %.

12.4.2. White tempering

The aim of white tempering – similarly to black tempering – is to decompose cementite, which causes the brittleness in white cast iron. However, in white tempering the decomposition of cementite is accomplished by burning out the carbon in an oxidizing atmosphere according to the reaction equation:

$$CO_2 + C = 2\,CO. \qquad (12.7)$$

White tempering is carried out according to the time – temperature diagram shown in Figure 12.12. The initial white cast iron (see composition in Table 12.4) is heated to the temperature interval $T = 950\text{-}1000\ ^\circ C$. It is held in an oxidizing atmosphere for about $t = 50\text{-}60$ hours. As a result of this, first the cementite constituent of ledeburite decomposes following Equation (12.6), then the free carbon burns out of the surface layers according to Equation (12.7).

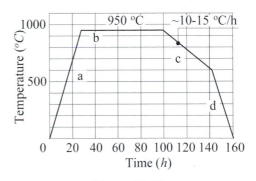

Figure 12.12

Temperature-time diagram of white tempering

The carbon from the inner layers diffuses toward the surface layers where it also burns out. The time required for white tempering is determined by the diffusion rate of carbon. Therefore, it is a more time-consuming process than black tempering. During slow cooling (v = 10-15 °C/h) the cementite decomposition occurs simultaneously with the burning out of the continuously formed free carbon according to the previously described processes. As the diffusion rate rapidly decreases with the lowering of the temperature, carbon cannot diffuse to the surface of thick-walled castings. Therefore, the characteristic microstructure of white tempered cast iron is purely ferritic only near the surfaces (to a depth of 4-8 mm), whilst the deeper layers consist of ferrite including graphite as temper carbon.

In Table 12.5, the mechanical properties of typical black- and white-tempered malleable cast irons are summarized.

Table 12.5 Mechanical properties of black and white malleable cast irons

ASTM Specification	R_m [MPa]	HB	A_5 %	Characteristic microstructure
A 220 Gr 40010	350	< 150	10	ferrite
A 220 Gr 45008	450	150-200	8	ferrite + pearlite
A 220 Gr 50005	550	180-230	5	pearlite
A 220 Gr 70003	600	210-260	3	pearlite
A 47 Gr 32510	350	< 150	10	ferrite + graphite
A 47 Gr 40005	400	150-180	5	ferrite + graphite

13. NON-FERROUS METALS AND THEIR ALLOYS

13.1. Introduction

Ferrous alloys are the most widely applied metallic materials in engineering practice. The widespread application of iron-based alloys is mainly due to their favorable mechanical properties that can be changed in a wide range by relatively inexpensive and simple production processes. Simultaneously, it is true that they also possess less favorable properties. Notably, the relatively high density ($\rho = 7.85$ **g/cm^3**), poor electric conductivity compared to certain non-ferrous metals and, in most cases, inadequate resistance to corrosion should be mentioned. Consequently, in many application fields, it is practical (in certain cases even unavoidable) to use non-ferrous metals or their alloys that have more favorable properties.

Non-ferrous metals may be classified according to several aspects. One of the most frequent classification methods of non-ferrous metals and alloys is based on the fundamental metals. Accordingly, aluminum-, copper-, magnesium-, titanium-based, etc., alloys can be distinguished. Another frequently used method classifies non-ferrous metals based on their properties. According to this classification, we can distinguish, for example, light metals, noble metals, refractory metals, superalloys, etc. The group of *light metals* includes metallic materials with low density (e.g. magnesium, beryllium, aluminum, titanium and their alloys). A certain group of non-ferrous metals has a characteristic color: copper and its alloys together with the noble metals, including silver, gold and platinum belong to this group. There are non-ferrous metals with a high melting point, which are classified as *refractory metals*: tungsten, molybdenum, tantalum, niobium belong to this group. A special group of non-ferrous metals – called *superalloys* – is based on their very special properties. For example, *W, Nb, Ta, Mo, Co, Ni* can be collected under the term superalloys due to their special properties, though some of them can be found in some of the previously mentioned groups as well.

In Table 13.1, some characteristic properties of non-ferrous metals are summarized. (In the table, in each group, the highest strength available commercially in the given group is shown. For comparison purposes, the relevant data of iron are included, too.)

Table 13.1 Characteristic properties of selected non-ferrous metals

Material	Density ρ g/cm^3	Young's modulus E, MPa	Tensile strength R_m, MPa	Specific strength R_m/ρ	Price USD/kg
Aluminum	2.75	69.000	570	211.000	1.30
Beryllium	1.85	290.000	380	205.000	660.00
Zinc	7.13	106.000	520	73.000	1.25
Magnesium	1.75	44.800	380	218.000	3.00
Nickel	8.90	207.000	1.360	153.000	9.00
Lead	11.36	13.800	70	6.000	0.80
Copper	8.93	124.800	1.300	146.000	2.45
Titanium	4.51	110.000	1.350	299.000	12.15
Tungsten	19.25	408.300	1.030	54.000	22.00
Iron	7.85	210.000	2.100	263.000	0.22

Considering these data, several important conclusions can be drawn. One of the most significant conclusions is that among the metals listed, various different sequences may be set up depending on what properties compose the basis of comparison.

Thus, for example, the sequence based on the ascending order of density will be *Mg, Be, Al, Ti, Zn, Fe, Ni, Cu, Pb, W*. However, the sequence is significantly changed if the strength is selected as the basic parameter of comparison. It can be clearly seen that iron has the highest strength, but significant strength can be achieved with the alloys of nickel, titanium, copper and tungsten as well. At the end of the strength series, magnesium and lead can be found.

The sequence is also significantly changed if the production costs and the world market prices are considered. (Although, an economic boom or prosperity occasionally may divert prices from the production costs, in a long-term comparison the prices listed can be regarded realistic.) As can be seen from the table, iron alloys are at the top of the list based on prices. The production cost of aluminum is 5-6 times that of iron. For titanium, it is more than 50 times, for beryllium roughly 3,000 times that of iron alloys.

In order to make further and more illustrative comparisons, a few specific parameters were derived from the properties given in Table 13.1. Thus, the term *specific strength* is introduced as the ratio of tensile strength (R_m) and density (ρ) as defined by the following expression:

$$r = \frac{R_m}{\rho} . \qquad (13.1)$$

Using the expression (13.1), the calculated specific strength parameters are given in Table 13.2. The table clearly shows that titanium possesses the highest specific strength followed by iron, magnesium, aluminum and beryllium. It is worth mentioning that, among the first five metals, four are light metals. However, it should be noted that iron, having a relatively high density is still ranked as second based on the specific strength.

Table 13.2 Specific properties and rank of selected non-ferrous metals

Material	R_m/ρ	R_m/price	(R_m/ρ)/price	Accumulated rank number	Final rank
Iron	263,000	9,545.45	1,195,454	12	1
Aluminum	211,000	438.46	162,307	14	2
Magnesium	218,000	126.67	72,666	28	3
Titanium	299,000	111.11	24,609	29	4
Copper	146,000	530.61	59,591	30	5
Zinc	73,000	416.00	58,400	32	6
Nickel	153,000	151.11	17,000	34	7
Beryllium	205,000	0.58	310	45	8
Lead	6,000	87.50	7,500	47	9
Tungsten	54,000	46.82	2,454	51	10

Similarly, an interesting comparison can be made using ratios based on the strength per unit price, as well as the specific strength per price. Both ratios are the best for iron due to its high strength and low price. If the lowest value obtained for beryllium is taken as a unit value, the strength of iron per unit price is 16,000 times that of beryllium.

Ranking the 10 metals according to all these analyzed parameters and summarizing the individual rank numbers, the final rank of metals shown in Table 13.2 can be obtained. This reflects the still valid industrial, practical tendency to apply the iron-based ferrous alloys in the largest quantity and in the widest range among all the metallic materials.

It is also confirmed by the data of Table 13.2 that aluminum and its alloys represent the most important group of light metals. Simultaneously, it also follows from the data of Table 13.1 and 13.2 that if one of the properties is of primary importance (for example, low density), the rank of metals may be significantly modified.

13.2. Aluminum and its alloys

13.2.1. Production and characteristics of pure aluminum

As we could see, aluminum belongs to the group of light metals. If we arrange the light metals according to their density in ascending order we find: magnesium (Mg: 1.7 g/cm^3), beryllium (Be: 1.8 g/cm^3), aluminum (Al: 2.7 g/cm^3) and titanium (Ti: 4.5 g/cm^3). From a practical application point of view, aluminum is the most important light metal. The aluminum ore, bauxite, is first processed to Al_2O_3 (aluminum oxide, often called as alumina) by a chemical process. The production of alumina requires a double amount of bauxite than the quantity of alumina produced. Alumina is electrolyzed first to so-called *primary (wrought) aluminum*. The purity of the wrought aluminum is in the range of 99.0-99.7 %. (This means that, besides Al, other elements are present in a quantity of 0.3-1.0 %. The most frequent contaminants of aluminum are Si, Fe, Cu, Zn, Mg. These are also the most important alloying elements provided they are added to aluminum on purpose). In order to produce 1 kg pure aluminum, roughly 2 kg alumina should be electrolyzed. Taking into consideration the ratio of bauxite to alumina, it follows that 4 kg bauxite is required to produce 1 kg pure aluminum.

The purest aluminum used in industrial practice is the so-called "*aluminum of four nines*". It contains 99.99 % aluminum. This is obtained by repeated electrolysis of wrought aluminum at about 1000 °C. Exactly twice as much electric energy is required to achieve this purity than for the primary electrolysis of aluminum. Therefore, this is used only in those applications where high purity is of utmost importance. These are the products for the electrical and chemical industry as well as for the production of mechanical instruments. Among all the aluminum alloys, high purity aluminum possesses the best electric conductivity, the highest resistance to corrosion, the highest ductility, but the lowest strength.

Aluminum crystallizes in the face-centered, cubic crystal system. This is the basis of its best formability among all the light metals. Besides low density, aluminum is also characterized by a low melting point (T_{melt} = 660 ^oC), excellent thermal and electrical conductivity, as well as good resistance to corrosion. The electrical conductivity of aluminum is the third best after silver and copper. The corrosion resistance of aluminum results from the high melting point of the compact oxide layer formed on its surface (T_{melt} = 2050 ^oC). This oxide layer has good adhesive properties and due to its passivating features high resistance to chemical effects. The higher the purity of aluminum, the better its resistance to corrosion. Unfortunately, the strength of pure aluminum is very low, the yield point and tensile strength of aluminum are less than one fourth compared to mild steel, but its elongation is higher by approximately 50 %. Although, the density of aluminum is about 1/3 that of steel, aluminum can be competitive, despite of its higher price, only when its strength is increased by alloying or in cases where the light weight of structures, the high conductivity and/or the high corrosion resistance are of decisive importance.

13.2.2. Alloys of aluminum

Because of the extremely low strength of pure aluminum, in engineering practice only its alloys are used. According to the effects of alloying elements, they are classified in the following main groups:

- Strength-increasing elements: *Cu, Mg, Si.*
- Corrosion resistance-increasing elements: *Mn, Sb.*
- Grain-refining elements: *Ti, Cr.*
- Thermal strength-increasing elements: *Ni.*
- Machinability-improving elements: *Co, Fe and Bi.*

Bi plays the same role in aluminum alloys machined with automatic equipment as *S* in steels, that is, it provides brittle chips during machining. In most cases iron (*Fe*) is considered contaminant in aluminum alloys. The upper limit of *Fe* content is about 0.7%. Aluminum forms hard and brittle metallic compounds with some alloying elements, such as Al_2Cu, Al_3Mg_2, Al_3Fe. Chemical compounds are never formed in aluminum by the elements *Si, Bi, Cd* and *Zn.* Alloying elements may also form compounds with each other. Certain elements can be regarded as alloying elements in some alloys and the same elements may be undesirable contaminants in other alloys. The best example of this is *Cu*, the most important alloying element of the *Al-Cu-Mg* alloy (which may be temper-graded to the highest strength). However, it is the most dangerous contaminant in the corrosion resistant *Al-Mg* alloy. Oxygen bonded in aluminum oxide and hydrogen absorbed during melting are always considered contaminants in aluminum alloys.

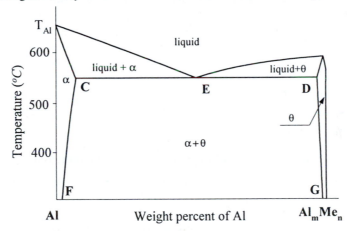

Figure 13.1

Characteristic phase diagram of aluminum alloys

Part of a phase diagram characteristic of aluminum alloys is shown in Figure 13.1. It can be seen from this diagram that the solubility of the *Me* component in the α-solid solution decreases with decreasing temperature. Therefore, the characteristic microstructure of aluminum alloys at room temperature consists of a mild, deformable α-solid solution and embedded hard metallic compounds. (These features of the phase diagrams provide the theoretical basis for the heat treatment of aluminum alloys to be discussed later.) On the right hand side of the phase diagram shown in Figure 13.1, the zone of the metallic compound formed by aluminum and the alloying element (for example, Al_2Cu) can be seen. It is also typical of aluminum alloys that at a defined composition a eutectic microstructure is formed consisting of an α-solid solution and the metallic compound of the aluminum and the alloying element.

Aluminum forms with its main alloying elements (*Cu, Mg* and *Si*) phase diagrams similar to that shown in Figure 13.1. The crystal structure of copper is identical to that of aluminum: *Cu* also has a face-centered cubic structure and the atomic radius of *Cu* is only 10 % smaller. However, the atomic and electron structure of copper is quite different. On the one hand, the atomic number of aluminum is 13 while that of the copper is 29, which thus has more than twice as much electrons. A further significant difference is that *Al* has three valence electrons, while copper has only one. Therefore, aluminum forms only a limited solid solution with copper. The saturated solid solution of aluminum at the eutectic temperature contains 5.65 % of *Cu*, but at room temperature it can dissolve only 0.5 % *Cu* (see points *C* and *F* in Figure 13.1).

The atomic structure of the other two main alloying elements of aluminum i.e. *Mg* and *Si* is almost identical to that of aluminum. The atomic number of *Mg* is 12 and that of *Si* is 14 - thus they are first neighbors of aluminum. Their atomic radii differ from that of aluminum by less than 10 %. However, their crystal structure is completely different. *Mg* has a hexagonal close-packed and *Si* has a diamond-type cubic crystal structure. Despite this, at the eutectic temperature $(T = 451^{\circ}C)$ aluminum can dissolve 13.4 % *Mg* which continuously decreases to 2.95 % at room temperature. However, precipitation is hindered by the hexagonal crystal structure of *Mg* to such an extent that in many cases it does not occur. Therefore, these types of *Al-Mg* alloys cannot be tempered. In the binary phase diagram of *Al-Si*, the saturated α-solid solution contains 1.6 % of *Si* and it decreases to 0.05% (virtually close to zero) at room temperature.

13.2.3. The theoretical basis of the tempering of aluminum alloys

A characteristic heat treatment procedure for aluminum alloys is called tempering. The main aim of tempering is to increase the strength of aluminum alloys by the precipitation of finely dispersed precipitates. The theoretical background of tempering of aluminum alloys is provided by the phase diagram.

Tempering is a precipitation hardening heat treatment process that results in the segregation of finely dispersed precipitates in the ductile α-solid solution matrix. These dispersed, uniformly distributed precipitates obstruct the motion of dislocations. Thus, the strength of aluminum alloys is increased by the *precipitation hardening mechanism* described in Chapter 11 (see Figure 11.15).

Precipitation hardening can be applied to increase the strength of *Al*-alloys, if they form a homogeneous single-phase solid solution at elevated temperatures and the solubility decreases with decreasing temperature. Consider the phase diagram shown in Figure 13.2 in order to analyze this.

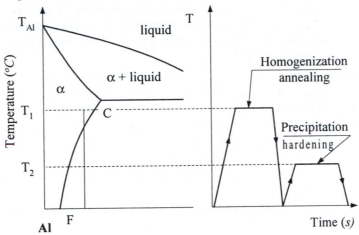

Figure 13.2

Theoretical diagram for analysis of the tempering of aluminum alloys

The first step of precipitation hardening is always a homogenization annealing aimed at providing homogeneous α-solid solution. The purpose is to completely dissolve the previously segregated precipitates. The homogeneous α-solid solution can be achieved by heating the alloy to the homogeneous α-zone and holding at this constant temperature until the whole amount of the alloy is transformed into a homogeneous α-solid solution. Since the homogenization is a diffusion process, the temperature should be selected high enough to reduce the time required to complete the process. In practice, this means a temperature (marked by T_1 in the figure) above the *CF* solubility limit, but below the eutectic temperature to avoid undesired side effects, such as melting and grain coarsening.

The homogenization is followed by rapid cooling (for example, quenching in water) to room temperature. The aim of the rapid cooling is to avoid segregation at this stage and to form an oversaturated α-solid solution. In this respect, the rapid cooling of aluminum alloys plays a role similar to the quenching applied in the hardening of steels. However, the consequences are quite different. The martensitic microstructure obtained by quenching in steels has the highest strength due to the lattice stresses caused by the excess carbon in the oversaturated α-solid solution. In aluminum alloys, the hardness and strength of the oversaturated α-solid solution produced by rapid cooling corresponds to a value between the annealed and the tempered states.

Precipitation hardening is the next stage of tempering. This stage is also called *aging*. The finely dispersed precipitates are regarded as a necessary condition for achieving excellent strength properties. If the cooling from the homogeneous α-solid solution zone occurred slowly through the solid solubility limit curve, the precipitates would be located along the grain boundaries and it would result in a coarser and less uniform distribution. The coarse grains of precipitates decrease the strength properties. This is the reason why the tempering of aluminum alloys is performed in the described three steps, i.e. applying the subsequent procedures of homogenization annealing, quenching and aging. In many cases, either natural or artificial aging may be applied.

Since the oversaturated α-solid solution is not an equilibrium phase, the precipitation of metallic compounds – and consequently, the increase in strength – in certain alloys can occur to a certain extent even at room temperature. This is called *natural aging*. However, the strength increase achieved by natural aging is a very slow process taking months or perhaps years to occur. Therefore, in practice, aging is performed at a higher temperature and is called *artificial aging*. The usual temperature of artificial aging (or simply *aging*) is about one fourth of the temperature of homogenization annealing.

During aging, complex metallographic processes occur that will be analyzed through the example of the *Al-Cu* alloy system. During artificial aging, the following partly sequential, partly overlapping procedures occur:

In the first stage of aging, the *Cu* atoms precipitate with a defined crystalline orientation from the initial oversaturated α-solid solution. The segregating copper atoms collect on the {100} crystal planes forming a coherent structure with the matrix. The thickness of the precipitates is up to the size of a few atoms (0.4-0.6 nm) while their extension in the planes is 8-10 nm. Such coherent precipitates can be seen in Figure 13.3. Since the diameter of copper atoms is about 11 % smaller than that of the aluminum atoms, the lattice in the neighborhood of precipitates is distorted to a tetragonal one. The previously described regions formed by the precipitates were first discovered by Guinier and Preston by X-ray diffractometric methods. Therefore, these are called *Guinier-Preston zones*. (Further precipitation is observed during aging performed at higher temperatures or for a longer period. Therefore, the coherent precipitates formed at relatively low temperatures are called

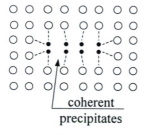

coherent
precipitates

Figure 13.3
Schematic view of coherent
precipitates formed in
aluminum alloys

Guinier-Preston I (or *GP I*) zones to distinguish them from those arising at a higher temperature.)

As aging proceeds, *Guinier-Preston II* zones are formed in those cases where the precipitate is not a pure metal or a solid solution, but rather a metallic compound in the equilibrium state. (As an example, the *Al₂Cu* – called θ-phase – may be mentioned. This is shown in Figure 13.1). Even in this case, first *GP I* zones are formed. However, after the development of the *GP I* zones, the copper atoms of the zone bond themselves to as many *Al* atoms as is required to form the metallic compound (thus two aluminum atoms are bonded to one copper atom). Initially, these atoms are located randomly, but as aging proceeds, they occupy the locations corresponding to the crystals of the metallic compound, but they do not currently form the metallic compound. The structure formed in this way – composed of at least two atoms – is also coherent with the {100} plane of the matrix. Its thickness is about 1-4 nm, and the extension in the plane may even reach the size of 10-100 nm. These structures are called *GP II* zones and they cause an additional increase in hardness and strength. The existence of GP-zones was successfully indicated by X-ray diffraction and electron microscopic examinations. Since these zones are extremely small, the magnification in electron microscopic examinations should be at least 100,000.

The development of the *GP-II* zone is followed by the formation of the θ'-phase, which is incoherent with the matrix. It possesses a tetragonal structure differing from the structure of the matrix and its thickness is in the range of 10-150 nm. This is already the precipitate of the metallic compound characteristic of the alloy. Its composition is still different from that of the equilibrium θ-phase.

The precipitate that can be seen under the microscope develops from the *GP*-zones as they start to grow by diffusion. The real process of precipitation occurs when the size of the growing particle is large enough so that the transition to the crystal structure – corresponding to the composition – dissipates at least as much energy as is required to compensate for the increase in the surface energy of new θ-phases. The θ-phase formed has its own (body-centered tetragonal) structure and possesses a boundary surface toward the matrix.

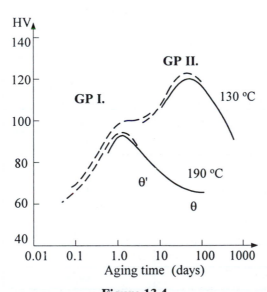

Figure 13.4

Characteristic aging curve of aluminum alloys

The previously described processes are manifested on the aging curves of aluminum alloys, indicating the increase in hardness as a function of aging time. Such aging curves are shown in Figure 13.4 for various temperatures of aging for the previously analyzed *Al-Cu* alloy.

The consequences of the development of phases as described above can be well followed on the aging diagram. When aging the alloy at *T* = 130 °C, first the *GP I* zones are formed resulting in a significant increase in hardness. On continued aging at this temperature, the *GP II* zones will be formed resulting in an even greater increase in hardness. With further holding at this temperature, formation of the θ'-phase is initiated followed by formation of the equilibrium θ-phase. Then growth and coagulation of precipitates occur, represented by a decreasing hardness compared to the maximum hardness achieved by the optimal hardening time.

It can be also seen from the diagram that above the optimal aging temperature (for example, at $T = 190\ ^{\circ}C$), *GP I.* zones are not formed since this temperature is already above the solution temperature of the *GP I.* zones. Thus, the maximum hardness remains much below the optimal value. At this higher temperature, formation of the equilibrium θ-phase as well as the growth of precipitates start after a shorter holding time, and together with this, a decrease in hardness can be observed. This phenomena is called overaging, which may occur if the holding time is longer than required to reach the maximum hardness at the optimal temperature or if the aging is performed above the optimal temperature.

Summarizing the main processes of the aging of aluminum alloys, it can be characterized by the following sequence of phases: α-oversaturated solid solution $\rightarrow$ *GP I* zones $\rightarrow$ *GP II* zones $\rightarrow$ θ'-phase $\rightarrow$ θ-phase.

13.2.4. Classification and main types of industrial aluminum alloys

Aluminum alloys can be classified in various ways. One of the most common classification methods is based on the major alloying elements. In the United States, a four-digit numerical designation is used to identify aluminum wrought alloys according to the major alloying elements. In this – widely accepted – designation system, the first digit indicates the alloy group, which contains specific alloying elements. The second digit indicates modification of the original alloy or impurity limits. The last two digits identify the aluminum alloy or indicate the purity. An overview of this four-digit classification system is shown in Table 13.3.

Table 13.3. The four-digit classification system of wrought aluminum alloys

Type of alloys grouped according to the major alloying element	Four-digit notation
Pure aluminum, min. 99.00% or higher purity	1xxx
Copper	2xxx
Manganese	3xxx
Silicon	4xxx
Magnesium	5xxx
Magnesium and silicon	6xxx
Zinc	7xxx
Other elements	8xxx

The basic four-digit designation is supplemented with further letters and digits (separated by a hyphen from the basic digits) indicating strain hardening, tempering, or other heat-treatment conditions. The letter *F* means *as fabricated, O –annealed, H – strain-hardened, T – heat-treated,* and *W – solution heat-treated.* The numbers following the above letters indicate further subdivisions. Thus, for example, *H1* means only *strain-hardened, H2 – strain-hardened and partially annealed, H3 – strain-hardened and stabilized.* (For further details, see the *ASM Metals Handbook, Volume 2: Properties and Selection of Non-ferrous Alloys*).

Classification of industrial aluminum alloys may also be performed on the basis of characteristic phase diagrams. Alloys containing a smaller quantity of alloying elements than the composition corresponding to the saturated α-solid solution (point *C* in Figure 13.5) - and consequently containing mainly solid solution – are the so-called *formable aluminum alloys* (region I). Alloys containing a larger quantity of alloying elements than those mentioned above are the so-called *aluminum casting alloys* (region II).

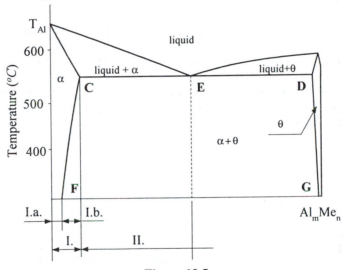

Figure 13.5
Classification of aluminum alloys based on a characteristic phase diagram

Formable aluminum alloys can be divided into two subgroups. The properties of alloys containing 100 % of α-solid solution at room temperature (region I. a) cannot be changed by heat treatment. Therefore, these are called *non-heat-treatable* alloys. The properties of alloys in the composition range defined by the line representing the solubility limit (region I. b) can be changed by heat treatment. Therefore, these alloys are considered *formable and heat-treatable.*

In the development of industrial aluminum alloys, the main aim is to increase the strength to the highest possible level. Based on the previous analyses, the possible methods to increase the strength may be summarized as follows.

One of the most important means of increasing the strength is *alloying*. The alloying elements used in aluminum alloys to increase strength are *Cu, Mg* and *Si*. Further increase in strength can be achieved by applying the previously described heat treatment processes, i.e. aging (obviously, it may be applied only to heat-treatable alloys). The strength-increasing effects of alloying and aging may be further improved by cold forming. Let us examine the effects of these three methods on the strength of a practical aluminum alloy. For this purpose, the *Al-Cu-Ni* three-component formable and heat-treatable alloy is selected.

The tensile strength of pure wrought aluminum (*Al*=99.5 %) used to produce the *Al-Cu-Ni* alloy is roughly $R_m = 100$ MPa. Adding 4% *Cu*, 2% *Ni* and 1.5% *Mg* to it, the tensile strength is increased by another 100 MPa, measured after alloying in "as cast" state. By aging this alloy, a further increase of 100 MPa is achieved, resulting in a tensile strength of $R_m = 300$ MPa. When this alloy is cast into a slab and is hot-forged and aged by heat treatment, the tensile strength of the finished product may even reach the value of $R_m = 400$ MPa. This means that the strength of the initial wrought aluminum can be increased fourfold by applying the above three methods together.

The *Al-Cu-Ni* alloy is equally produced in the form of castings, extruded or rolled semi-finished products. The above analysis indicates that the highest strength of the alloy can be achieved if plastic deformation is also applied during production. This is also valid for other aluminum alloys. This is the reason why aluminum alloys are distinguished as formable and casting alloys. An outstanding subgroup of the casting alloys includes the materials of pistons for internal combustion engines. These are called *piston alloys* and will be discussed in detail later.

13.2.4.1. Formable aluminum alloys

The main alloying elements of formable aluminum alloys are *Cu, Mg, Si* and *Zn*. The percentage quantities of these alloying elements usually do not exceed the values corresponding to the saturated solid solution at the eutectic temperature. Therefore, these alloys may contain maximum 5 % *Cu*, 10 % *Mg*, 1.5 % *Si* and 4 % *Zn*. By alloying further additive elements to these alloys, certain properties may be further improved. Thus, for example, the addition of about 2% *Ni* increases thermal strength, 1-2 % of *Mn* improves corrosion resistance, particularly against seawater.

Among formable aluminum alloys, the *non heat-treatable alloys* form a separate group. The typically two-component *Al-Mg* alloys called *hydronalum* are known for their good corrosion resistance against seawater. Their relatively low strength ($R_m = 200$-300 MPa) can be increased by cold forming. Similarly good corrosion resistance is characteristic of the non-heat-treatable *Al-Mn* aluminum alloys. These are used primarily in the food industry (for example, as milk transporting vessels). These aluminum alloys have the lowest tensile strength, which does not exceed the value of $R_m = 150$ MPa.

Most heat-treatable aluminum alloys typically contain three or more components. One of their basic types is the so-called *Dural-series* alloy containing *Al-Cu-Mg* components. The alloy containing 4 % *Cu* and 2 % *Mg* can be hardened to the maximum strength (R_m = 500 MPa). Similarly, the *Al-Cu-Ni* alloy containing 4 % *Cu* and 2 % *Ni* can be hardened to high strength. Applying artificial aging, its strength may be increased to R_m = 400 MPa. This alloy has a particularly high thermal strength too. The disadvantage of both above-mentioned heat-treatable alloys is that the presence of the copper alloying element, added to increase strength, makes them susceptible to corrosion.

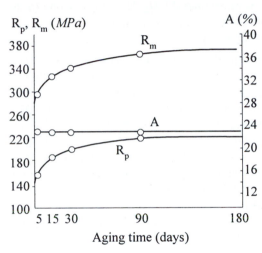

Figure 13.6
The aging diagram of *AlZnMgTi* alloy

The typical representative of naturally aging, heat-treatable aluminum alloys is the weldable *AlZnMgTi* alloy. Its high strength is achieved by hardening. Natural aging occurring during rest following quenching from about 400 °C is illustrated in Figure 13.6. 1-2 days after the quenching, the alloy is still mild and well formable. The final strength can be achieved after 90-180 days. Its tensile strength is increased to nearly 400 MPa and its yield point can reach 200 MPa. Even with these high strength parameters, its specific elongation is still significant, being about 20-22 %. A further important feature of this alloy is that the heat-affected zone of welding also hardens by natural aging after welding. As a drawback, its susceptibility to corrosion due to its *Zn* content should be mentioned.

A typical composition of the heat-treatable *Al-Si-Mg* alloys contains 1.5 % *Si* and 1.5 % *Mg*. Its strength can be increased only to about R_m = 300 MPa, but at the same time its corrosion resistance – due to the absence of copper – is very good. It is equally applied to produce safety equipment for mines and uninsulated wires for high-voltage power lines.

13.2.4.2. Aluminum casting alloys

The aluminum casting alloys can be classified into three subgroups. These are the so-called *silicon group* (4000 series), the *magnesium group* (5000 series) and the *cuprous group* (2000 series). The most outstanding aluminum casting alloys belong to the silicon group. Typical representatives of this group are the *Al-Si* and *Al-Si-Mg* alloys. Both are eutectic alloys having a definite melting point. In the *Al-Si* phase diagram, the composition at the eutectic point is *Si* = 12 %. The melting point of the eutectic composition is $T_{melt} = 578$ °C. Consequently, they exhibit the lowest degree of shrinkage. When cast into sand moulds the value of shrinkage is about 1-1.15 % compared to that of other aluminum alloys, which is in the range of 1.25-1.5 %. In metal mould casting, their shrinkage is even lower. It is about 0.5-0.8 %. Therefore, these are the most suitable alloys among all the aluminum casting alloys for metal mould casting. Besides these, their strength properties are the best. It is worth noting that the *Al-Si-Mg* alloys can also be heat-treated.

The second group of aluminum castings is made up of aluminum alloys containing magnesium. These alloys represent the casting versions of the formable alloys known as *hydronalum*. The *Al-Si-Mg* and *Al-Mg-Mn* alloys belong to this group as well. All these alloys can be characterized by good corrosion resistance. The best among them is the hydronalum group.

The third group of aluminum casting alloys is the cuprous one. The main types are the *Al-Cu* and *Al-Cu-Ni* alloys. These alloys are the least susceptible to shrinkage cavity formation. They possess good thermal conductivity and to some extent good thermal resistance. These alloys can be worked quite well, but in machining, they tend to "stick". Sticking can be decreased by adding 0.2 % *Si* and 0,3 % *Mg*. The thermal strength within this group can be increased by alloying with *Ni*. This alloy – often called "*Y*" *alloy* – is used for castings of engine cylinder heads and pistons working in hot conditions (for example, pistons of Diesel engines). In this alloy, we can find 4 % *Cu* besides 2 % *Ni*.

Obviously, the strength parameters of casting alloys are lower than those of formable alloys. Tensile strength is also influenced by the casting technology. When the alloy is cast into a sand mould, its strength is lower than in the case of a metallic mould. For example, the strength of *Al-Si* cast in sand is about 150-200 MPa compared to the value of 180-260 MPa ensured in a metallic mould. The tensile strength of the heat-treatable *Al-Cu-Ni* and *Al-Si-Mg* alloys may be increased to 300 MPa by hardening and subsequent aging.

The so-called *piston alloys* are a separate group of aluminum casting alloys. For these alloys, the operating conditions of the piston are decisive. The temperature on the surface of the piston in the combustion chamber during operation is in the range of 200-280°C, thus it is subjected to a considerable thermal load. A further problem is that the thermal expansion of the cast iron sleeves is about half of the expansion of aluminum alloys. Therefore, the piston has to be used with the proper gap sealed by the piston rings. The smaller the gap, the more silent is the engine and the less its fuel consumption. However, this increases the danger of sticking of the piston. For this purpose, several piston alloys have been developed. In the United States, mainly copper-based alloys, in Europe silicon- based alloys are more widely used. In Japan, carmakers use silicon-based hypereutectic alloys called *hypersil*.

Al-Cu-Ni and *Al-Cu-Si* alloys are two characteristic representatives of *copper-based* piston alloys. The *Al-Cu-Ni* (the so-called "Y" alloy) contains 4 % *Cu*, 2 % *Ni* and 1.5 % *Mg*. It has a hardness of $HB = 90-120$ and a high thermal expansion coefficient ($\alpha = 24 \times 10^{-6}$ 1/°C). It has also good thermal conductivity; therefore, it is used also in Diesel engines. The alloy *Al-Cu-Si* is produced with 10 % *Cu* and 2 % *Si*. It has a hardness of *HB* = 90-100 and its thermal expansion coefficient is still relatively high ($\alpha = 22 \times 10^{-6}$ 1/°C). Both alloys are frequently used in engines subjected to a high thermal load.

The *Al-Si* eutectic alloy is the characteristic representative of *silicon-based* piston alloys. Besides 12 % *Si*, it contains *Cu, Ni* and *Mg* alloying elements in a quantity of about 1 % each. (Therefore, it is sometimes referred to as the "1-1-1" alloy.) Its Brinell hardness is *HB* = 90-110 and it has a lower thermal expansion coefficient ($\alpha = 20 \times 10^{-6}$ 1/°C). Because of its lower thermal expansion coefficient, it may be fitted with a smaller gap, thus the engines assembled with this type of pistons are more silent even when running cold.

In Japan, the essentially *silicon-based,* but hypereutectic piston alloy is widely used. It has a high *Si* content (*Si* = 22 %) and besides this, it contains *Cu, Ni* and *Co* alloying elements in the amount of 1.5 % each. Its hardness is *HB* = 130 and it has the lowest thermal expansion coefficient of all piston alloys ($\alpha = 17 \times 10^{-6}$ 1/°C). This enables fitting the pistons with the smallest gaps, resulting in silent and low-consumption cars.

13.3. Magnesium and its alloys

Magnesium is the lightest metal used in engineering practice. Its density is 1.7 g/cm^3. Its melting point is almost the same as that of aluminum (*T* = 650 °C). *Mg* has a hexagonal close packed crystal structure. This means that *Mg* has only one plane favorable for plastic deformation, i.e. the (0001) base plane. This is the reason for its low formability at room temperature, which is somewhat better at elevated temperatures. The corrosion resistance of magnesium is lower than that of aluminum. It has a high affinity to oxygen, but the oxide layer formed on its surface is not as compact as that of aluminum. Therefore, corrosive media may easily break through it, causing general surface corrosion. It is especially sensitive to humidity, salt water, inorganic acids and salts. This is the reason why magnesium and its alloys have to be protected from the environment with a protective coating.

The strength of pure magnesium is poor. Its yield strength is about 70 MPa, but it can be increased by alloying. The main alloying element of magnesium is alumi-num. Further additional alloying elements are *Zn* and *Mn*. However, its strength properties are lower than those of the hardened aluminum alloys even in the alloyed state, its specific strength is high due to its low density (see Table 13.1). Therefore, *Mg* alloys are important materials in aviation and aerospace industry.

Magnesium forms a solid solution with aluminum, as shown in Figure 13.7 on the binary phase diagram of *Al-Mg*. The upper limit of solubility at the eutectic temperature (*T*=437 °C) is 12.7 %. Solubility decreases to about 3 % at room temperature. Therefore, in magnesium alloys containing 3-12 % *Al* during cooling, the metallic compound of *Al$_{12}$Mg$_{17}$* precipitates, increasing the hardness.

The practical *Mg-Al* alloys usually contain 3-9 % aluminum and they are called "*electron alloys*". The electron alloys containing 3-6 % *Al* are formable, while the *Mg*-alloys containing 4-9 % *Al* are typically used for castings. The hardness of electron alloys can be improved by alloying with 0.5-1 % *Zn* and 0.1-0.5 % *Mn*.

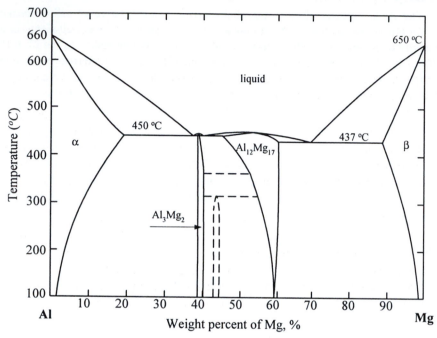

Figure 13.7
The *Al-Mg* phase diagram

Due to the flammability of magnesium, the casting of electron alloys is very difficult. The molten alloy is protected by a salt layer. The sand mould is prepared by adding 3-10 % sulfur and the surface of the mould is coated with sulfur powder prior to casting. The tensile strength of electron castings is in the range of $R_m =$ 100-200 MPa, the elongation is about 2-10 % and the hardness is between $HB =$ 50-70. The formable electron alloy is extruded or rolled between T=300-400 ºC. Prior to hot forming, electron alloys are heated in airtight furnaces or in a salt bath. The mechanical properties of extruded electron alloys are roughly twice as high as those of cast magnesium (R_m = 200-300 MPa, A = 10-20 %). The hardness value is approximately the same.

Two development tendencies can be observed concerning up-to-date *Mg* alloys. One of these tendencies is represented by the development of alloys containing extremely low amounts of contaminants. The other trend is characterized by adding more than 5 % rare earth metals as alloying elements (for example, cerium). In both cases, a corrosion-resistant film of *MgO* is formed. A similarly interesting development is represented by the reinforcing of magnesium with ceramic materials, for example, silicon carbide fibers. This is especially advantageous in structures operating at elevated temperatures.

13.4. Beryllium and its alloys

Beryllium is even lighter than aluminum (its density is only 1.85 g/cm^3). At the same time, its modulus of elasticity is higher than that of steels ($E = 290,000$ MPa). Due to its relatively high yield strength ($R_p = 200$-350 MPa), its specific strength is also significant (see Table 13.1). Beryllium maintains its strength and rigidity up to relatively high temperatures.

The ore of beryllium is beryllium sulfate, but it can be found only in very small quantities in the Earth's crust. Pure beryllium is extracted from this ore with a complicated and expensive process. This is the reason why beryllium is one of the most expensive metals. Its price is roughly 3000 times that of plain carbon steel.

Beryllium has a hexagonal close-packed crystal structure. Therefore, it is very difficult to form plastically, particularly at room temperature. At higher temperatures it is quickly oxidized, and therefore, it can be further processed only in vacuum. This means another additional cost item in most of its production processes. *BeO* is carcinogenic so that its processing requires special care.

Due to its low density and its fourfold modulus of elasticity, it is an essential material in aeronautics and aerospace industry. It is applied as the material of space aerial and special mirrors with surfaces of several square meters.

13.5. Titanium and its alloys

Although titanium is the ninth most common element in Earth's crust, its practical application is quite recent. Its wide spread application has been promoted mainly by the aerospace industry and astronautical developments. Titanium is extracted from its two main ores, i.e. *ilmenite (FeO-TiO$_2$)* and *rutile (TiO$_2$)*, by the so-called Kroll process. In the metallurgical process, first *TiCl$_4$* is produced. Pure titanium is obtained by reducing it with *Na* or *Mg*. Since its affinity to oxygen is exceptionally high, its production can take place only in vacuum.

At room temperature, titanium has a hexagonal close-packed crystal lattice. This is the so-called α-titanium. α-titanium transforms to body-centered cubic crystal of β-titanium at $T = 882$ °C. Despite its hexagonal crystal lattice, pure titanium has relatively good formability even at room temperature. The good formability of titanium is decreased by *C, O* and *N* contaminants. Even a very small quantity of these elements makes titanium rigid and brittle. Its melting point is high compared to other light metals ($T = 1670$ °C). Melting of titanium is usually performed in water-cooled copper ladles with an electric arc. The electrodes are made of high-strength titanium and the strongly reactive metal bath is protected by vacuum or argon gas from picking up *O* and *N*.

The density of titanium is 4.5 g/cm^3. Its strength parameters are significantly improved with the degree of purity. Strength can also be increased by alloying and it depends very much on its microstructure (see Table 13.4). The yield strength of titanium of commercial purity is much higher than that of the plain carbon steels. However, its modulus of elasticity (E = 110,000 MPa) is only half that of steels; the strength of structures made of titanium is equivalent to that of steel structures. Its rigidity is only about half that of steels because of the lower modulus of elasticity. The yield strength of titanium can be increased by about 50 % by cold forming (up to 800 MPa), but at the same time the elongation is decreased to about one tenth of its original value, i.e. to 5 %.

Table 13.4 Mechanical properties of titanium and its alloys

Name	Yield strength R_p, MPa	Tensile strength R_m, MPa	Specific elongation A, %
High purity titanium	120	170	55
Commercial titanium (99.0 %)	485	550	25
α-Ti alloy (5 % Al-2.5 % Sn)	780	860	15
α–β Ti alloy (6 % Al-4 % V)	970	1030	8
β-Ti alloy (13 % V-11 % Cr)	1210	1290	5

Due to its low density and excellent mechanical properties, titanium became one of the most important strategic materials of the aerospace industry. It is applied as the construction material of both fuselage and jet engines.

The corrosion resistance of pure titanium is equivalent to that of the austenitic stainless steels containing 18 % of Cr and 8 % of Ni. It is attacked only by haloid, sulfuric and phosphoric acids. The corrosion resistance of titanium is provided by the thin, compact oxide layer (TiO_2) on its surface. This property of titanium makes it useful in the manufacturing of equipment for the chemical industry, as well as for the production of surgical implants. Titanium of commercial purity having excellent corrosion resistance is used for the production of heat exchangers, reactor vessels, pumps and various items of chemical and oil processing equipment. Contaminants increase its strength, but significantly decrease its corrosion resistance. In the applications of Ti, it should always be considered that it reacts strongly with its contaminants (O, N and C) and the reaction rate increases with increasing temperature.

The diameter of the titanium atom (0.291 nm) differs from that of many metallic elements by less than 15%. Therefore, Ti can form a substitutional solid solution with many elements. The alloying elements increase the strength of titanium alloys mainly by the solid solution hardening mechanism and they significantly modify its allotropic transformation temperature. From this aspect, the alloying elements may be classified into four groups as shown in Figure 13.8.

Tin and *zirconium* belonging to the first group of alloying elements, increase the strength of the solid solution without changing the allotropic temperature (part a. in Figure 13.8). However, *aluminum* stabilizes the α-phase, thus it increases the temperature of the $\alpha \rightarrow \beta$ phase transformation (part b. in Figure 13.8). *Oxygen* and *hydrogen* have similar effects; however, they are not considered as alloying elements, but rather as contaminants. The elements *V, Ta, Mo* and *Nb* stabilize the ductile β-phase. They also lower the allotropic transformation temperature and in certain cases the β-phase remains even at room temperature above a certain content of the above alloying element (see part c. in Figure 13.8). Alloying titanium with *Cr, Mn* and *Fe* results in eutectoid reactions. These elements also lower the temperature of the $\alpha \rightarrow \beta$ phase transformation and result in $\alpha+\beta$, the dual-phase microstructure (part d. in Figure 13.8).

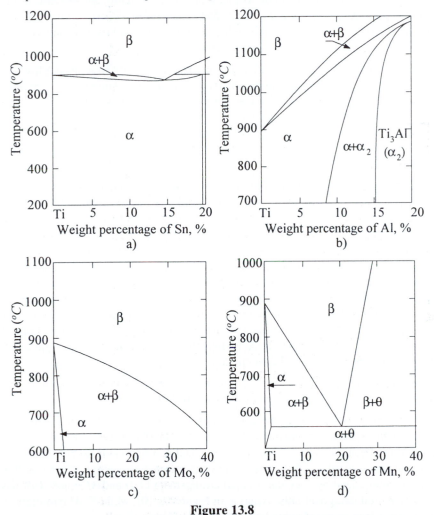

Figure 13.8
The effect of alloying elements on the phase diagrams of titanium alloys

One of the most characteristic types of α-*titanium* contains 5 % *Al* and 2.5 % *Sn*. These alloying elements increase the strength of the alloy by the solid solution hardening mechanism. Such alloys are annealed at high temperatures in the β-zone. During rapid cooling, a characteristic Widmanstätten-type α-titanium microstructure is obtained. This microstructure possesses high fatigue strength. If the alloy is cooled in the furnace, lamellar α-solid solution can be obtained with good creep resistance.

Though the β-*titanium* phase can be stabilized down to room temperature by a large amount of *V, Ta, Mo* and *Nb* alloying elements, commercial alloys usually contain less β-stabilizers than this amount. Rapid cooling results in a metastable β-phase in the whole volume of the alloy. The transformation of the metastable β-phase by segregation of the stable α-phase results in a significant increase in the strength through a precipitation hardening mechanism. These alloys are the favorite materials of the aerospace industry, being the fundamental materials of high strength frames and supporting elements.

By adjusting the ratio of stabilizing elements, a dual-phase microstructure consisting of α and β phases can be obtained even at room temperature. The most characteristic representative of these alloys contains 6 % *Al* and 4 % *V*. The properties of this alloy may be varied in a wide range by heat treatment. By selecting the proper temperature and deformation rate, this alloy can produce more than one thousand percent uniform elongation. This alloy is one of the most characteristic *superplastic alloys*.

By alloying titanium with niobium, an excellent conductive alloy is obtained called *superconductor*. Alloying with nickel produces the so-called *shape memory alloys*. The substance of this unique property may be summarized as follows. The microstructure of titanium alloy containing about 50 % nickel is transformed into martensite. At the end of this heat treatment, the component is formed to its required shape by a special thermo-mechanical treatment. Following this, it can be reshaped to any form, however, when the temperature is increased, the original shape (i.e. the shape it had at the end of the thermo-mechanical treatment) is regained. This means that it behaves as if the material "*remembered*" its former shape. This property can be utilized in many applications. For example, control devices starting to operate at a certain temperatures or special medical equipment, various implants, etc., should be mentioned.

Aluminum is one of the most important alloying elements of titanium. The complete *Ti-Al* binary phase diagram is shown in Figure 13.9.

Besides the α-stabilizing effect, the most important role of *Al* in *Ti*-alloys is that it forms special *intermetallic compounds* with titanium. These are the so-called γ- (*TiAl*) and α$_2$-alloys (*Ti$_3$Al*). These metallic compounds do not follow the chemical valence rules. They rather represent ordered crystalline structures that can be characterized by a regular ordered arrangement of *Ti* and *Al* atoms. *TiAl* has a face-centered tetragonal lattice (part a. in Figure 13.10), while *Ti$_3$Al* (similarly to *Ni$_3$Al*) has a face-centered cubic lattice (part b. in Figure 13.10).

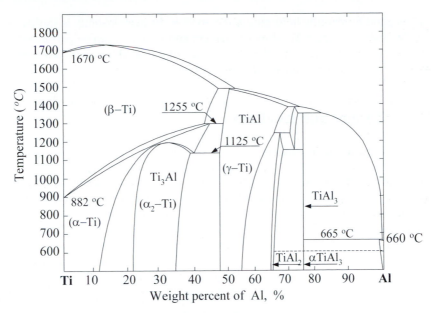

Figure 13.9
The *Ti-Al* binary phase diagram

The ordered crystal structure shown in Figure 13.10 significantly obstructs the movement of dislocations. However, at room temperature it results in a weaker formability, while at high temperature these alloys show excellent creep resistance. Therefore, these alloys are favorite materials in supersonic aircraft and jet engines.

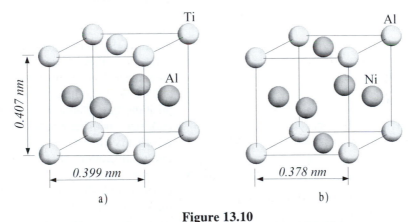

Figure 13.10
Ordered crystal structure of intermetallic compounds in *Ti*-alloys
a) *TiAl* – face centered tetragonal, b) *Ti$_3$Al* and *Ni$_3$Al* face centered cubic lattice

The *Ti*-alloys and titanium itself are castable; however, they should be protected from harmful oxidation, usually by applying vacuum casting. With the application of appropriate forming processes, they may be rolled into very thin sheets.

It has good formability in forging and extrusion. Its machinability is similar to that of the austenitic steels. Welding of titanium and its alloys can be performed under a protective argon gas atmosphere. Each side of the weld has to be protected from oxygen and nitrogen by an ample flow of argon.

13.6. Copper and its alloys

13.6.1. Production of copper

Copper is extracted from its ores (containing copper and sulfides) by metallurgical processes. The sulfide ores of copper are oxidized and the oxide of copper is reduced by carbon. Primary copper is never pure, it usually contains about 1 % contaminants (e.g. *Pb, Bi, Sb, As, Ni, Fe* and noble metals). Pure copper is produced by electrolysis of wrought copper. The positive pole (the anode) is made of wrought copper cast into slabs. The negative pole (the cathode) is made of thin sheets of rolled pure copper, while the electrolyte consists of copper sulfate and the aqueous solution of sulfuric acid. During electrolysis, pure copper precipitates on the cathode, the contaminants are partly dissolved and partly deposited on the bottom of the bath. The purity of the copper obtained by this method is 99.9 % and it is called *cathode copper* or *electrolytic copper*. It is mainly used in electrical engineering, e.g. for the production of high-voltage power lines. Wrought copper always contains O. This is hard to remove because the copper oxide (Cu_2O) can form an alloy with copper as shown in the Cu - Cu_2O phase diagram (Figure 13.11). As can be seen from this figure, oxygen and copper can dissolve each other in the liquid state. In the solid state, they cannot dissolve each other. They form eutectic a at 3.45 % Cu_2O. Below this composition, the solidification of the melt starts with the crystallization of primary copper and ends with the solidification of a *copper – copper oxide eutectic*. The solidified Cu-Cu_2O eutectic can be recognized along the grain boundaries by its characteristic blue color on the unetched micrographs.

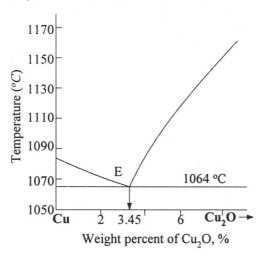

Figure 13.11
The Cu-Cu_2O phase diagram

The oxygen content of copper may cause the so-called *hydrogen illness*. When copper containing oxygen is subjected to a high-temperature reducing atmosphere – which may occur during the annealing of copper plates - then H_2 reduces Cu_2O

according to the reaction equation $Cu_2O + H_2 = 2\ Cu + H_2O$ and water vapor is generated. At high temperatures, water vapor can cause cracks in copper sheets or fine hairline cracks, which may lead to fracture during loading in subsequent applications. Therefore, copper must not be heated in a reducing atmosphere because all copper contains more or less oxygen. An oxygen content exceeding 0.1 % – equivalent to 0.9 % Cu_2O content – is extremely dangerous.

13.6.2. Properties of copper

Copper crystallizes in the face-centered cubic crystal system. Thus, it has 12 slip systems – including four slip planes (the {111} family planes) and three slip directions (the <110> family directions) – providing good formability. Copper has a unique combination of properties from the point of view of engineering applications. High electrical and thermal conductivity, good corrosion resistance, relatively good strength properties are the most important ones.

Alloying elements forming limited solid solutions with copper can increase the strength of copper. However, copper also forms metallic compounds, mostly in the form of electron compounds. The latter usually appear as a second phase together with the solid solution in copper alloys rich in alloying elements. Due to their hardness and brittleness, electron compounds significantly decrease the strength properties of copper alloys.

The electrical conductivity of pure copper is ranked second after silver. It equals 58 m/Ω mm^2. The electrical conductivity is significantly decreased by even a small amount of contaminants forming a solid solution. *P, Al, As, Fe, Sb, Sn, Zn* are the most frequent contaminants in the sequence of their effects. *Pb, Ag, S, O* also decrease conductivity but to a lesser degree. Copper also possesses good corrosion resistance. Humid air attacks its surface, but it reacts with the CO_2-content of the air forming "patina" (copper carbonate) as a protective layer. Some important characteristics of pure copper are summarized in Table 13.5.

Table 13.5 Selected properties of copper

Name of Property	Value of Property
Density	$\rho = 8.9$ g/cm^3
Melting point	T = 1083 ^{o}C
Coefficient of thermal expansion	$\alpha = 17 \times 10^{-6}$ mm/m ^{o}C
Electrical conductivity	e = 58 m/Ω mm^2
Tensile strength	$R_m = 200$ MPa
Specific elongation	A = 40 %
Hardness	HB = 60

Copper may be hardened by cold forming. Cold forming resulting in 80 % reduction of the cross section increases the strength by 90 %, while elongation is decreased by 75 %.

The main alloying elements of copper are: *Zn, Sn, Al*. Additional alloying elements are: *Pb, Ni, Mn*. The most important alloys of copper are the following:

- *Cu-Zn* alloys known as *brass*,
- *Cu-Sn* alloys named *tin bronze*,
- *Cu-Al* alloys named *aluminum bronze* and
- *Cu-Sn-Zn* alloys known as the so-called *red alloy*.

In practice, the name *"phosphorus bronze"* is also often used. It does not represent an individual alloy, rather a degree of purity. Phosphorus can be dissolved in copper to a certain amount, but its main purpose is deoxidization. Phosphorus is a good deoxidizing element for copper. Therefore, the name *phosphorus bronze* means a bronze that is deoxidized by phosphorus.

13.6.3. Classification of copper alloys

In the United States, copper alloys are classified according to a designation system administered by the *Copper Development Association (CDA)*. In this designation system, the numbers C10100 to C79900 designate wrought alloys, and the numbers from C80000 to C99900 designate casting alloys. In Table 13.6., the main alloy groups of copper are listed with the appropriate designation numbers.

Table 13.6 Designation system of copper alloys

Type of alloys grouped according to the major alloying element	Designation number
Wrought alloys	
pure copper and high copper content alloys	C1xxxx
copper-zinc alloys (brasses)	C2xxxx
copper-zinc-lead alloys (lead brasses)	C3xxxx
copper-zinc-tin alloys (tin brasses)	C4xxxx
copper-tin alloys (phosphor bronzes)	C5xxxx
copper-aluminum alloys (aluminum bronzes)	C6xxxx
copper-nickel and copper-nickel-zinc alloys (nickel silver)	C7xxxx
Cast alloys	
cast pure copper, high copper content alloys, brasses	C8xxxx
cast copper-tin, copper-tin-lead, copper-aluminum	C9xxxx

13.6.3.1. Cu-Zn alloys (brasses)

The *Cu-Zn* alloys (i.e. the brasses) consist of a series of alloys with a copper content of 90, 80, 72, 67, 63 and 60 % (i.e. 10-40 % of *Zn*). The first two alloys, possessing extremely good formability, are mainly used to produce decorative artifacts. The rest of the alloys are known as *brass*. The phase diagram of *Cu-Zn* alloys is shown in Figure 13.12.

In this alloy system, α-solid solution crystallizes primarily along the *AD* liquidus curve. This is the substitutional solid solution of *Zn* and *Cu*. It has good formability, similarly to pure copper. Along the *BCD* horizontal line, a β-phase having a composition corresponding to point *C* is formed as a result of the peritectic reaction. In this reaction, α-solid solution having a composition corresponding to point *B* and the liquid having a composition corresponding to point *D* react with each other. The β-phase primarily crystallizes along the *DH*-liquidus curve as a single-phase electron compound. This electron compound has an atom/electron ratio of $N_A:N_E$ = 2:3. The concentration range of the β-phase decreases with decreasing temperature as shown in Figure 13.12. However, the β-phase is an electron compound, but it has good formability at the temperatures of the homogeneous zone.

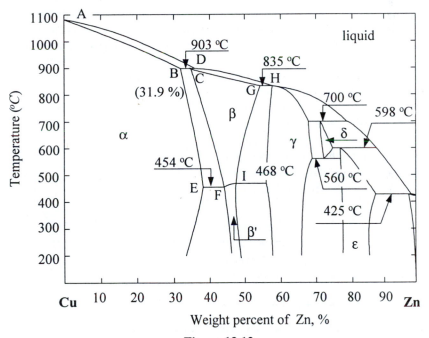

Figure 13.12
Part of the *Cu-Zn* phase diagram

The *γ-phase*, appearing above 50 % zinc content, is an electron compound (Cu_5Zn_8) too. It is a rigid, brittle phase that can be formed neither by hot nor by cold forming. It causes a sudden decrease in the strength and ductility properties. Therefore, these copper alloys are not used in engineering practice.

Originally, the β-phase has a body-centered cubic lattice with random atomic arrangement. In the temperature range of 454-468 °C (line *EFI* in Figure 13.12), it is transformed into an ordered lattice such that copper atoms are located at the corner points and zinc atoms at the body centers. This phase is denoted by β'.

Figure 13.13

Microstructure of brass having 67 % *Cu* content
Etching agent: 5 g *FeCl*+ 26 cm³ *HCl*+10 cm³
water, magnification: N=200×

The *BE* solubility limit curve is the unique feature of the *Cu-Zn* phase diagram. It indicates that the zinc solubility of the α-phase increases with decreasing temperature. From this feature, the following may be concluded. The brass having 63 % *Cu* contains $\alpha+\beta$ phases in the temperature range of 900-650 °C. If this alloy is cooled rapidly, its microstructure contains $\alpha+\beta'$ phases even at room temperature. The reason is that the $\beta \rightarrow \alpha$ phase transformation may be hindered by rapid cooling. When such two-phase alloys (for example, the *CuZn67* alloy) are heated to the temperature range of T = 500-600 °C, the β'-phase recrystallizes into an α-phase and the microstructure of the alloy becomes homogeneous with multi-angular grains intersected by twin crystals. This is illustrated in Figure 13.13. The grain containing twin crystals is the α-phase, the black stripes on the borders are the β'-phase.

The change in the strength properties of the *Cu-Zn* alloys is shown in Figure 13.14. The sudden increase in tensile strength (R_m), as well as the sudden drop of specific *elongation* (*A*) are caused by the appearance of the β'-phase.

The mechanical properties are the most favorable when the zinc content is in the range of 33-40 %. This explains why, in practice, these brasses are used most widely. Their characteristic compositions and mechanical properties are

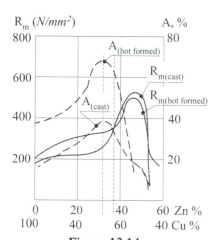

Figure 13.14
Mechanical properties of *Cu-Zn*
alloys vs. composition

summarized in Table 13.7. The common feature of the first two alloys is that their microstructure is a homogeneous α-solid solution. They are utilized in the production of castings, cold-rolled, forged or deep-drawn components.

Table 13.7 Chemical compositions and mechanical properties of Cu-Zn alloys

Code	Cu, %	Zn, %	R_m, MPa	A, %
CuZn33	67	33	320	45
CuZn37	63	37	350	45
CuZn40	60	40	400	40

The *CuZn33* alloy has a lower strength than the *CuZn37*. The microstructure of brass having a copper content of 60 % contains an α-solid solution and a β'-electron compound with ordered crystal structure. It can be hot-forged, and is therefore, also called *forgeable brass*. It is also known as *Muntz* or *Weiss*-metal. Table 13.7 shows its characteristic mechanical properties. The *Zn* content is 40 %, the tensile strength R_m = 400 MPa, (equivalent to 40 daN/mm^2), the elongation A =40 %. Therefore, it is also called 40/40/40 brass.

Besides the previously mentioned alloys of copper, a so-called *noble brass* is used for manufacturing of high-strength machine components. It contains at least 50 % *Cu*, as well as together at least 88 % *Cu+Zn*. This brass is alloyed with 12 % of other components, e.g. *Mn, Al* and *Fe*. Thus, this is a multi-component alloy. It has a tensile strength of R_m = 300-650 MPa, an elongation of A = 10-40 % and a hardness of HB = 90-150.

13.6.3.2. The *Cu-Sn* alloy (*tin bronze*)

The *Cu-Sn* alloy (*tin bronze*) has been used for a very long time. The Bronze Age (about 3000 years *BC*) was named after this alloy. It has been used by mankind to make utensils and decorative artifacts. The binary phase diagram of *Cu-Sn* alloys shown in Figure 13.15 is quite complicated. (However, it should be noted that the *Cu-Sn* alloys used in the industry have a *Sn* content below 40 %.) A typical feature of the diagram is that the first sections of the liquidus and solidus curves show high temperature ranges. It follows from this that the primarily crystallizing dendrites of industrial tin bronze castings are always laminated. The diagram contains five phases. At practical cooling rates, alloys containing 0-14 % *Sn* consist of a stable α-solid solution of tin and copper (this is marked by the dotted line drawn in the diagram at 14 % *Sn* content). This is a well formable alloy. Actually, the β and γ-phases are electron compounds with a disordered body-centered crystal structure. However, their properties, similarly to the β-phase of brass, differ from that characteristic of regular compounds. The δ and ε-phases are brittle and hard metallic compounds. Alloys containing these compounds in larger quantities are not suitable as structural materials.

Three eutectoid microstructures can be found in the *Cu-Sn* phase diagram, denoted by the letters E_1, E_2 and E_3 . The first two of them are always decomposed. However, the E_3 eutectoid does not decompose in practice because of the low diffusion rate corresponding to the low temperature of its formation. The δ-phase constituent of the E_3 eutectoid can be decomposed to $\alpha+\varepsilon$ phases only after prolonged annealing in the temperature range of 300-350 °C, below the eutectoid temperature.

Bronzes with a 5-10 % *Sn* content have the best strength properties. In practice, bronze containing 6 % *Sn* is used to manufacture sheets, wires and rods. The tensile strength is in the range of R_m = 400-500 MPa with a specific elongation of A = 50-70 %. Bronzes containing 10-14 % *Sn* are used to produce bearings because of their good sliding properties. Bronze containing 10 % *Sn* is used to manufacture bearings loaded with a low surface pressure and a high rpm. In contrast, bronze having a 14 % *Sn*-content is used for bearings loaded with a high surface pressure and a low rpm. The tin bronze containing 20 % *Sn* consists of α + decomposed E_3 eutectoid ($\alpha+\varepsilon$) in equilibrium conditions. This latter phase is a hard and brittle microstructure, which radically reduces elongation. This bronze can only be manufactured by casting. It is used occasionally because of its hardness and corrosion resistance. The sliding property of bronze bearings can be improved by alloying with 15-30 % *Pb*. This ternary alloy containing *Cu-Sn-Pb* is called *lead bronze*.

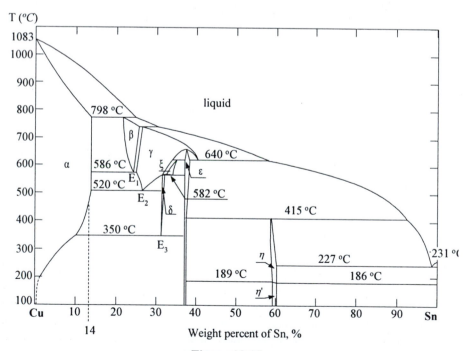

Figure 13.15
The binary phase diagram of *Cu-Sn*

13.6.3.3. The Cu-Al alloy (*aluminum bronze*)

The binary phase diagram of *Cu-Al* alloys is shown in Figure 13.16. It has practical significance up to 9 % of *Al*-content. In this range, *Cu-Al* alloys contain pure α-solid solution. The tensile strength of bronze containing 9 % *Al* is $R_m = 350$ MPa. Its specific elongation is $A = 12$ % that is relatively favorable, however its yield point is low $R_{p0.2} = 70\text{-}80$ MPa.

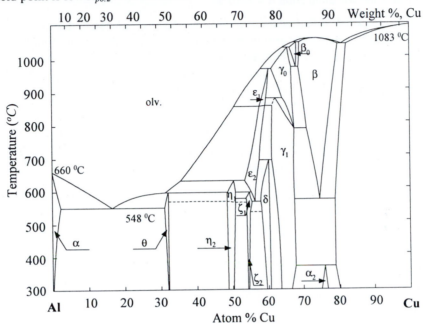

Figure 13.16
Equilibrium phase diagram of the Cu-Al two-component system

The aluminum bronze has good casting properties in a permanent mould and good formability with rolling. The alloys used for these purposes typically contain 4, 5 and 9 % *Al*. The mechanical properties of these alloys are summarized in Table 13.8.

Table 13.8. Mechanical properties of aluminum bronze alloys

Al-content	R_m, MPa	A, %	HB
4	350	50	60
5	450	30	70
9	500	30	70

The hardness of aluminum bronze can be increased by alloying it with *Mn, Ni, Si, Sn*. Besides the 4-9 % *Al*-content, the maximum amount of all the other components together should not exceed 15 % in the bronzes listed in the table.

These are the so-called *complex Al-bronzes*, differing from the usual aluminum bronzes by their higher hardness. The typical mechanical properties in the "*as cast*" condition are the following: hardness $HB = 130$, tensile strength $R_m = 450$ MPa and specific elongation $A = 8\text{-}10$ %.

The so-called *rolled aluminum bronzes* deserve separately mention. Their mechanical parameters are as follows: hardness $HB = 80\text{-}160$, tensile strength $R_m = 450\text{-}650$ MPa and specific elongation $A = 8\text{-}30$ %. A well-known representative of these bronzes is the so-called "*steel bronze*", which contains 9 % *Al* and 13 % *Mn*. This is the preferred material for fruit knives since besides hardness, resistance to acids is also of great importance. This alloy is especially resistant to oxalic acid. Other important applications of these alloys are in the chemical industry as handles, levers, valve seats, screw spindles and hand wheels. These alloys are frequently applied where both chemical resistance and high strength are required. The multi-component aluminum bronzes can be hardened.

13.6.3.4. The *Cu-Sn-Zn* alloy (*red alloy*)

The red alloy may be considered a special version of bronze. The bronze containing 4-10 % *Sn* is alloyed with 4-7 % *Zn* and in some cases 0.7 % *Pb*. This alloying provides good mould filling ability. It is the preferred material for railway fittings, wheels of centrifugal pumps and bearing shells. To improve machinability, lead is added to the alloy. The *statue bronze or monumental metal* is a characteristic type of red alloy. It is mainly used for statue casting.

13.7. Zinc (*Zn*) and its alloys

Zinc is extracted from its sulfide or carbonate ores by metallurgical processes or by a cold electrolytic procedure. In the metallurgical processes, the ores are first melted and the obtained zinc oxide is reduced to primary or wrought zinc. The melting point of zinc is 419 °C and at 950 °C it evaporates. Since the reducing temperature is in the range of 950-1100 °C, zinc is in the vapor state during the reduction process. Therefore, the reducing process should be carried out in a hermetically sealed, airtight furnace. The wrought zinc is contaminated by lead and other metals; therefore, it has to be further refined by remelting. Even refined zinc contains about 1 % *Pb*. Pure zinc containing 99.99 % *Zn* (the so-called "*four nines*" zinc) may be obtained by the electrolysis of $ZnSO_4$.

Zinc has very good casting properties. It can also be easily formed by rolling at a temperature of 140-170 °C. In the temperature range of 120-140 °C, zinc is less formable and above 200 °C it is brittle. *Zn* crystallizes in the hexagonal system, thus it has only one slip plane (0001). Due to the texture developing during plastic deformation, the mechanical properties perpendicular to the direction of rolling are better than in the rolling direction.

In industrial practice, pure zinc is used for the production of sheets and for alloying (for example, in brass), as well as for the *galvanizing* of steels to provide a protective coating with good corrosion resistance. The anti-rust protection of galvanized steel parts remains, even if the zinc coat is scratched. This is because in electrochemical corrosion zinc is the anode (positive pole) where oxygen is generated, and steel is the cathode (negative pole) where hydrogen is generated. It is well known that in electrochemical corrosion, the anode is suffering loss.

The mechanical characteristics of sheets made of pure zinc are as follows: tensile strength R_m = 150 MPa and specific elongation A = 15 %. The hot form-ability of zinc is reduced by *Sn* and is especially reduced by the simultaneous presence of *Sn* and *Pb* due to the formation of a low melting point eutectic.

The main alloying elements of zinc are *Al* and *Cu*. The casting alloys produced with these alloying elements are suitable for pressure die-casting and permanent mould casting. The mechanical characteristics of casting zinc alloys are as follows: tensile strength R_m = 150-250 MPa, hardness HB = 70-80 and specific elongation A = 1-2 %. The drawback of the *Zn-Al* alloys is that the objects produced from these alloys are not dimensionally stable, because their solid solution first shrinks, then expands during the allotropic transformations. This effect can be reduced by alloying with copper.

13.8. Lead and its alloys

Lead has a bluish (gray) color. In open air, it is oxidized in the form of a thin protective film having a dark gray color. The density of lead is 11.3 g/cm^3. The melting point of lead is T = 327°C. Its mechanical characteristics are the following: tensile strength R_m = 15-18 MPa, hardness HB = 7-8 and specific elongation is in the range of A = 30-50 %. Pure lead is the softest metal crystallizing in the face-centered cubic system, which provides perfect formability. Lead is classified as an ideally plastic material because of its low strength, low hardness, considerable elongation and low temperature of recrystallization (20-200°C). Because of its low recrystallization temperature, the forming of lead at room temperature is considered hot forming from the metallurgical point of view. For industrial purposes, lead is hardened by alloying. The most important alloying element of lead for hardening is antimony.

Lead possesses good resistance to sulfuric acid, concentrated and dilute hydrochloric acid and it is the most resistant to sulfurous acid among all metals. Therefore, it is an important material in the production of kettle linings and pipes operating in sulfuric acid or chlorinated lime. Due to its corrosion resistance, lead is used as a protective coating for cables.

13.8.1. The *Pb-Sb* alloys

Antimony (*Sb*) is a silver-white metal having a melting point of 630 °C and a density of 6.7 g/cm^3. Pure antimony is a very brittle metal. Antimony belongs to the group of those rare metals that expand upon cooling instead of shrinking Practically, it is used only for alloying. *Sb* hardens lead and, therefore, the lead-antimony alloy (14-25 % *Sb*) is called *hard lead*. The material of hard shot contains 2 % *Sb*. The strength of hard lead containing 8 % *Sb* may be increased by heat treatment: on quenching from the temperature of 235°C after subsequent tempering its hardness may be increased to R_m = 80 MPa.

13.8.2. The *Pb-Sn* alloy

The binary phase diagram of the Pb-Sn two-component alloy is shown in Figure 13.17. Lead with 64% tin forms a eutectic having a melting point of 183 °C. In alloys having higher lead content than the eutectic composition, the α-phase (rich in lead) is surrounded by a eutectic network. In the production of battery cells, the alloy is heated above the eutectic temperature (T=183°C) and the eutectic is removed by spinning.

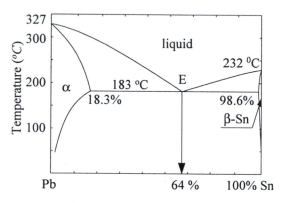

Figure 13.17
The *Pb-Sn* binary phase diagram

The *Pb-Sn* alloys are also used as soft soldering materials. Changing the quantity of *Sn*, the melting point of the soldering material can be changed in a relatively wide range. The melting point of the alloy containing 30 % *Sn* is 260 °C and decreases by about 20 °C with every addition of another 10 % *Sn*. The lowest melting point alloy contains 64 % of *Sn* (this is the pure eutectic composition) and has a melting point of 183 °C.

The *Pb-Sn-Sb* alloy is one of the best filling metals in the manufacture of bearings due to its good sliding properties, provided by *Pb* and *Sn*, and its wear resistance, ensured by *Sb*. The *Pb*-based bearing metals contain 5-10 % *Sn* and 14-16 % *Sb*. In Figure 13.18, the microstructure of white bearing metal is shown.

The alloy containing 25 % *Sn*, 25 % *Sb* and 50 % *Pb* is used as *letter metal* in the linotype industry. It has good casting properties at 240 °C: it fills the mould easily, has a low shrinkage and is sufficiently hard. A less expensive version of this alloy contains 2.5-3 % *Sn*, 10-12 % *Sb* and 85-87.5 *Pb*. It has a melting point of 280 °C. This is the *letter metal* of *linotype* composing machines.

Figure 13.18
Microstructure of white metal
Etching agent: 2 % solution of HNO_3, N=250×

The *Pb-Sn-Bi* alloys have low melting points. Their melting point can be readily controlled by their composition. The three-component eutectic of *Pb-Sn-Bi* (15.5 % *Sn*, 52 % *Bi*, 32.5 % *Pb*) has a melting point of 100 °C. Therefore, on combining the three metals in various compositions, melting fuses are produced with a melting point in the range of 100-327 °C. The melting point of these metals can be further reduced by adding cadmium. For example, the alloy containing 13.4 % *Sn*, 50 % *Bi*, 26.6 % *Pb* and 10 % *Cd* has a melting point of 60 °C. It is commonly called *Wood metal*.

13.9. Tin and its alloys

Tin exists in two allotrope modifications: i.e. *white* and *gray tin*. The white tin has a silver-white color and above the temperature of 13.2 °C it has a tetragonal lattice. On cooling it below 13.2 °C for a longer period, it is transformed into gray tin (possessing a cubic diamond lattice) and decomposes into powder. However, the mild and formable white tin may be undercooled and thus its allotropic transformation can occur only after a long holding time.

The tensile strength of tin is very low, in the range of $R_m = 30\text{-}40$ MPa, and its specific elongation is $A = 40$ %. At 200 °C, tin is so brittle that it can be crushed to powder. Tin has good cold formability. It is usually rolled to very thin sheets and foils. The rolling is performed in bands containing many layers of tin sheets separated by an oil film.

Manufacturing of tin cans and tinning of iron sheets are the main fields of industrial applications of tin. Tinning is a coating process to protect iron sheets from rusting. An excellent property of tin is that none of its compounds is harmful to health. Therefore, it is widely used in the food processing industry. The tin foils and tubes have already been substituted by aluminum foils and tubes, but the complete replacement of tin has not been accomplished yet. Tin coating protects only as long as it is scratch-free since, in the electrochemical corrosion process, oxygen develops on iron. Therefore, mild steel sheets coated with tin in an electrolytic process are treated with a further protective layer of lacquer baked in a furnace. Manufacturing of bearing and soft soldering materials is the most important application field of tin. The requirement of the good bearing is that it should contain hard, load-bearing crystals embedded in a soft matrix. Therefore, bearing metals should always have a heterogeneous microstructure.

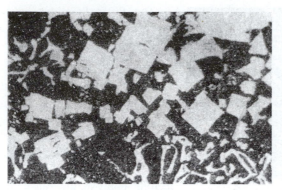

Figure 13.19
Microstructure of bearing metal *Pb* 67
Etching agent: 2 % *HNO₃*, *N*= 400×

The microstructure of a bearing metal is shown in Figure 13.19. The crystals with a regular shape in the figure are *SnSb* metallic compounds (these are the so-called load-bearing crystals) and the embedding matrix is a eutectic of *Pb* and *SnSb*. The hard crystals protruding from the soft matrix provide a place for the lubricating film. When the load on the bearing increases, the protruding crystals are pressed inside the matrix and thus further crystals can take part in load bearing. The good bearing metals have good thermal conductivity and a not too high melting point in order to melt when a problem occurs and to protect the more expensive shaft. All these requirements are fulfilled by the bearing metal containing 80 % *Sn*, 10 % *Sb*, 5-7 % *Cu*-t and 1-3 % *Pb*. The bearing crystals are the Cu_3Sb and Cu_3Sn metallic compounds.

A decrease in tin content decreases the gliding properties. Because of the shortage of tin, it is often substituted by lead. The tin content was gradually decreased from 80 % to 10 %; however, it was found that tin-antimony based bearing metals having only 5 % of tin have equivalent sliding properties to the high tin content additives. The tin-based bearing metals are usually cast into steel or bronze sleeves to provide the required strength properties. Prior to casting, the sleeves have to be tinned and casting has to be performed at the lowest possible temperature.

13.10. Noble metals: Au, Ag, Pt

Gold (Au) is a noble metal with a melting point of 1063 °C and a characteristic yellow color. Its density is ρ = 19.3 g/cm³. Gold, similarly to its alloying elements (*Ag* and *Cu*), crystallizes in the face-centered cubic system. Its tensile strength is R_m = 145 MPa, the specific elongation *A* = 50 % and the hardness *HB* = 18. Gold is a soft material with good formability. The strength of gold can be increased by alloying *Ag* and *Cu* into it. The purity of gold is measured in carats. The number of carats means the pure gold content of the alloy expressed in units of 1/24 of the total amount. The usual coin alloy contains 90 % *Au* and 10 % *Cu*.

Gold is the most important jewelry material. Nowadays, the gold content of the alloy is expressed in thousandths, which is stamped into jewelry instead of the carat number. Thus, for example, 14-carat gold means 585 thousandths of *Au* content.

Gold plays an important role in electronics and in the semiconductor industry as well. The most widely used alloys of gold are summarized in Table 13.9, indicating the carat number, the content of alloying elements and the tensile strength.

Table 13.9 The composition and tensile strength of various alloys of gold

Alloy	Carat	Au, %	Ag, %	Cu, %	R_m, MPa
Pure gold	24	100	-	-	145
22-carat	22	90	5	5	400
18-carat	18	75	15	10	700
14-carat	14	58	10	32	700

Silver (Ag) is used, besides in jewelry, for the manufacture of wires and electrical contacts due to its good electrical conductivity. Alloys of silver are used as soldering materials.

The melting point of platinum *(Pt)* is very high $(T = 1771\ ^oC)$ and, therefore, it is suitable for the production of thermometers for elevated temperatures. *Pt-PtRh* alloys are most frequently used as thermometers. Due to its high chemical resistance, it can be used as melting and heating crucibles in chemical laboratories.

13.11. Alloys of nickel (Ni) and cobalt (Co)

Both nickel and cobalt are mainly used as alloying elements because of their good corrosion and thermal resistance.

Nickel crystallizes in the face-centered cubic system and, therefore, it has good formability. Nickel possesses a high melting point $(T = 1453\ ^oC)$. It is characterized by a relatively low yield point $(R_p = 60\ MPa)$ and a high hardening capability in agreement with its large uniform elongation $(\varphi_m = 30\ \%)$. Therefore, compared to its yield point, it has a relatively high tensile strength $(R_m = 320\ MPa)$. The modulus of elasticity is almost equal to that of steel $(E = 207,000\ MPa)$.

Cobalt has a hexagonal close-packed lattice up to $471\ ^oC$. Above this temperature, it crystallizes in the face-centered cubic system. It has a high melting point $(T = 1495\ ^oC)$. The tensile strength of cobalt is only 234 MPa in the as-cast condition and its purity amounts to 99.9 %. On increasing the amount of contaminants by 0.1 % (thus reaching a purity of only 99.8 %), the tensile strength increases to $R_m = 950\ MPa$ after applying a so-called zone remelting. Its modulus of elasticity is practically equal to that of steel $(E = 211,000\ MPa)$.

Nickel is the fundamental metal in alloys having excellent corrosion resistance. It forms a complete series of alloys with copper, named *Monel.* Besides good corrosion resistance, this alloy has good formability. It is the preferred material for coin making. Nickel affects a whole set of physical properties. It is used to manufacture permanent magnets due to its excellent magnetic properties. The alloy containing 50 % *Ti* and 50 % *Ni* is used to produce so-called *shape memory alloys*. The thermal expansion of the iron alloy containing 36 % *Ni* is the lowest. This so-called invar steel, which is used in the instruments industry as well as for manufacturing of bimetal.

The alloys containing *Fe* besides high contents of *Ni* and *Co* are the so called *superalloys* which are the preferred materials of parts requiring high strength even at elevated temperatures. The main application fields are parts of turbines and aircraft jet engines. Their high strength remaining even at elevated temperatures, and excellent creep resistance, are due to the stability of microstructure even at high temperatures.

13.12. Refractory metals

The group of refractory metals consists of those metals that have extremely high melting points, exceeding even 2000 °C. Therefore, they possess high thermal resistance. Tungsten, molybdenum, tantalum and niobium belong to this group. The application fields of these metals and their alloys cover the production of lighting filaments, equipment for the chemical industry, generators for nuclear power plants and missile technology. Some characteristic properties of these metals are summarized in Table 13.10.

Table 13.10 Characteristic properties of refractory metals

Metal	T_{melt} °C	ρ g/cm^3	$R_{p0.2}$, MPa	R_m MPa	TTKV °C
Nb	2468	8.57	55	117	-140
Mo	2610	10.22	207	345	30
Ta	2996	16.60	165	186	-270
W	3410	19.25	103	455	300

From Table 13.10, it can be seen that besides the outstandingly high melting points, refractory metals possess typically high density and, therefore, their specific strength is not very high. Another feature of these metals is that their oxidation starts in the temperature range of 200-450 °C. Following this, they become quickly contaminated and the increase in the contamination content leads to their increased brittleness. Therefore, special measures have to be taken during the processing of these metals (i.e. casting, hot forming, welding or powder metallurgical processing). These usually mean vacuum processing technologies.

In certain applications, these metals are specially coated. The coatings have to fulfill the requirement of thermal resistance and compatibility with the base metal. Additional requirements are protection against contamination and a thermal expansion coefficient close to that of the base metal. Silicon and aluminum coatings satisfy these requirements up to a temperature of 1650 °C.

The metals listed in Table 13.10 possess a body-centered cubic crystalline structure; consequently, they have a transient temperature characterizing the ductile to brittle transitions (see the last column of Table 13.10). The transient temperatures of *Nb* and *Ta* are well below room temperature, therefore, these metals have good formability even at room temperature. In contrast, *Mo* and especially *W* possess high transient temperatures, consequently, they are brittle and hard to be formed plastically. If the structure of these metals is transformed into a fibrous one by hot forming, the transient temperature is decreased and the formability is improved.

The mechanical properties of refractory metals can be improved by alloying. When alloying *W* with hafnium, rhenium or carbon, their thermal resistance can be increased to 2100 $^{\circ}$C. Their alloys can be hardened by the characteristic solid solution hardening mechanism. *Mo* and *W* form a series of solid solutions similarly to *Cu* and *Ni*. The strength of some of their alloys can be increased by the dispersion hardening mechanism as well. Their composite versions, as for example *W*-fiber-reinforced *Nb*, possesses excellent properties even at elevated temperatures.

BIBLIOGRAPHY

Askeland, D. R.: The Science and Engineering of Materials
Chapman & Hall, London, 1996.

Berecz, E.: Chemistry for Engineers
Tankönyvkiadó, Budapest. 1991.

Cahn, R. W.: Physical Metallurgy
North Holland Publ. Co., Amsterdam, 1998.

Callister, W. D.: Materials Science and Engineering
John Wiley, New York, 1994.

Chalmers, B.: Physical Metallurgy
Chapman & Hall, London, 1969.

Cottrell, A. H.: Theoretical Physical Metallurgy
Edward Arnold, London, 1960.

Crawford, R. J.: Plastics Engineering
Pergamon Press, Oxford, 1987.

Elliott, R. P.: Constitution of Binary Alloys
Mc-Graw Hill, New York, 1965.

Éles, L. - Szőke, L.: Quality steels
Műszaki Könyvkiadó, Budapest, 1981.

Esterling, K.: Tomorrow's Materials
The Institute of Metals, London, 1988.

Farag, M. M.: Selection of Materials and Manufacturing Processes for Engineering
Prentice Hall, New York, 1989.

Guy, A. G.: Introduction to Materials Science
McGraw-Hill Company, New York, 1978.

Hume-Rothery, W.: The Structure of Metals and Alloys
The Institute of Metals, London, 1956.

Lakhtin, J.: Engineering Physical Metallurgy and Heat Treatment
Mir, Moscow, 1990.

Metals Handbook. vol.2.: Non-ferrous Alloys and Pure Metals
        American Society for Metals, Ohio, 1979.

Metals Handbook. vol.8.: Metallography, Structures and Phase Diagrams
        American Society for Metals, Ohio, 1979.

Metals Handbook. vol.9.: Metallography and Microstructures
        American Society for Metals, Ohio, 1979.

Porter, D. A. - Easterling, K. E.: Phase Transformations in Metals and Alloys
        Chapman & Hall, London, 1992.

Prohászka, J.: Anyagszerkezeti ismeretek
        Tankönyvkiadó, Budapest, 1964.

Prohászka, J.: Introduction to Materials Science
        Tankönyvkiadó, Budapest, 1988.

Prohászka, J.: Anyagtechnológia
        Tankönyvkiadó, Budapest, 1989.

Prohászka, J.: A szilárdtest kutatás újabb eredményei
        Akadémiai Kiadó, Budapest, 1988.

Rollason, E. C.: Metallurgy for Engineers
        Edward Arnold, London, 1984.

Smith, F. W.: Principles of Materials Science and Engineering
        McGraw Hill, New York, 1990.

Tamman, G.: Lehrbuch der Metallographie
        L.Voss Verlag, Leipzig, 1923.

Tisza, M.: Metallography
        University Publisher, University of Miskolc, Miskolc, 1985.

Tisza, M.: Materials Science for Mechanical Engineers
        University Publisher, University of Miskolc, Miskolc, 1998.

Van Vlack, L. H.: Elements of Materials Science and Engineering
        Addison-Wesley Co., New York, 1989.

Verhoeven, J. D.: Fundamentals of Physical Metallurgy
        John Wiley & Sons, New York, 1975.

Verő, J.-Káldor, M.: Physical Metallurgy
        Tankönyvkiadó, Budapest, 1977.

Verő, J. - Káldor, M.: Physical Metallurgy for Metallurgists
        Műszaki Könyvkiadó, Budapest,1971.

Vinson, J. R.: Advanced Composite Materials
        ASTM Technical Publication, Philadelphia, 1977.

Zorkóczy, B.: Metallography
        Tankönyvkiadó, Budapest. 1974.

APPENDIX – SI UNITS

Both in science and in technology, the *International System of Units* (*Systeme International d'Unites*) is accepted. These units are commonly referred to as *SI Units*. Though, SI Units are widely used, it is worthwhile to present a brief summary here. For a more detailed presentation, the reader is referred to ASTM publications (e.g. *Metric Practice Guide*).

In the International System of Units, the major basic units are as follows:

Quantity	Unit	SI Symbol
Length	meter	m
Mass	kilogram	kg
Time	second	s
Electric current	ampere	A
Temperature	Kelvin	K
Amount of substance	mole	mol

Derived SI Units commonly encountered in Physical Metallurgy are listed below.

Quantity	Unit	SI Symbol
Area	square meter	m^2
Density	kilogram/cubic meter	kg/m^3
Force	Newton	$N\ (kg \cdot m/s^2)$
Energy	Joule	$J\ (N \cdot m)$
Power	Watt	$W\ (J/s)$
Stress	Pascal	$Pa\ (N/m^2)$

Conversion between US and appropriate SI Units

$$1 \text{ dyne} = 10^{-5} \text{ N}$$
$$1 \text{ erg} = 10^{-7} \text{ J}$$
$$1 \text{ calorie} = 4.184 \text{ J}$$
$$1 \text{ psi} = 6.8948 \text{ kN/m}^2 \text{ (kPa)}$$
$$1 \text{ ksi} = 6.8948 \text{ MN/ m}^2 \text{ (MPa)}$$

A system of prefixes for fractions and multiples

Fraction	Prefix	Symbol		Multiple	Prefix	Symbol
10^{-1}	deci	d		10^{1}	deka	da
10^{-2}	centi	c		10^{2}	hecto	h
10^{-3}	milli	m		10^{3}	kilo	k
10^{-6}	micro	μ		10^{6}	mega	M
10^{-9}	nano	n		10^{9}	giga	G
10^{-12}	pico	p		10^{12}	tera	T

INDEX

AUTHOR INDEX